Faust, Miklos.

Physiology of temperate zone fruit trees

$44.95

DATE			

PHYSIOLOGY OF TEMPERATE ZONE FRUIT TREES

PHYSIOLOGY OF TEMPERATE ZONE FRUIT TREES

Miklos Faust
Fruit Laboratory
Beltsville Agricultural Research Center
Agricultural Research Service
United States Department of Agriculture
Beltsville, Maryland

A WILEY-INTERSCIENCE PUBLICATION

JOHN WILEY & SONS

New York / Chichester / Brisbane / Toronto / Singapore

Library of Congress Cataloging in Publication Data:

Faust, Miklos.
Physiology of temperate zone fruit trees / Miklos Faust.
p. cm.
Bibliography: p.
1. Fruit trees—Physiology. I. Title. II. Title: Temperate zone fruit trees.
SB357.28.F38 1989 89–32062
634–dc20 CIP
ISBN 0–471–81781–3

Printed in the United States of America

10 9 8 7 6 5 4 3 2 1

To Professor Robert M. Smock,
the person who introduced me
to the physiology of fruit trees,

and to my wife, Maria A. Faust,
who encouraged me and endured the procedure
of my writing this book

PREFACE

The physiology of fruit trees is complicated. Fruit trees are perennial woody plants that are taxed every year to produce large fruit crops. Often, as much as 70% of the fruit tree's carbohydrates are harvested as fruit, yet the tree must have sufficient carbohydrate stores to maintain basic life processes—to grow, form flower buds for the next season, and survive the cold stress of winter. In addition, with the on-going shift of fruit industry orchards from colder to warmer climates, additional physiological demands such as increased transpiration and respiration are put on the trees. All these facts must be taken into account in orchard design.

Although much is known about the various physiological functions of plants in general, less is known about the same functions in trees. Even less is known about such processes in temperate zone fruit trees. The last book that summarized the physiology of temperate zone fruit trees was written by Kobel in 1954 more than 30 years ago. I hope that this book will familiarize a wide circle of readers with the present knowledge on the applied physiology of fruit trees and be of help to students, growers, and specialists.

During the writing of this book, I deliberately selected information developed on temperate zone fruit trees. Occasionally this may not be the most up-to-date knowledge on the particular subject, but it is the state of the art with regard to temperate zone fruit trees. Readers should use this book in conjunction with other basic writings on the physiology of plants.

I should like to thank the publishers, John Wiley & Sons, for their understanding and cooperation and all my colleagues and collaborators who have helped or advised me. Particular thanks are due to R. Korcak, G. Steffens, S. Miller, A. N. Miller, D. Swietlik, R. Scorza, and S. Y. Wang for reviewing parts of the book and Xiu-Ping Sun for the illustrations.

MIKLOS FAUST

Beltsville, Maryland
June 1989

CONTENTS

PHYSIOLOGY OF TEMPERATE ZONE FRUIT TREES

1

PHOTOSYNTHETIC PRODUCTIVITY

Photosynthetic productivity in fruit trees depends on many factors. Some of the factors are similar to those identified for plants in general; others are relatively unique for woody perennials and for fruit trees in particular. The factors determining photosynthetic productivity of trees can be divided into internal and environmental groups. The internal factors range from leaf structure and chlorophyll content of the leaf to water-conducting ability and osmotic adjustment by the tree and the presence of strong sinks such as the fruit. The environmental factors include the availability of light, the high temperature that could influence photosynthesis but more likely influences dark respiration, which results in a substantial loss of carbohydrates, and the availability of water, which together with air humidity (vapor pressure deficit) determines stomatal opening and has a major influence on CO_2 exchange and consequently on photosynthesis. In fruit trees, in addition to these complex effects, one has to consider the previous history of the plant, which acts almost like a 'memory' and influences the photosynthetic productivity.

About a half century ago in 1935, in a historical series of experiments, A. Heinicke and N. Childers determined the total photosynthesis of a young bearing apple tree throughout the entire year. Their investigation clearly shows that at the beginning of the season the tree loses energy; then as the leaf area develops, the photosynthates (Pn) slowly increase, and as the leaves senesce, the rate of photosynthesis declines. Throughout the season the single most important factor governing Pn is light level. On days with high levels of irradiation the Pn are high, and conversely on days of low irradiance the Pn could be only 25% of that occurring on bright days. The daily fluctuation in Pn, light, temperature, and transpiration in Heinicke and Childers' experiment is shown in Figure 1.1. The experiment clearly indicated fluctuation in photosynthesis throughout the season. This is an important indication that photosynthesis productivity for obtaining maximum yields should be considered on the yearly basis.

As has been said before, light is the most important factor in photosynthesis. When fruit trees are exposed to solar radiation, the radiation interacts with the trees through absorption, scattering, and transmission through the canopy. About 30% of the global radiation is absorbed by the leaves in a modern orchard. The photosynthetically active component of the absorbed radiation, up to 28% in terms of solar energy, is used for carbohydrate production and stored chemically in the tree in the form of high-energy compounds. More than 70% of solar radiation absorbed by the trees is converted into heat and used as energy for transpiration and convective heat exchanges with the surrounding air. These processes determine the water use of the trees and the temperature of leaves and fruit. One continually needs to consider both of these interrelated aspects of utilization of light energy. For high photosynthetic productivity, light

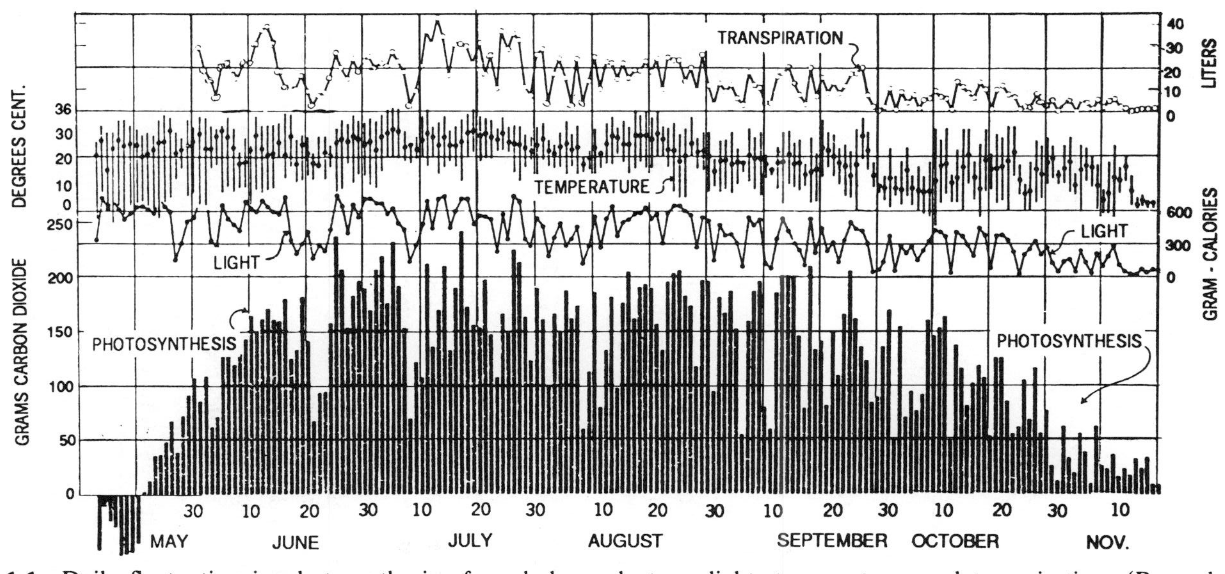

Figure 1.1 Daily fluctuation in photosynthesis of a whole apple tree, light, temperature, and transpiration. (Reproduced by permission from Heinicke and Childers, 1937; chart prepared by Norman F. Childers.)

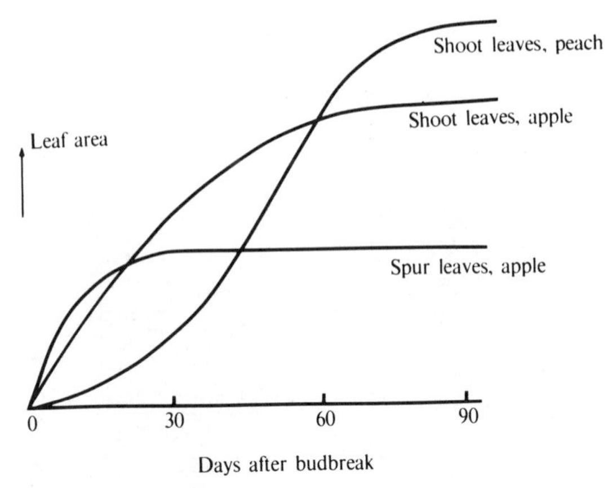

Figure 1.2 Development of leaf area in apple and peach (data from Cain, 1973, and Chalmers et al., 1983). Shoot leaves of apple usually start to grow when spur leaves complete their development.

interception needs to be maximized, which requires that the tree canopy cover the available ground space in the orchard and is designed in such a way that it allows for maximum light interception and the desired distribution of light within the canopy.

LIGHT INTERCEPTION AND UTILIZATION

Development of Leaf Area

The leaf area of trees develops rapidly during the spring up to a maximum value; then it becomes stable during midseason and finally decreases when the leaves start to fall during autumn. In fruit-producing species, there is some variation from this scheme. In apple, leaves are located on spurs and on shoots. The spur leaves develop very rapidly, usually by the end of petal fall (Caine, 1973), and the leaves located on the shoots develop as the shoots develop, which may take an additional 60 days. In apple, a high degree of spur formation is desirable to increase productivity; thus the leaf area at the beginning of the season is relatively high in this species as compared with other species that develop leaves only on shoots. In peach, there are no spurs and the trees are usually heavily pruned for reasons advanced in another chapter. In this species, the development of the leaf area begins with low values during spring and increases for about 90 days until shoot growth ceases (Figure 1.2).

As the leaves of a tree develop during the spring, interception of light by the tree increases. It is generally assumed that a larger canopy intercepts more light. This may not be so. If too many leaves are crowded into the tree, they shade each other, and that shading is detrimental to photosynthesis.

A uniform measure to quantify the leaf area has been developed and used for many crops. It is called the *leaf area index* (LAI). This is a ratio of leaf area to ground area and is used in conjunction with light interception studies and transpirational measurements as a basis of canopy efficiency.

The leaf area of a given branch or a tree can be determined by measuring the actual area of the leaves. However, it was necessary to develop some kind of estimation of the leaf area without actually determining the leaf area itself in every instance. Obviously, the larger the tree is, the larger the leaf area associated with it will be. This has focused attention to the determination of leaf area associated with a given branch or trunk diameter. Holland (1968) developed an equation for apple relating the total leaf area of a branch (A) to its girth (G). He used the apple cultivar 'Cox's Orange Pippin' on M2 rootstock for his studies. His equation is

$$\log A = \log K + b \log G$$

The coefficient b varies with time because the leaf area increases during the season to a plateau and then decreases, whereas the girth remains the same. The July value for the total leaf area in Holland's studies was

$$\log A = 3.159 + 2.215 \log G$$

The K value indicates a coefficient that is dependent on a particular system used for growing the tree and a particular cultivar, and it needs to be established first from a sample of 15–20 branches and is applicable to that system or cultivar only. Barlow (1969) and Verheij (1972) developed similar relationships relating leaf area and trunk diameter of entire trees of apples (Figure 1.3). These relationships were $\log A = 0.329 + 2.586 \log D$ (Barlow) and $\log A = -2.050 + 2.230 \log D$ (Verheij). In these equations, D denotes trunk diameter. The coefficient b in all three studies is very similar, but differences in the intercept are great. The relationships are sufficiently different to not allow absolute comparisons, but the relationship within a given set of circumstances is well established and could be used to establish the LAI in a relatively simple way. A detailed description of the establishment of LAI can be found in Jackson's review (1980a).

The LAI in apple ranges from 1.5 to 5. It depends on rootstock and scion cultivars, pruning and fertilization, and other cultural practices (Jackson, 1980). Verheij and Verver (1973) concluded that a LAI of 2.45 for apple was too dense for maximum fruit yield and fruit quality. In modern apple orchards, LAI is approximately 1.5 for trees on M9 rootstock, 1.7 for palmettes, and 2.2 for pillars (Jackson, 1980a). Peach leaf area index is generally higher than that of apple and ranges between 7

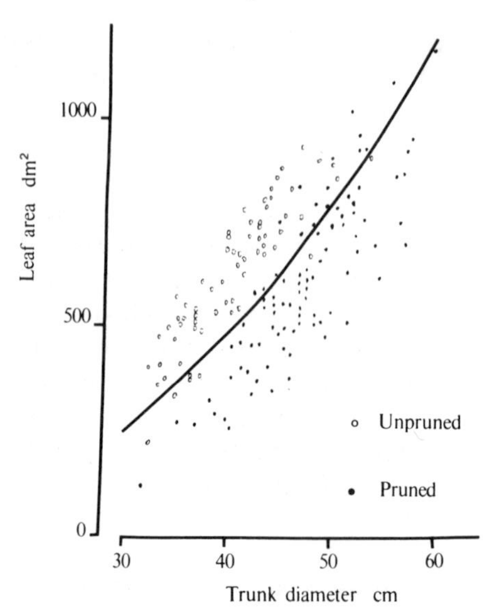

Figure 1.3 Relationship between leaf area and trunk diameter of apple trees. 'Golden Delicious' on M9 rootstock. (Reproduced by permission from Jackson, 1980a.)

and 10. For peach most of the leaf area is close to the top of the tree. Only 5% of the leaf area is in the lowest 10% of the tree, 14% is in the next 50% from the ground, and 80% of the leaves are in the top 40% of the tree (Chalmers et al., 1983).

Relationship between LAI and Light Interception

The LAI is useful in conjunction with light interception records as a basis to analyze the canopy productivity. In general, the interception of light by the canopy can be expressed as

$$I_L/I_0 = e^{-KL}$$

where I_0 is incident light energy, I_L is light intercepted by the LAI of L, and K is the extinction coefficient of short-wave radiation. Light intensity declines logarithmically with LAI, and total light interception is a logarithmic function of LAI. The K values were calculated by Cain (1973), Proctor (1978), and Jackson (1978) with averages of 0.56, 0.43, and 0.60, respectively. Each set of data represent a certain set of circumstances. Jackson's data appears to be the most generally useful because of the tree type he used for the measurements. Therefore the average value of 0.60 for K is generally accepted for most purposes.

TABLE 1.1 Percentage of Full Radiation Needed for Various Quality Factors in Apples

Character	Development Satisfactory	Development Unsatisfactory
Fruit size	>50%	<50%
Red color	>70%	<40%
Development of spurs	>30%	<25%

Light Penetration into the Tree

Light penetration into the tree is usually determined by the size and the shape of the tree. Heinicke (1963, 1964, 1966) reported that light intensity within large 'Delicious' apple trees decreased by 58% within about 2 m from the top of the tree. Jackson (1967, 1970) found even more rapid decline in radiation within the canopy of a 4-m-high 'Cox's Orange Pippin' apple tree. Light decreased to 34% of the above-the-tree levels within 1 m from the top. Lakso and Musselman (1976) measured the penetration of *photosynthetically active radiation* (PAR) into large 'McIntosh' apple trees and found that only 15% of the available light penetrated the interior of the tree. Light penetration was even less for 'Wayne', where only 7% reached the interior of the tree. The best red color of the fruit is associated with light levels of more than 70% of full radiation. Apples receiving less than 40% of the radiation develop inadequate red color (Heinicke, 1966). The size of the fruit is similarly affected. Shaded fruit is always smaller than that exposed to full radiation. Light intensity needed for various quality factors in apple is given in Table 1.1. At low LAIs, tree height and tree design has little influence on light penetration into apple trees. At high LAIs, above 1.5, tree height and canopy type greatly influence light penetration into the canopy. At LAI 3.0, 20% less light penetrates the interior of standard apple trees compared to dwarf trees (Heinicke, 1964) (Figure 1.4).

Shading could occur as intertree or intratree shading. Sansavini (1982) studied the degree of both types of shading for a northern latitude of 44.29°. Obviously, with increasing tree size, the intratree shading increases. By comparison, intertree shading is a function of the distance between trees and the elevation of the sun. The magnitude of intertree and intratree shading is illustrated in Figure 1.5.

Jackson (1980) calculated the portion of the canopy that receives more than 30% of incident radiation so that $I_L/I_0 = 0.3$. He used this value because it is the lowest value at which acceptable production of apples is possible. From the equation $I_L/I_0 = e^{-KL}$ one can calculate

$$0.3 = e^{-KL_{0.3}}$$
$$\ln 0.3 = KL_{0.3}$$
$$L_{0.3} = \ln 0.3/-K$$

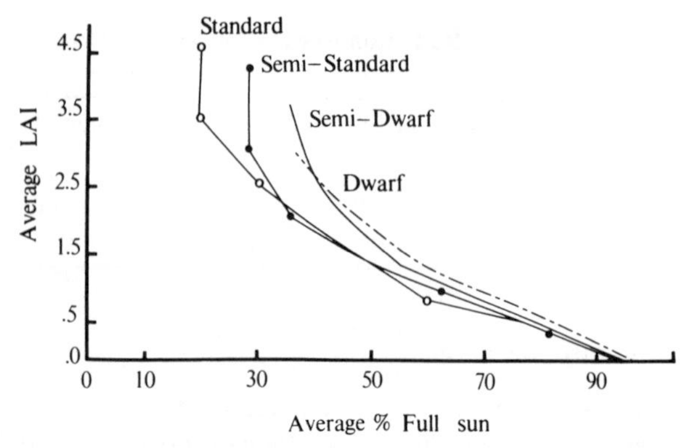

Figure 1.4 Relationship between LAI, tree size, and light penetration into 'Delicious' apple trees as percentage of full sun. Tree size was influenced by rootstock and not by short-internode bunching of leaves. (Reproduced by permission from Heinicke, 1964.)

where $L_{0.3}$ is the leaf area that receives at least 30% light. The same equation is applied to discontinuous canopies where only a part of the ground ($1 - T_f$; see next section) is covered by the trees. Thus the leaf area exposed to more than 30% incident radiation is expressed as

$$L_{0.3} = \ln 0.3/-K(1 - T_f)$$

where T_f is the transmission of light depending on the orchard form. Calculations of this type can be carried out for any chosen value of the illuminated percentage of the canopy (L_I). The general equation is

$$L_I = \ln I-K(1 - T_f)$$

Interception of light and penetration into the canopy depends on the canopy type and the distance between the trees. Palmer (1981) calculated light interception by trees of three different apple-planting systems: palmette, spindle, and full-field (bed system) types. Calculated profiles within individual sections of trees for two LAIs are given in Figure 1.6.

The part of each day the canopy or portion of canopy receives light depends on the orientation of rows in the orchard. If rows are oriented in a north–south direction, both sides of the row receive the same number of hours of light. In contrast, if rows are oriented east to west, the north side of the tree receives two-thirds fewer hours of direct radiation than the south side (Gyuro, 1974) (Figure 1.7).

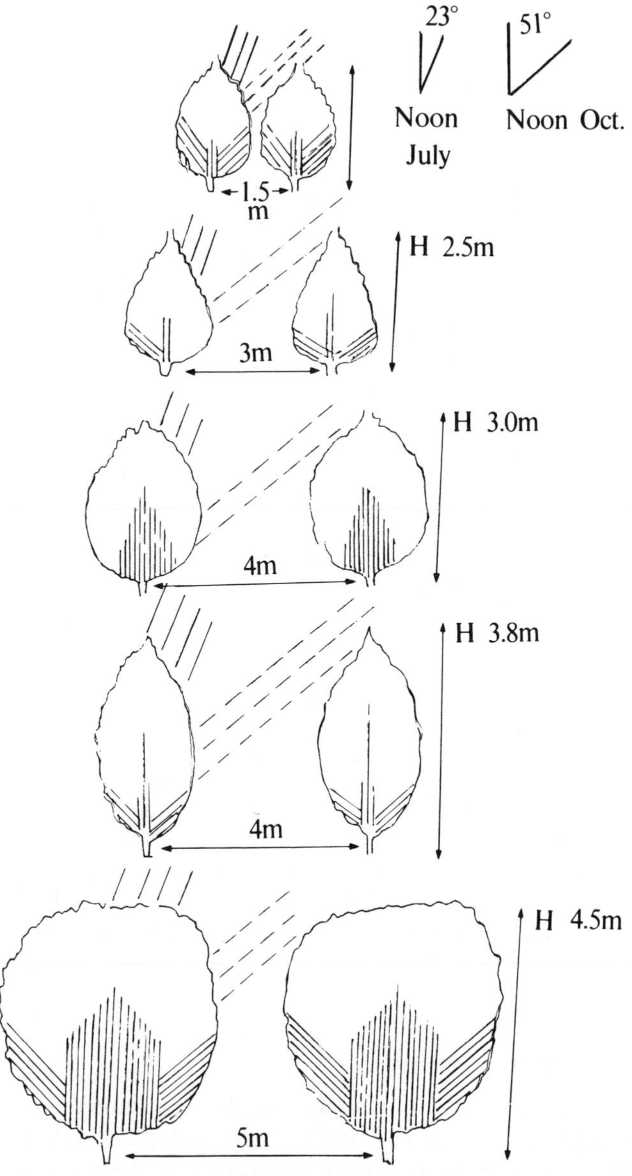

Figure 1.5 Influence of tree shape and planting distance in intratree and intertree shading in apple orchards. Vertical lines indicate intratree shading. Slanted lines indicate intertree shading apparent in October. Little intertree shading occurs when sun angle is high (July). Comparable shading to that occurring in October occurs daily at lower sun angles. (Reproduced by permission from Sansavini, 1982.)

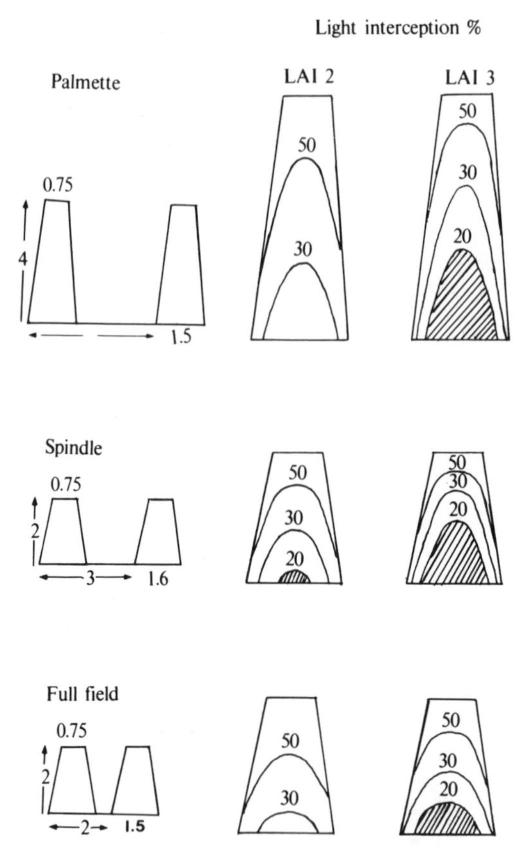

Figure 1.6 Calculated light profiles within hedge sections of apple plantings of given dimensions for two LAIs and three orchard systems: palmette, spindle, and full field. Shaded area receives too low light for satisfactory productivity. (Reproduced by permission from Palmer, 1981.)

Light Interception by Orchards

Orchards are discontinuous canopies. Light transmission through the orchard canopy occurs in two ways. Light reaches the orchard floor between trees of a given geometry regardless of the LAI, and it also reaches the orchard floor after transmission through the canopy, that is, through the gaps between the leaves and through the leaves themselves. These two types of transmission were expressed by Jackson (1980a) as

$$T = T_f + T_c$$

where T is total transmission to the ground, T_f is transmission dependent on orchard form, and T_c is transmission through the canopy.

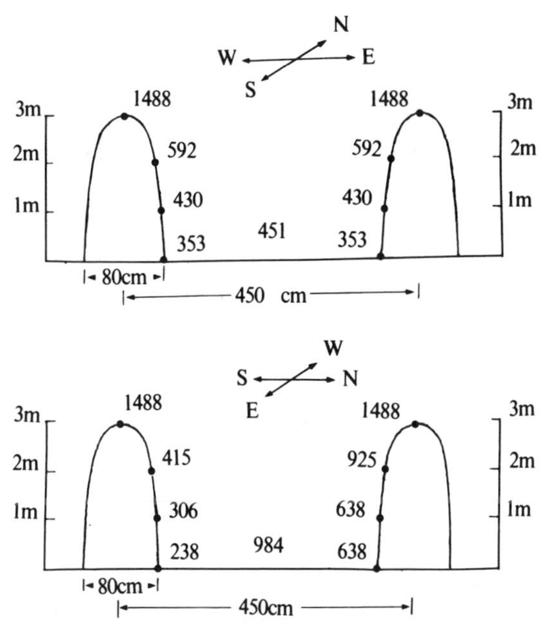

Figure 1.7 Hours of sunshine received during the growing period in apple hedgerow by various parts of the tree in north–south or east–west row orientations. (Reproduced by permission from Gyuro, 1974.)

Usually more light is transmitted to the ground if tree height is low and/or the trees are far apart. Conversely, the light interception of an orchard can be increased by planting the trees closer together and/or by increasing the tree height. Jackson (1980a) calculated T_f values for a variety of orchard forms commonly used in high-density apple orchards. His values are for direct or diffuse light conditions (Figure 1.8). These values were developed with nonreflecting, nontransmitting models (Jackson, 1980b). Jackson assumed that ignoring the reflection introduced relatively little error and the accuracy of using the solid model was satisfactory. Transmittance through the canopy depends on the LAI. Trees with high LAI are virtually solid and transmit little light. Orchards with low LAIs are likely to transmit through the trees (T_c). Even in such cases, the solid models are useful because the shape of the tree (its height and form) is a determining factor that can be influenced by orchard design and developed by orchard practices. Thus light interception that depends on tree shape (T_f) can be considered when orchards are in the planning stage.

Transmission through the tree is related to the density of the canopy. More light is transmitted with few leaves on the tree than through a similar size tree with a dense canopy. Transmission through the canopy

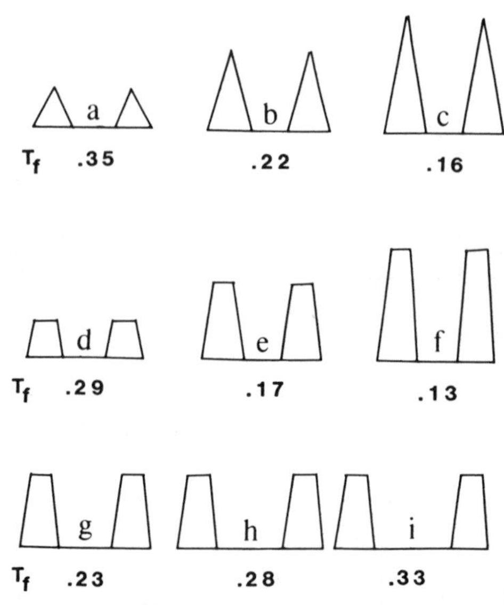

Figure 1.8 Transmission of light to orchard floor depending on tree form and on planting distance. Transmissibility T_f expressed as the fraction of global radiation reaching the orchard floor, where T_f values are averages obtained in diffuse and direct radiation. (Reproduced by permission from Jackson, 1980b.)

decreases as leaves develop during the spring. Hemispherical (fisheye) photography can be used to determine light transmission through the canopy (Lakso, 1984). Photographs taken from the base of the tree upward demonstrate the percentage of sky visible through the canopy. The percentage of sky visible also indicates the light transmission through the canopy to the ground. As the leaf area increases, the percentage of sky visible decreases. Correlation between development of leaf area and openings in the canopy through which light can penetrate is illustrated in Figure 1.9.

Leaves may be randomly distributed in the canopy of a tree or they may be grouped together on relatively dwarf branches with only short internodes between the leaves. This is important in the various 'spur-type' apples that have somewhat reduced internode length or in a variety of dwarf, short-internode trees of several fruit-producing species. Oikawa and Saeki (1977) developed a model that considers light interception by plants that have their leaves grouped on the stems. They assumed the same LAI that was distributed among fewer stems with more leaves per stem or among more stems with fewer leaves per stem. The ground area was the same; thus the model could distinguish between light penetration between few stems with many leaves compared with many stems with few leaves. In fruit trees, the situation is not that clear-cut. Some trees have

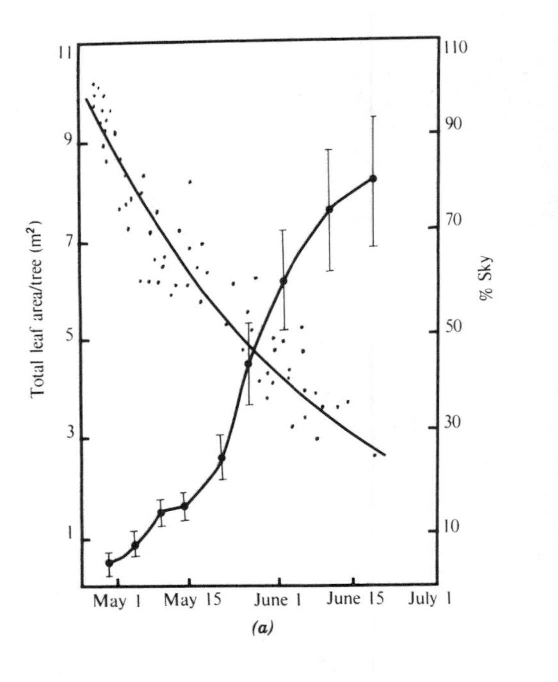

(a)

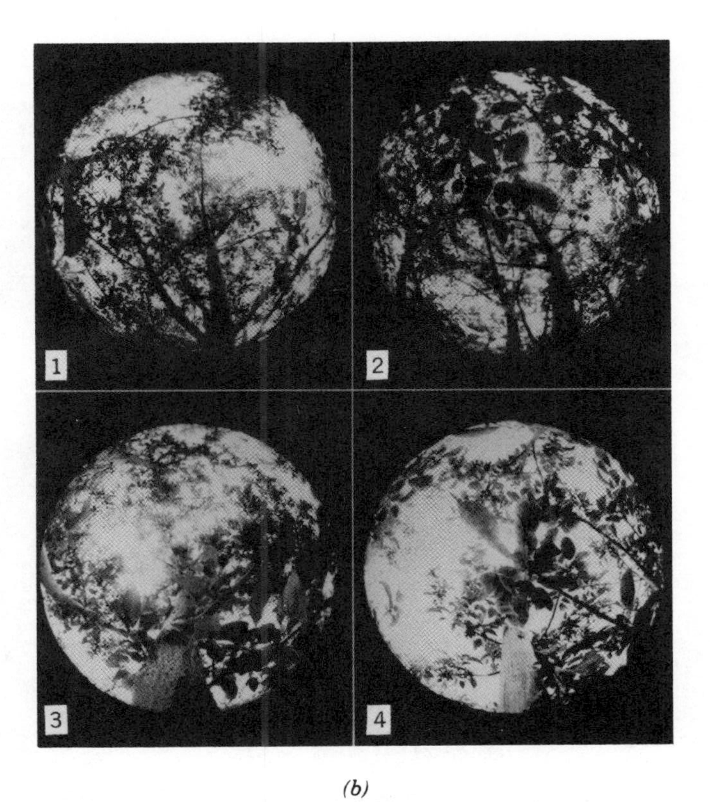

(b)

Figure 1.9*(a)* Relationship between leaf area and percentage of visible sky from the base of the tree, indicating light transmission through the tree. As leaf area increases, percentage of sky decreases. (*b*) Light penetration (% sky) in various apple trees. Lower right is a dwarf tree on M9 rootstock; other are semidwarf trees. (Reproduced by permission from Lakso, 1975, 1984).

Figure 1.10 Light penetration into unpruned naturally developing peach canopy types: *A*, compact; *B*, standard; *C*, semidwarf; *D*, dwarf with short internodes. Sizes are proportional. (Reproduced by permission from Scorza, 1984).

short internodes, but the number of shoots is not reduced from that of standard trees. Other trees, often called *compact*, have a low rate of apical dominance and most of their buds break, developing into secondary shoots. Canopies of such plants are very crowded even though the length of their internodes is not reduced. Scorza (1984) measured light penetration into standard, compact, semidwarf (cross of compact and dwarf) and dwarf (short-internode) peach trees. Light penetration into the center of compact and dwarf trees was much less than into standard or semidwarf trees (Figure 1.10). The least amount of light penetrated

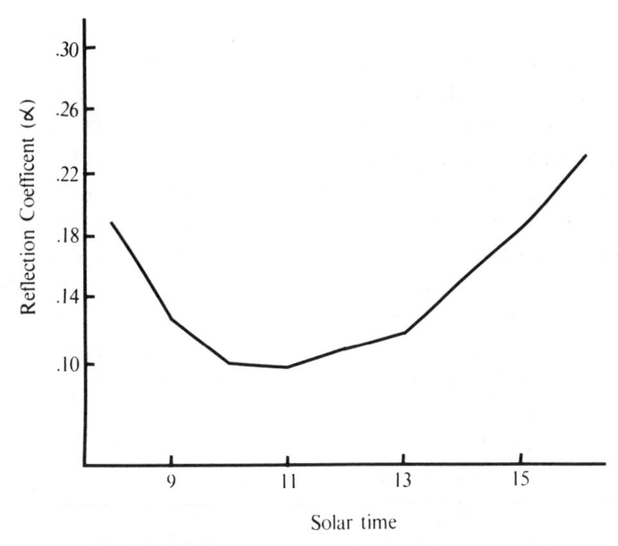

Figure 1.11 Diurnal variation of reflection coefficient α of a mature 'McIntosh' apple tree. (Reproduced by permission from Proctor et al., 1972.)

into the short-internode trees even though these trees are very small compared to the standard or even compact trees.

Absorption and Reflectance by Leaves and Transmittance of Light through Leaves

Trees receive incident solar radiation, which consists of two components: direct solar radiation as a beam of parallel rays reaching the orchard surface and diffuse solar radiation scattered in the earth's atmosphere and reaching the trees from all directions. Radiation interacts with the tree canopy through absorption and scattering. These processes vary widely in various parts of the spectrum and depend on leaf structure, leaf age, spectral distribution, and angle of incident radiation. Radiative scattering can be divided into reflection and transmission. The upward-scattered radiation is defined as reflected radiation and is characterized by the reflection coefficient α. The part of the incident radiation transmitted below the surface of the canopy is characterized by the transmission coefficient Γ. The reflection coefficient is the average reflectivity over a specific waveband. The transmission coefficient can be defined in the same way. The reflection coefficient in the older literature was often called *albedo*, derived from the Greek for whiteness.

The radiation balance of an apple tree has been determined by Proctor et al. (1972). The mean seasonal diurnal variation in the radiation components over the apple tree are similar to those for other tree crops (Figure 1.11). Reflection of solar radiation by apples is lowest at solar noon and increases almost linearly with the zenith angle. The strong dependence of reflectance on the zenith angle is expected, considering the canopy design of apple trees. Orchards are discontinuous surfaces. Therefore, greater penetration of solar radiation should be expected into the orchard at solar noon with correspondingly reduced reflection coefficient than at lower zenith angles (Figure 1.11). Values for net long-wave radiation (L_n) for apple orchards are negative. Each day the net long-wave radiation loss rises gradually to solar noon and decreases afterward. There is no seasonal trend in daily long-wave radiation. The net long-wave radiation loss from the canopy is about the same as the short-wave loss by reflection. Since the net outgoing long-wave radiationn (L_o) depends on surface radiative temperatures, a good relationship exists between the long-wave exchange coefficient and canopy temperature.

Fruit tree leaves reflect and transmit light over all visible wavelengths. The reflection coefficient of a canopy depends on its geometry and the angle of the sun as well as the radiative properties of its components. Maximum values of 0.25 were recorded over relatively smooth surfaces such as cut grass. Reflection coefficients decrease with the height of the crop. Values as low as 0.10 have been recorded for forests (Standhill et al., 1966). In general and also over apple orchards, reflectance is relatively large for green light near 550 nm, but the absolute reflectivity is larger in the near-infrared spectrum (Palmer, 1977). Reflectance is greatest early in the season (May) and decreases as the season advances. For apple, reflectance over the range of 400–700 nm was greatest in May, at 10–11% of radiation received, and averaged around 9% for the rest of the season. Similar values were reported for peach. Mean daily reflectance values obtained were 0.18 at the beginning of the season and 0.14 at the end of the season (Chalmers et al., 1983).

In the visible spectrum, most of the radiation penetrating the leaf epidermis is absorbed by the pigments of the chloroplast. Absorption of green light is less intense (500–540 nm) than that of blue light (400–540 nm) or red light (600–700 nm) so that the reflected and transmitted light is strongly green. The incident radiation is scattered in all directions by multiple reflections at cell walls. About the same light is scattered forward and backward so the fluxes of the transmitted and reflected portions of the visible radiation are often nearly equal (Monteith, 1973). For apple leaves, transmittance between 400 and 700 nm declined from 7% in May to 1.5% for sun leaves and 3.5% for shade leaves by the end of September (Palmer, 1977).

Palmer (1977) reviewed canopy transmittance data of different wavelengths of light through apple canopies. Transmission of infrared through the canopy is high. Approximately five times as much radiation is transmitted between 750 and 1400 nm than between 400 and 680 nm. Suckling et al. (1975) reported that for a dwarf apple orchard canopy, the transmission, absorption, and reflectivity values were 0.53, 0.19, and 0.28 of global radiation including infrared, compared with 0.42, 0.07, and 0.51 for PAR. The processes of absorptivity and transmissivity for either global or PAR radiation are influenced by the LAI and, therefore, are expected to change during the season as LAI changes. Suckling et al. (1975) reported such changes for a young apple orchard (Figures 1.12 and 1.13). Canopy absorptivity, reflectivity, and transmissivity all differ with direct or diffuse radiation, as illustrated from data obtained on bright, sunny days (Figures 1.14 and 1.15). In peach, the transmissivity of the canopy reflects the different nature of the development of the leaf area. Transmissivity of the canopy to the incoming short-wave flux was 0.8 at the beginning of the season and decreased progressively as the leaf area developed to a low value of 0.15 by the end of the season (Chalmers et al., 1983).

Proctor et al. (1972) measured the components of radiation balance of a 10-year-old apple tree in Ontario and found an average reflection coefficient of 0.17 (Figure 1.16). Net short-wave radiation (Q_n) radiation was linearly related to short-wave solar radiation (Q_s):

$$Q_n = a + bQ_s$$

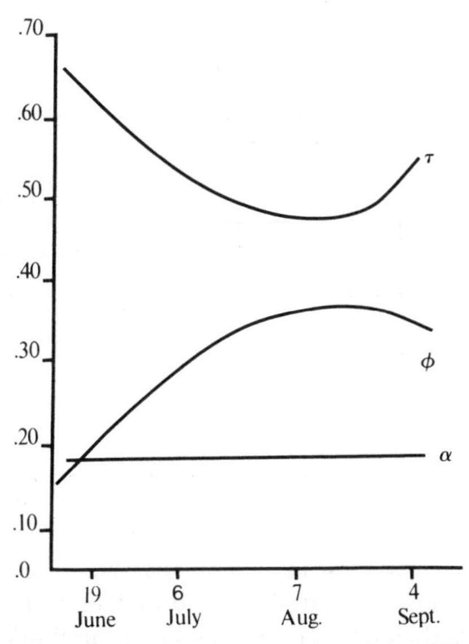

Figure 1.12 Seasonal variability of canopy absorbtivity ϕ, reflectivity α, and transmissivity τ of global radiation of a young apple orchard. (Reproduced by permission from Suckling et al., 1975.)

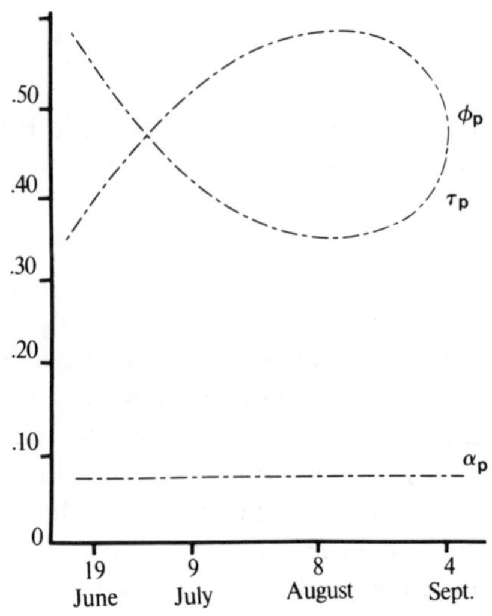

Figure 1.13 Seasonal variability of canopy absorbtivity ϕ_p, reflectability α_p, and transmissivity τ_p of photosynthetically active radiation of young dwarf apple orchards. (Reproduced by permission from Suckling et al., 1975.)

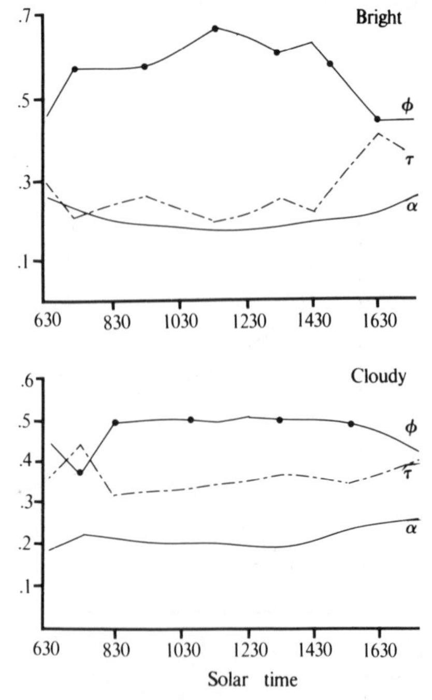

Figure 1.14 Diurnal variation in canopy absorbtivity ϕ_p, reflectivity α_p, and transmissibility τ_p of photosynthetically active radiation of a young dwarf apple orchard on a bright and cloudy day, representing the difference between direct and diffuse radiation in the photosynthetically active range. (Reproduced by permission from Suckling et al., 1975.)

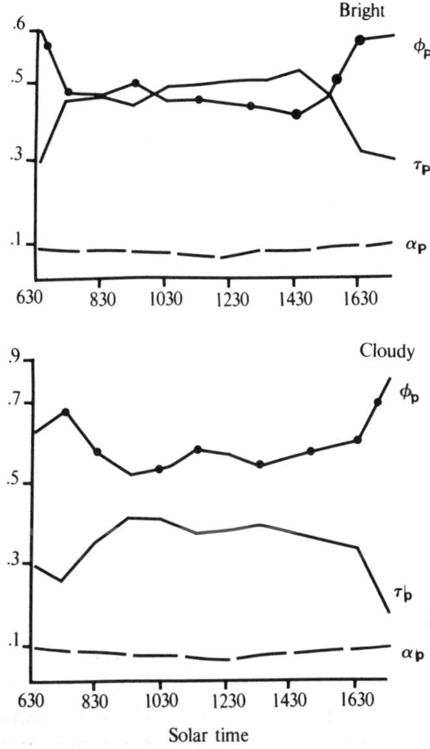

Figure 1.15 Diurnal variation in canopy absorbtivity ϕ, reflectivity α, and transmissivity τ of global radiation of young dwarf apple orchard on a bright and cloudy day, representing the difference between direct and diffuse radiation. (Reproduced by permission from Suckling et al., 1975.)

His values were $a = -98\ \mathrm{W\,m^{-2}}$ and $b = 0.80$. Landsberg et al. (1973) obtained comparable results with reflection coefficients ranging from 0.13 to 0.19. From their results $a = -14.7\ \mathrm{W\,m^{-2}}$ and $b = 0.67$. Both values closely approximated $Q_n - 0.6\,Q_s$. Considering that a fraction of about 0.15 is reflected and about 0.17 is transmitted ot the orchard floor between the trees, this is a reasonable value. Thorpe (1978) estimated that a radiation exchange (Q_e) was about half of net radiation (Q_n). Thus about $0.30\,Q_s$ is exchanged by the leaves and calculated an empirical formula for the relationship between net radiation exchange per unit of leaf area (Q_l) and solar short-wave radiation (Q_s):

$$Q_l = 0.34\,Q_s - 5$$

Landsberg and Jones (1981), using the data of Suckling et al. (1975) for the absorption of short-wave radiation of dwarf young orchard, stated that $Q_l = 0.3\,Q_s$ is a useful approximation for most modern orchards. This indicates that a useful estimate of the available energy in the leaves

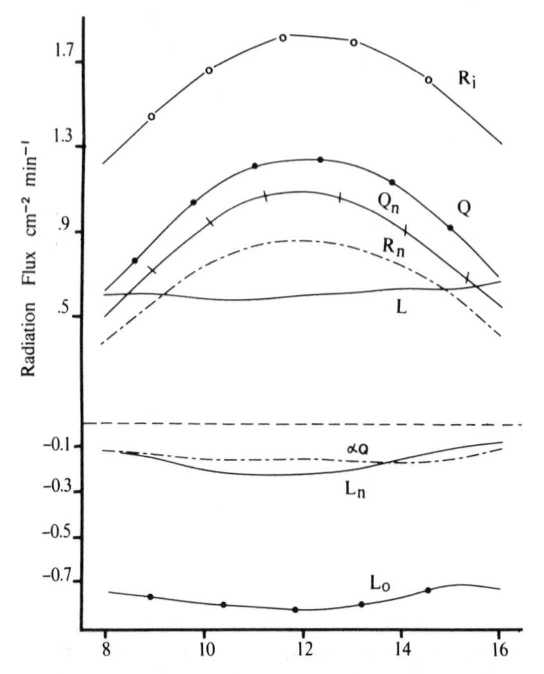

Figure 1.16 Radiation balance of a mature 'McIntosh' apple tree: Q, global short-wave radiation; Q_n, net short-wave radiation; αQ, reflected short-wave radiation; R_i, total global radiation; R_n, net global radiation; L, incoming long-wave radiation; L_n, net long-wave radiation; L_o, outgoing long-wave radiation. (Reproduced by permission from Proctor et al., 1972.)

for major light-energy-related processes is about 30% of the available solar radiation. With $Q_s = 800\ \mathrm{W\,m^{-2}}$, net radiation in orchards is estimated as $Q_n = 560\ \mathrm{W\,m^{-3}}$ and energy utilized by the leaves as $Q_l = 280\ \mathrm{W\,m^{-2}}$.

PHOTOSYNTHETIC POTENTIAL

Structure of the Leaf

The structure of the leaf of fruit trees is typical of a C3 plant. However, the structure of the leaf varies with location, whether located on the shoot or on a spur, and with exposure to light. It is difficult to describe leaves of all fruit tree species; therefore, apple will be used as an example. In general, leaf thickness increases from the base of the shoot toward the apex (Table 1.2). As the leaf position approaches the apex, the palisade cells become more elongate and the palisade layers become more com-

TABLE 1.2 Leaf Characteristics of Apple[a] 'McIntosh'

	Leaf Position from Base		
	3rd	8th	17th
Palisade thickness, (μm)	100	107	123
Spongy mesophyll thickness, (μm)	124	88	84
Intercellular space in mesophyll, %	28	21	14
Leaf thickness, (μm)	252	220	232
Leaf area, cm^2	12	36	25
Stomatal frequency, mm^{-2}	284	396	556

[a] Reproduced by permission from Cowart, 1936.

pact and well defined and occupy a greater percentage of the mesophyll tissue. The decreasing percentage of the intercellular space with leaf position indicates the relative density of the leaf. Stomatal frequency increases with leaf position toward the apex (Table 1.2). However, stomatal frequency is also characteristic of the cultivar. For example, 'Northern Spy' has almost 30% higher stomatal frequency than 'McIntosh', 'Cortland', and 'Delicious.'

The general leaf structure of apple is illustrated in Figure 1.17. The leaf structure also changes with exposure to light and type of wood on which the leaf is located (Table 1.3).

As leaves develop in shade, their chlorophyll content increases (Table 1.4; dw = dry weight). This is especially important in peach, where the LAI is high and many of the leaves are shaded.

Chloroplasts of shade-grown leaves have thicker grana than sun-grown counterparts (Skene, 1974). Grana of shade-grown leaves do not change when the leaf is exposed to sun. In contrast, grana of sun-grown leaves are enlarged when the leaf is exposed to shade. Photosynthetic responses of the shade leaves at low light levels are similar to those developed in full sun (Barden, 1977).

Another way to characterize leaves is by the *specific leaf weight* (SLW). The SLW is the dry weight per unit area. Several reports indicate the SLW is affected by shade (Barden, 1978). This is consistent whether leaves are grown in controlled studies or in the orchard. As shading increases, the SLW decreases (Barden, 1978). In general, as SLW increases the Pn rate also increases. This relationship is even closer in orchard conditions than in controlled studies.

Effect of Leaf Age on Photosynthesis

Maximum Pn rates rise rapidly during the expansion of young leaves of apple, and the peak rate occurs a few days after the leaves are fully

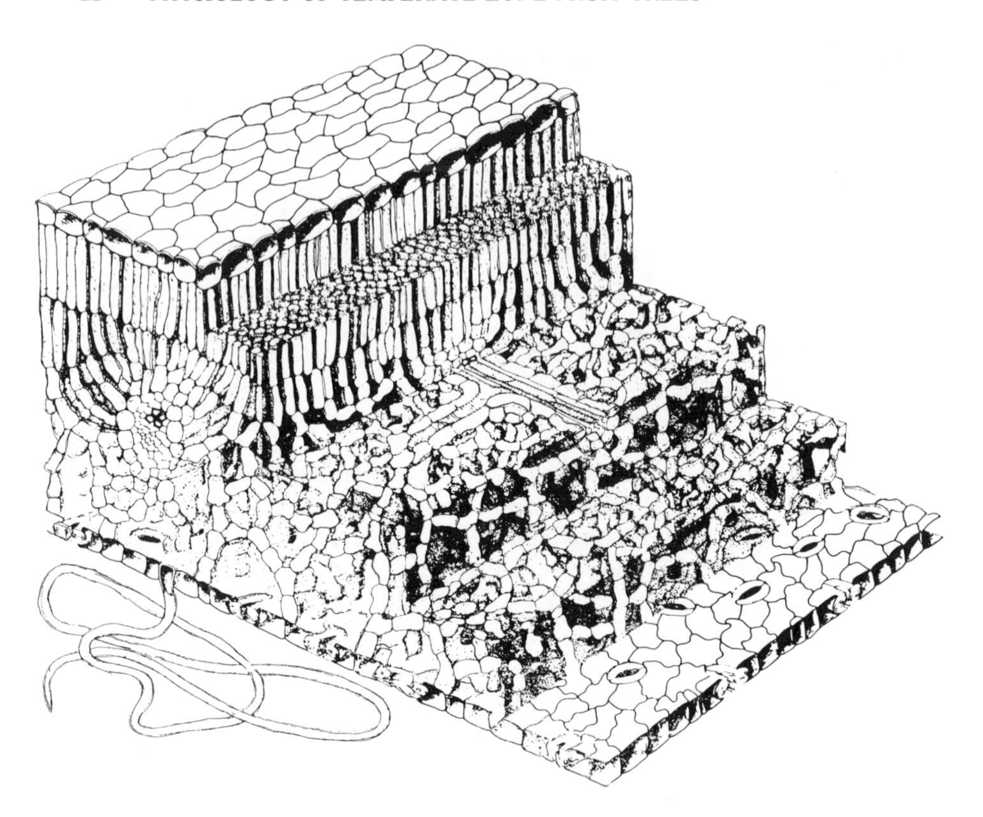

Figure 1.17 Leaf structure of an apple leaf. (Drawn by Rita Eames, Cornell University, 1943. Reproduced by permission.)

TABLE 1.3 Leaf Structure of Apples as Affected by Exposure and Location[a]

	Thickness (μm)		
	Leaf	Palisade	Spongy Mesophyll
Exposed shoot	204	93	86
Exposed nonfruiting spur	195	87	81
Exposed fruiting spur	189	85	79
Shaded nonfruiting spur	148	54	68

[a] Reproduced by permission from Cowart, 1936.

expanded (Barden, 1978; Kennedy and Fujii, 1986). The limited photosynthesis in the young leaves may be explained by immature stomata (Slack, 1974) in these leaves. The stomata mature about 6 weeks after emergence of the leaf. An alternative explanation for the low Pn in young leaves is the low N content of the expanding leaf. DeJong (1982) was able to show a direct correlation between N content and Pn.

TABLE 1.4 Effect of Shade on Chlorophyll Content of Peach Leaves[a]

Sunlight (%)	Average Leaf Area (cm^2 leaf)	Chlorophyll a ($mg\ dm^{-2}$)	Chlorophyll b ($mg\ dm^{-2}$)	Chlorophyll a ($\mu g\ mg^{-1}$ dw)	Chlorophyll b ($\mu g\ mg^{-1}$ dw)
100	29	2.8	1.3	2.6	1.2
36	34	4.9	1.7	6.1	2.2
21	38	4.8	2.2	7.1	3.4
9	35	5.2	2.5	12.1	5.7

[a] Reproduced by permission from Kappel and Flore, 1983.

Photosynthetic rates of the spur leaves of apple remain constant for a period of several weeks after the leaf is fully expanded and is followed by a downward trend when the leaf becomes senescent (Kennedy and Fujii, 1986). The downward trend in photosynthesis can be delayed if the trees are given additional N during early fall. Apparently the late N treatment delays leaf senescence and Pn rate remains higher (Heinicke, 1934). The annual senescence of apple leaves also can be stopped and the chlorophyll content preserved if the temperature is increased to 18°C during the day and 10°C at night (Lakso and Lenz, 1986).

The Pn rate of extension shoot leaves of apple remains higher than the Pn rate of spur leaves, and during August and September this rate can be three times higher than the spur leaf rate (Palmer, 1986).

Photosynthetic Potential of Fruit Tree Leaves

It is difficult to determine the maximum net photosynthetic potential for leaves of fruit tree species because the measurements to determine Pn were made at various conditions and the units used for reporting the results are also different.

Values were reported of 11–34.9 mg $CO_2\ dm^{-2}\ hr^{-1}$ for apples (Dozier and Barden, 1971; Ferree and Barden, 1971; Heinicke, 1966; Looney, 1968), 12.7–23.2 mg $CO_2\ dm^{-2}\ hr^{-1}$ for peach (Marini and Marini, 1983), 23.4 mg $CO_2\ dm^{-2}\ hr^{-1}$ for cherry, and 20.5 mg $CO_2\ dm^{-2}\ hr^{-1}$ for plum (DeJong, 1983). Leaves of spur-type apples trees have approximately 12% higher photosynthetic rate than their nonspur counterparts (Looney, 1968).

Effect of Light on Photosynthesis

The response of Pn to increasing irradiance is a hyperbolic response characteristic of C3 plants. Single-leaf photosynthesis is saturated approx-

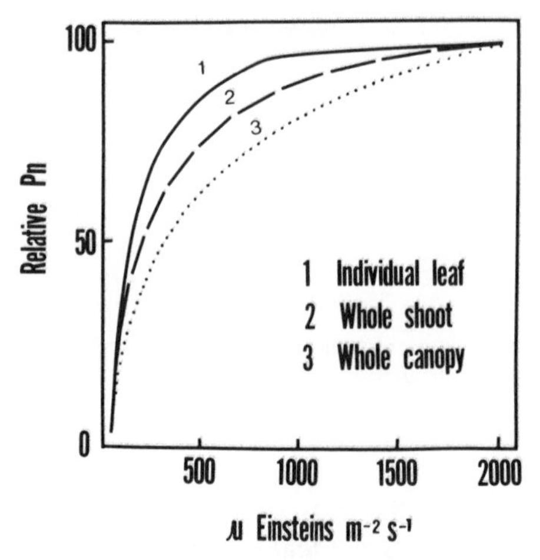

Figure 1.18 Light saturation of Pn in apple. Individual leaf, whole shoot, and entire canopy sun levels. (Reproduced by permission from Lakso, 1986.)

imately at 20–40% of full sunlight (Lakso, 1986). Light saturation is somewhat lower for a group of leaves and approaches full sunlight when the entire canopy is considered (Figure 1.18) (Lakso, 1986; Lakso and Seeley, 1978). In general, photosynthesis is saturated between 400 and 600 $\mu E\, m^{-2} s^{-1}$ in individual apple leaves (Landsberg et al., 1975). In peach, cherry, almond, plum, and apricot this value is slightly higher (Figure 1.19) and is between 400 and 700 $\mu E\, m^{-2} s^{-1}$ (DeJong, 1983; Marini and Marini, 1983). On a clear day with high total incident radiation whole-tree photosynthesis may be lower than on hazy or partially cloudy days, which give higher levels of diffuse light penetrating into the canopy (Lakso and Musselman, 1976). Leaves within the interior of the canopy are often exposed to a short period of PAR as the sun moves and sunflecks penetrate the canopy. Lakso and Barnes (1978) were able to show that apple leaf photosynthesis responds to changing irradiance in 5–15 s. High and low light alternating in 1-s cycles produces 85–95% of Pn rates of continuous high level of light. This suggests that interior leaves could be relatively efficient or at least instantaneously respond to incoming light when a sunfleck strikes them.

Stomatal Control of Photosynthesis

Because photosynthetic rate is an important factor in determining fruit yield, it is important to identify those components that contribute most to the control rate of photosynthesis. Since stomata play a central role in gas exchange in plants, it is important to evaluate stomatal versus nonstoma-

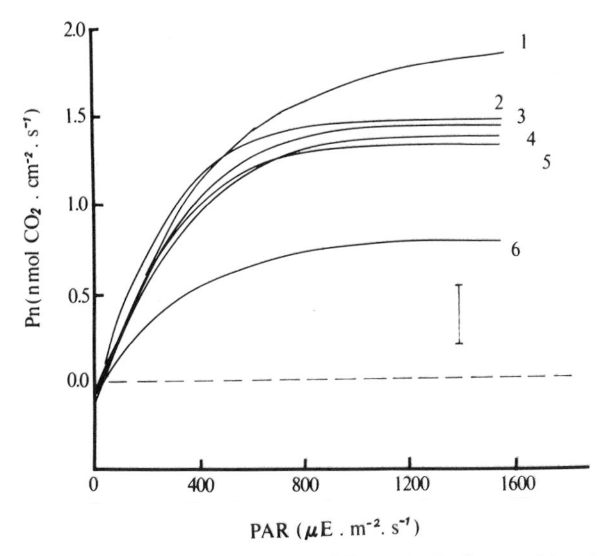

Figure 1.19 Relationship between net CO_2 assimilation and level of PAR for several fruit species. (1) Almond; (2) Cherry; (3) Apple; (4) Peach; (5) Plum; (6) Apricot. (Reproduced by permission from DeJong, 1983, and Proctor et al., 1976.)

tal limitations of photosynthesis in order to identify those processes where improvements can be made.

Stomatal conductance is under a variety of controls that interact. The maintenance of turgor is essential for stomatal opening. It can be maintained by a variety of processes discussed in Chapter 3. Stomata are also influenced by diurnal patterns and internal CO_2 concentration, both discussed in other sections of this chapter.

The similarity in stomatal behavior and photosynthetic rates suggests a limiting role of stomatal control on photosynthesis. Lakso (1986) calculated stomatal limitations of photosynthesis in fruit trees. His calculations indicate that in low-stress conditions stomata constitute only about 20–30% of the limitation to photosynthesis (Table 1.5).

The limitation was calculated by Lakso (1986) by using Farquhar and Sharkey's (1982) equation $I = A_0 - A/A_0$, where I is the limitation, A is the measured photosynthetic rate, and A_0 is the calculated rate at infinite stomatal conductance (zero resistance).

Jones (1986) claimed that the estimation of stomatal limitation based on the preceding equation is arbitrary and attributes rather little importance to the stomata. He proposed a method that does not involve extrapolation to unnatural conditions. This method involves a comparison of slopes of the photosynthetic supply-and-demand functions at the actual CO_2 concentration and stomatal conductance.

TABLE 1.5 Stomatal Limitation to Photosynthesis in Fruit Tree Leaves

Species	Experiment	Source	Stomatal Limitation
Plum	Nitrogen	DeJong, 1983	0.30
Peach	Nitrogen	DeJong, 1983	0.27
Cherry	Nitrogen	DeJong, 1983	0.28
Apricot	Nitrogen	DeJong, 1983	0.31
	VPD[a] stress low	Schulze et al. 1972	0.18
	VPD stress high	Schulze et al. 1972	0.26
Peach	Soil water		
	Stress low	Tan and Buttery, 1982	0.34
	Stress high	Tan and Buttery, 1982	0.57
Apple	Soil water		
	Stress		
	Low	Lakso, 1986	0.28
	High	Lakso, 1986	0.30
	Ringing		
	Control	Lakso, 1986	0.31
	Ringed	Lakso, 1986	0.40

[a] VPD = water pressure deficit.

Diurnal Variation in Net Photosynthesis

Diurnal variation has been observed in several fruit tree species. The maximum photosynthetic rates tend to occur in the morning with lower rates occurring at equivalent PAR levels in the afternoon. Lakso and Seeley (1978) reviewed the possible reasons for diurnal variation. The low water potential, high VPD, and high temperature, all occurring in the afternoon, are expected to close stomata and decrease Pn. Yet Pn rates do not follow this trend. In apricot in desert climate the stomatal opening greatly depends on the water deficit of the surrounding air (Schulze et al., 1972). Sweet cherries do not show higher Pn midday depression in stage III or stage II of fruit growth (Roper et al., 1986). Apples have shown a slight midday depression in Pn in the orchard (Landsberg et al., 1975) but not in a closed-flow assimilation system with constant temperature and humidity (Lakso and Seeley, 1978). A typical diurnal response in the field maximum stomatal opening (conductance) occurring in the midmorning followed by a gradual decline in the afternoon can be interpreted as a humidity response. However, evaluation of diurnal patterns of stomatal behavior of leaves differing in their time of exposure to light showed a poor correlation with VPD, but stomatal opening was well correlated with time elapsed after exposure to full light (Lakso, 1986).

Effect of Carbon Dioxide Concentration on Photosynthesis

The photosynthesis of fruit trees responds positively to increased CO_2 concentrations. The linear response of CO_2 assimilation to increasing internal CO_2 concentrations between 50 and 250 μl liter^{-1} is typical of C3 species (DeJong, 1983). The CO_2 compensation point for almond, plum, peach, cherry, and apricot has been reported to be between 55 and 65 μl liter^{-1} and is not significantly different among these species. However, for apple the compensation point has been determined as 20 μl liter^{-1} (Watson et al., 1978), indicating a higher rate of efficiency. All the stone fruit species listed above maintained intercellular CO_2 concentrations near 230–260 μl liter^{-1} when ambient CO_2 concentration was 320 μl liter^{-1} and water, light, and temperature were not limiting. For C3 leaves the CO_2 assimilation response usually begins to level off at intercellular concentrations above 250–270 μl liter^{-1}. Controlling CO_2 concentrations at 230–260 μl liter^{-1} maximizes the use of leaf carboxylation potential and minimizes water loss while maintaining reasonable rates of CO_2 assimilation (DeJong, 1983). Apricot leaves tend to maintain lower CO_2 concentrations. DeJong (1983) expressed the opinion that this may indicate the adaptation of this plant to arid conditions. Mesophyll conductance of apricot (0.09 cm s^{-1}) is considerably lower than the other stone fruit species (0.18–0.25 cm s^{-1}).

The mechanism that couples stomatal conductance to photosynthesis appears to be the stomatal sensitivity to CO_2, especially internal CO_2. Field leaves of apple show a tight coupling of stomata to photosynthesis, whereas greenhouse-grown trees show less tight coupling. Lakso (1986) interpreted this by high sensitivity to internal CO_2 in the field-grown leaves. Sensitivity of stomata of greenhouse-grown trees can be increased by exposure to evaporative stress for a few days (Lakso, 1986).

A close correlation between stomatal conductance and mesophyll conductance occurs in fruit trees. Lakso (1986) reported a stomatal conductance – mesophyll conductance ratio of 1.5–2 when both values were calculated on the CO_2 basis. A similar close correlation was found in stone fruits.

The importance of CO_2 control on stomatal opening is that it can modify stomatal response to VPD. If stomata respond strongly to internal CO_2, they tend not to open more than a given conductance to maintain a certain level of internal CO_2. Therefore, the natural opening response of stomata to low VPD would be overridden by the internal CO_2 control. Lakso (1986) recalculated West and Graff's (1976) data relating transpiration of apple leaves to VPD in ambient and CO_2-free air. Estimations of stomatal conductance show a full humidity response of the stomata in CO_2-free air, but in ambient CO_2 concentration the opening was inhibited at low VPDs (Figure 1.20).

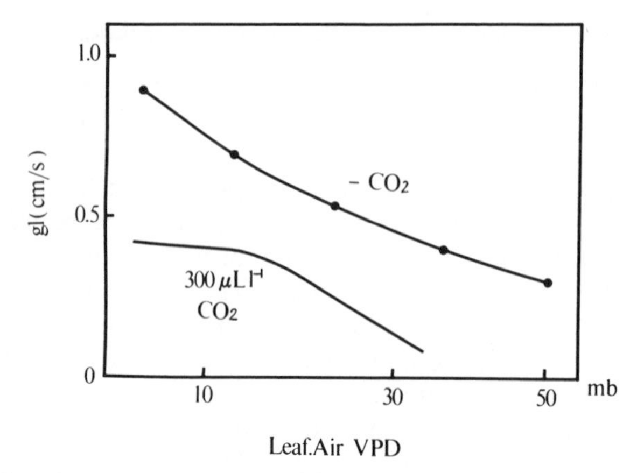

Figure 1.20 Stomatal opening in CO_2-free and CO_2-containing air. (Reproduced by permission from Lakso, 1986.)

Effect of Leaf Nitrogen on Photosynthesis

Leaf N content appears to be a major factor determining the photosynthetic rate per unit leaf area of the leaf. A good linear correlation has been obtained in peach (DeJong, 1982) and other stone fruit species (DeJong, 1983) between leaf N content and leaf photosynthetic rate. The higher the leaf N content is, the higher the leaf photosynthetic rate is. Detailed analysis revealed that photosynthetic assimilation rates expressed on a leaf area basis, mesophyll conductance, and ratio of leaf conductance to water vapor were all linearly related to leaf N content (R_2 = 0.908, 0.921, 0.685, respectively).

Among the *Prunus* species Pn rates for peach, plum, and cherry were all very similar on a leaf area, N, and water use basis. In comparison the Pn of almond was higher on the leaf area basis, the same on the leaf N basis, and lower on the water use basis than the other species. Opposite differences were found with apricot. In apricot the Pn rate was lower than the other species on the leaf area and leaf N basis, but it was the same on the water use basis (Table 1.6).

The Pn rates based on N in leaves exposed to low light intensities were the same in peach regardless of N status. However, Pn rates based on N were at least 30% higher in high-N-fertilized trees than in leaves of trees not receiving N fertilization (DeJong, 1986). Although high rate of N fertilization may increase Pn rate, the lush growth resulting from high-N fertilization increases shading within the tree. The shoot growth on high-N trees also increases. Both of these factors decrease the available carbohydrates for the activities of the tree connected to fruit production. Thus N nutrition would have to be used in moderation and certainly not for the sole purpose of increasing Pn rates.

TABLE 1.6 Photosynthetic Rates Expressed on Various Bases[a]

Species	Leaf Area Basis ($nmol\ CO_2\ cm^{-2}\ s^{-1}$)	Nitrogen Basis ($nmol\ CO_2\ mg^{-1}\ N\ s^{-1}$)	Water Use Basis ($nmol\ CO_2\ \mu g^{-1}\ H_2O$)
Plum, peach, cherry	1.33–1.37	6.17–6.75	0.19–0.21
Almond	1.80	6.47	0.16
Apricot	0.67	3.76	0.19

[a] Reproduced by permission from DeJong, 1983.

Effect of Temperature on Photosynthesis

The photosynthesis of apple leaves responds to temperature, the maximum being between 20 and 30°C under saturating PAR conditions. A decline in Pn rates occurs above 30°C. Generally, field-grown foliage shows the same temperature response as experimental material grown in controlled conditions. Above 35°C, the Pn rate of apple decreases very rapidly, especially at high irradiance levels (Lakso and Seeley, 1978; Landsberg et al., 1975). Apple trees assimilate substantial amounts of CO_2 even at high temperatures. At 40°C, 20 mg CO_2 dm^{-2} hr^{-1} Pn rates were recorded at high irradiance in leaves of well-watered apple trees (Lakso and Seeley, 1978). Heinicke and Childers (1937) reported that the mean temperature of 28°C was associated with a decrease of Pn rates. They worked with an entire caged tree in the orchard. Thus their values are the closest to the response of an entire tree in orchard conditions.

Temperature below the optimum also result in decreased Pn. In addition to the immediate effect of low temperature, leaf temperatures below −1.3°C result in a physiological shock with decreased Pn rates up to 2 weeks after exposure (Seeley and Kammereck, 1977). This is important when thinning sprays are applied. Thinning sprays are usually more effective under poor photosynthetic conditions. Applications of thinning sprays after a frost are well known to have increased effectiveness. It is very likely that decreased levels of Pn after frost are responsible for this.

High temperature directly inhibits photosynthesis by stimulating photorespiration in apple and pear (Tormann, 1978). There is a correlation between light levels and temperature optimum for photosynthesis. If light changes to a higher level, it also establishes a higher temperature optimum (Seeley and Kammereck, 1977).

Temperature can influence photosynthesis indirectly through a variety of effects. Temperature effect is rarely independent of VPDs, which can influence Pn by influencing stomatal opening. Temperature influences dark respiration rates, thus influencing photosynthetic productivity. Temperature may have an important effect on fruit set, and fruit number directly influences Pn as an increased sink effect. During the fall, decreasing Pn in aging leaves could be halted by maintaining a higher temperature regime for the trees (Lakso and Lenz, 1986).

PHOTOSYNTHETIC PRODUCTIVITY

Photosynthetic Efficiency

Photosynthetic efficiency can be determined in many ways. The most useful method for fruit trees is the determination of the dry-weight increment in relation to leaf area.

Using growth analysis data and dividing the total dry-weight increments by leaf area, Proctor et al. (1976) obtained 1.07 and 0.62 kg m^{-2} for fruit-carrying and defruited apple trees for the growing season. This is very close to values obtained for apple by Avery (1969), who obtained values of 0.81 and 0.60 for fruiting and defruited trees, but higher than those obtained by Forshey et al. (1983), who obtained a value of 0.40 kg m^{-2}. The productivity of peaches appears to be lower. Miller (1986) determined values of 0.33 and 0.27 kg m^{-2} for unthinned and thinned peach trees.

The maximum photosynthetic efficiency is determined by the presence of sinks, which is the most important factor. Priestley (1986) recalculated Magness's data from 1931, which clearly indicated that at a high leaf–fruit ratio (30) the dry-weight production was 0.63–0.94 kgm^{-2}. At the same time leaves with a low leaf–fruit ratio (20) produced more dry matter, 0.94–1.46 kg m^{-2}. Thus photosynthetic efficiency must be determined in conjunction with determining sink size.

Effect of Fruit on Photosynthesis

Heavy fruit load in apple trees can strongly reduce leaf area as compared to those having no fruit, but the total dry matter at harvest is usually higher in fruiting trees (Hansen, 1971; Lenz, 1986b). This indicates a higher photosynthetic efficiency of leaf or Pn by the fruit themselves. Early studies provided two lines of investigation that indicate increased photosynthetic efficiency in the presence of the fruit. Productivity studies indicated that carbohydrates transported into the fruit were more than the suppression of growth induced by fruiting (Avery, 1969; Chandler and Heinicke, 1926). Direct determinations using $^{14}CO_2$ provided evidence that the fruit acts as a strong sink and causes a very efficient transfer of photosynthates from the leaves to the fruit. Hansen (1982) estimated the net assimilation rate at varying fruit–leaf ratios. As leaf area increased from 200 to 800 cm^2 per fruit, the net assimilation rate decreased by about 40% and the leaves accumulated starch with a concomitant increase in SLW. The response of leaf to the pressure of the fruit as sink is rapid. Removal of fruitlets of apple reduced translocation from the surrounding leaves within 24–48 h (Hansen, 1973). The sink strength of the fruit is not uniform during the growing period. The growth of stone fruit has three distinct phases. The fruit effect on Pn occurs during the third, or final, swell period, when most of the carbohydrates accumulate in the fruit (Chalmers et al., 1975; Crews et al., 1975). The fruit may affect the leaf Pn up to 45 cm away in peach (Crews et al., 1975).

Lenz (1986a) summarized existing data on the photosynthesis of attached apple fruit. Carbon dioxide fixation (photosynthesis) during light exposure was high enough in the fruit to reach the compensation point in

'Golden Delicious' but not in 'Cox Orange' apples. Consequently, it is obvious that the fruit Pn can compensate for dark respiration in the light especially during the early growing period of the fruit but not in the dark or during the later stages of fruit growth when the stomata are not functional. The fact that apple fruit is able to fix CO_2 by dark fixation through phosphoenol pyruvate (PEP) carbohydrates complicates the measurements (Blanke, 1985) and requires a careful sorting of CO_2 fixation due to dark fixation or photosynthesis.

Leaf Area Requirement for Productivity

From the preceding discussion it is clear that the photosynthetic capacity is quite responsive to the demand for photosynthate in other parts of the tree. Hansen and Stoyanor (1972) estimated that 200 cm^2 of leaf area is needed to grow a 100-g [fresh-weight (fw)] apple. An additional 75 cm^2 leaf area is needed for each increment of 25-g fruit. Using labeled C, Hansen and Stoyanor were able to show that with 300 cm^2 leaf area per fruit, 55–75% of C was transported into the fruit within the same branch. When the leaf area increased to 500 cm^2 per fruit, only 35–45% of C was transported into the fruit, and the remaining portion was in the woody part of the tree. It was mentioned before that the presence of fruit limits shoot growth. A leaf area twice the size that required to saturate the fruit is needed to allow growth elsewhere in the tree.

Changing Nature of Sink Strength

In some cases sink strength changes or a source may become a sink in its early development or later during the growing season. It is also possible that the source changes as the growing season advances. Johnson and Lakso (1986) estimated that apple shoots early in their life act as sinks and import carbohydrates from the reserves of the tree for at least 3 weeks. Shoots that eventually grow to only 2 cm become exporters of carbohydrates sooner than shoots whose final growth will be 50 cm (Figure 1.21).

Early in its life apple fruit depends completely on the Pn transported from the spur leaves. Partial or complete removal of spur leaves decreases or eliminates fruit set completely (Ferree and Palmer, 1982). The presence of a shoot on the spur is detrimental to fruit set. Since the shoot is an importer of carbohydrates in its early days, it is not surprising that it successfully competes with the fruit for carbohydrates. The availability of carbohydrates soon after fruit set is also crucial for peach fruit set. Inhibiting Pn by herbicide sprays is an effective way to thin peaches (Barden et al., 1986). Removal of spur leaves 55–117 days after petal fall has no effect on eventual fruit size. However, shading of shoot leaves

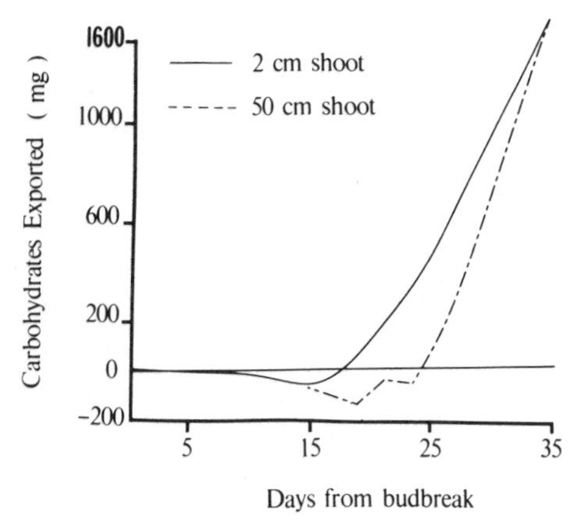

Figure 1.21 Carbohydrate import and export by shoots reaching short (2 cm) and long (50 cm) final length. (Reproduced by permission from Johnson and Lakso, 1986.)

later in the season reduces the size of the fruit (Rom, 1985; Rom and Ferree, 1986). This is further collaborated by ringing experiments that prevented carbohydrate transport from the shoot leaves but not from the spur leaves to the fruit. Such ringing did not influence fruit set but decreased fruit size (Ferree and Palmer, 1982). It appears that later in the season a substantial quantity of carbohydrates are provided by the shoot leaves to fruit growth and quality. This further explains the observation of Ferree and Palmer (1982) that the presence of spur shoot is detrimental to fruit set but results in a higher yield through its effect on fruit size.

Sorbitol Metabolism

Sorbitol has a unique role as an end product of photosynthesis in most temperate zone fruit trees. Sorbitol, a polyol, is a six-carbon alcohol. Plants of the Rosaceae family, where many of the fruit trees belong, produce sorbitol. It is estimated that up to 30% of all primary production goes through polyols rather than sugars (Bieleski, 1982). In fruit trees, of course, the percentage of sorbitol production is much higher. Labeling studies in apple revealed that in the ethanol soluble fraction of the leaves 80% of the ^{14}C activity was in sorbitol and only 17% in sucrose, 1.5% in glucose, and 1.2% in fructose (Bieleski and Redgewell, 1977). Plum leaves fix about 33% of CO_2 as sorbitol (Anderson et al., 1959, 1961).

Polyols are synthetized in the cytoplasm much like sugars (Rumpho et al., 1983). Models on synthesis of sorbitol in fruit are incomplete, but it is assumed that all polyols are synthetized in the same manner. Glucose-P is reduced to sorbitol-P by an enzyme, aldose-6-P-reductase (A6PR) and the P is split from sorbitol-P by a phosphatase. The reduction step requires nicotinamide adenine dinucleotide phosphate (NADPH). It is believed that NADPH is generated in the chloroplast along with adenosine triphosphate (ATP) and subsequently utilized by a shuttle system to carry it out from the chloroplast envelope. It is generally agreed that neither NADPH nor ATP can penetrate the chloroplast envelope. Therefore, the shuttle system is the only possibility that can carry the reducing power and energy from the chloroplast to the cytoplasm where sorbitol synthesis takes place. The shuttle system involves the conversion of 3-P-glycerate to glyceraldehyde-3-P, which is transported from the chloroplast. When glyceraldehyde-3-P is in the cytoplasm, it is converted by a nonreversible enzyme, glyceraldehyde-3-P dehydrogenase, to 3-P-glycerate regaining the reducing power in the form of NADPH (Kelly and Gibbs, 1973), which is in turn available to reduce glucose-6-P to sorbitol-6-P. To complete the shuttle, 3-P-glycerate is retransported into the chloroplast to be converted again using ATP and NADPH to glyceraldehyde-3-P. The scheme is represented in Figure 1.22.

The existence of A6PR has been demonstrated in apple, peach, pear, and apricot (Negm and Loescher, 1981). The specific phosphatases were shown in apple leaves (Grant and ApRees, 1981). The involvement of the shuttle mechanism has only been speculated by Rumpho et al. (1983) based on data obtained for spinach (Kelly and Gibbs, 1973). The involvement of glucose in sorbitol synthesis has been confirmed by tracer studies. When ^{14}C glucose is presented to the leaves of apple (Priestley and Murphy, 1980), pear (Bieleski, 1976, 1977), or plum (Anderson et al., 1961), most of the label appears in sorbitol. The reverse reaction apparently does not take place. Thus sorbitol emerges as an end product that is ready for transport from mature leaves of fruit trees.

Several types of information indicate that sorbitol is the major translocating substance from the leaves in fruit trees:

(1) There is a concentration gradient of sorbitol from source to sink, that is, from the leaves to the roots, throughout the season (Chong, 1971); the sorbitol concentration is the greatest in the leaves, and it is the least in the fruits;
(2) Sorbitol is the major carbohydrate in the conducting tissues of the apple (Hansen, 1970), and it is very responsive to sink demand (Hansen and Graslund, 1978);
(3) Apple phloem would load sorbitol, but sorbitol concentrations need ot be high to overcome the preference of phloem loading to sucrose (Bieleski, 1969). Labeling data indicate (Hansen, 1967a,b) that sorbitol concentration in leaves is high, thus this prerequisite to phloem loading does exist in fruit tree leaves;
(4) Once in the phloem sorbitol is not metabolized (Bieleski, 1969).

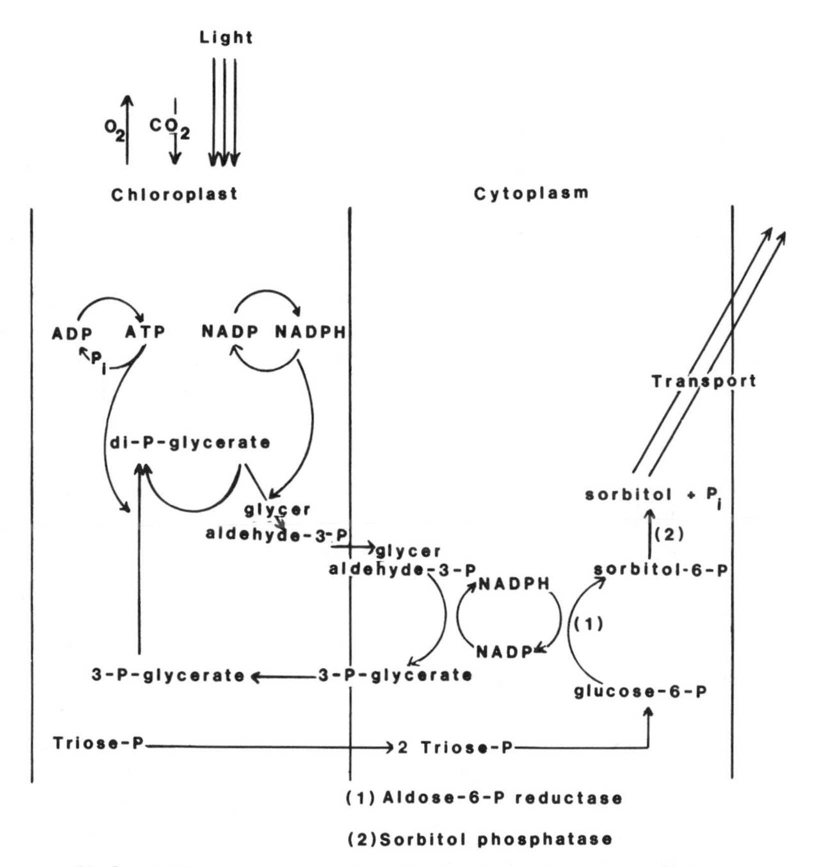

Figure 1.22 Production of sorbitol, schematic representation.

It comprises 65–80% of the soluble carbohydrates in the phloem of apple (Bieleski, 1969; Hansen, 1970, 1975; Priestley, 1980). Sorbitol is a major translocating substance in apricot (Reid and Bieleski, 1974), plum (Villiers et al., 1974) and peach (Manolow et al., 1977). In contrast, transport from the roots and the trunk during spring usually involves sucrose (Brown et al., 1985).

In the receiving sink tissues sorbitol is metabolized by an NAD-dependent sorbitol dehydrogenase (SDH) that oxidizes sorbitol to fructose (Negm and Loescher, 1979). This enzyme has been isolated from a variety of the photosynthetic tissues including apple and pear callus tissues (Negm and Loescher, 1979), apple fruits (Marlow and Loescher, 1980; Yamaki, 1980), and immature leaves that are still importers of carbohydrates (Loescher et al., 1982; Negm and Loescher, 1981). The enzyme data is consistent with labeling data in which sorbitol was converted primarily to fructose (Hansen, 1970). Plum fruit also readily metabolizes sorbitol (Villiers, 1978), but apricot fruit does not seem to have this ability (Reid and Bieleski, 1974).

Loescher et al. (1982) recognized that as the young leaves develop, they possess the SDH enzyme until they are importers of carbohydrates. The activity of this enzyme slowly decreases as the photosynthetic capability develops and the leaves undergo sink-to-source transition. With the increasing photosynthetic capability the A6PR enzyme activity involved in producing sorbitol develops. During autumn, when photosynthesis decreases again, the reverse process takes place: A6PR decreases and SDH increases. The apparent exception is the plum, for which sorbitol metabolism activity remains in the leaves throughout the season (Villiers 1978; Villiers et al., 1974). Whether the metabolism proceeds through SDH in plum leaves is not known.

Sorbitol metabolism has considerable interest in apple fruit. Certain cultivars such as 'Delicious' accumulate sorbitol in the intercellular spaces in advanced stages of maturity. Because of the high osmoticum in the intercellular spaces, they retain water and the surrounding tissues become anaerobic during storage and eventually break down. The condition is called *watercore* because most often the core tissue is the one most affected. However, in some cultivars wateriness occurs under the skin and earlier in the season. Labeling studies (Williams, 1966) indicated that conversion of sorbitol to fructose is impaired in the fruit disposed to watercore. However, the SDH activity is very similar in fruit susceptible and resistant to watercore (Marlow, 1982). In fact, SDH activity generally increases in apple when in some cultivars watercore develops (Marlow and Loescher, 1982). Thus a sudden decrease in SDH activity cannot explain the development of watercore. We must note, however, that SDH activity has been measured in extracts with added cofactors, whereas labeling studies were done in situ. Thus the two types of information are not directly comparable.

Calcium and the maturation process have central roles in the development of watercore. Calcium prevents whereas maturation enhances watercore development. Calcium is known to preserve and maturation to destroy the membrane integrity of apple cells. Thus a relationship between membrane integrity and the development of this disorder may exist. No other fruits have this disorder, yet membrane changes are essentially the same in all fruits. Thus the cause of watercore remains largely unknown.

Most environmental data involving sorbitol is available only in connection to watercore development. Temperature seems to play a major role in the development of watercore. The fruit temperature seems to make more difference than the level of radiation (reviewed by Marlow and Loescher, 1982). From the data available one cannot conclude that sorbitol levels are higher when temperature (i.e., radiation level) is higher.

TABLE 1.7 Dark Respiration Rates of Various Tissues of Apple Trees

	Respiration ($\mu g\, CO_2\, hr^{-1}$)	
Tissue	grams/dry weight	grams/fresh weight
Leaf	930	366
Stem	910	335
Fruit	75	10
Roots	394	352

[a] Reproduced by permission from Proctor et al. (1976).

However, it has been speculated that polyol synthesis could present a mechanism for dissipating reduction and possibly reduce the need for photorespiration (Fox et al., 1986).

Dark Respiration

Dark respiration of tissues of fruit trees is very important and a major determining factor in tree size and yield. Proctor et al. (1976) estimated the dark respiration rates of component tissues of apple. On both the dry- and fresh-weight bases the dark respiration of fruit was low; respiration of the leaves, stems, and roots were several magnitudes higher (Table 1.7).

Dark respiration can quite often utilize a substantial part of the Pn produced during the light period. Respiration rates of 1–1.6 $mg\, CO_2\, dm^{-2}\, hr^{-1}$ was reported for normal conditions (Dosier and Barden, 1971; Ferree and Barden, 1971) for leaf tissues. The effect of night temperature on dark respiration is very important. Proctor et al. (1976) determined the effect of temperature (Figure 1.23) on the dark respiration of leaves. Dark respiration at about 15°C is around 1 $mg\, CO_2\, dm^{-2}\, hr^{-1}$, which can increase at 25°C to 3.8 $mg\, CO_2\, dm^{-2}\, hr^{-1}$. Good apple-growing areas have cold nights either because the climate is generally cold or because the climate is dry and the nights cool. In contrast, in climates with warm nights tree growth and crop size are usually limited but the fruit size is normal. This could be easily explained by the high rate of night respiration. Under these circumstances the carbohydrate supply is usually not enough for tree growth and flower bud formation (thus the tree is small and crops are low), but carbohydrates are sufficient to satisfy the growth of the fruit (thus fruit size is normal).

The overall rate of dark respiration may require close to half of all carbohydrates produced by photosynthesis. Schulze et al. (1977) estimated the overall yearly carbohydrate loss due to respiration in apricot in a cool-night climate. The total use of carbohydrates in a nonirrigated apricot tree for dark respiration was 23 kg C of 57 kg C fixed; this was similar to the irrigated tree, which respired 32 kg C of the total 75 kg C

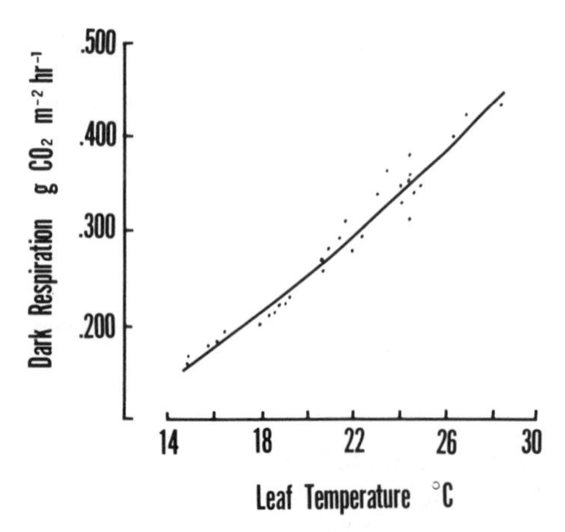

Figure 1.23 Dark respiration of leaf tissue of apple at various temperatures. (Reproduced by permission from Proctor et al., 1976.)

fixed. Although the irrigated tree fixed more carbon compared to the nonirrigated tree, the proportions used for dark respiration (42 vs. 40%) were very similar (Figure 1.24).

Carbohydrate Utilization in the Tree

The relationship between sources and sinks plays a dominant role in carbohydrate utilization in fruit trees. *Source strength* may be defined as the capacity of assimilatory tissues to produce Pn for export and *sink strength* as the potential capacity of tissues to accumulate or utilize metabolites. It is possible that a source is converted to a sink during the season (e.g., fruit) and a sink becomes a source (leaf). The major aim of horticulturists is to influence both sink and source for the advantage of producing sustained yield. In general, the sink strength of the growing limb-twig structure is discouraged and the sink strength of the fruit is encouraged.

Most studies are concerned with the partitioning of carbohydrates between fruit, new growth, leaves, and roots. In general, the sink strength is from high to low: fruit, shoot, root. Landsberg (1980) outlined a simple model that defines the dry-weight gain and partitioning of an apple tree.

$$\frac{\Delta W}{\Delta t} = \frac{\Delta W}{\Delta t}(P_L + P_S + P_R + P_F)$$

where ΔW is defined as dry-weight gain as photosynthates (P) in leaves (P_L) plus that in shoots (P_S) plus in that roots (P_R) plus that in fruits

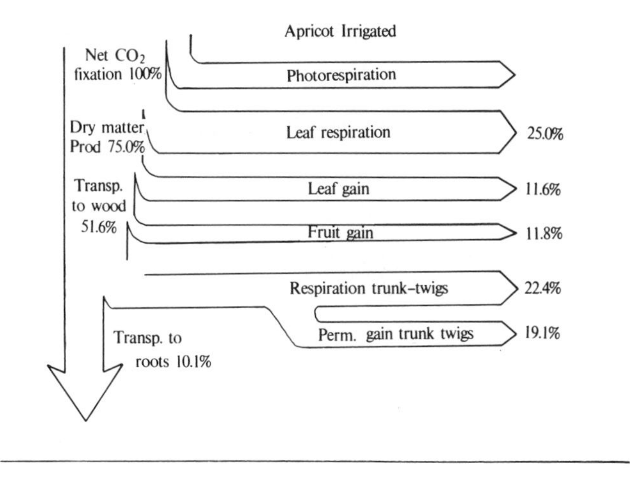

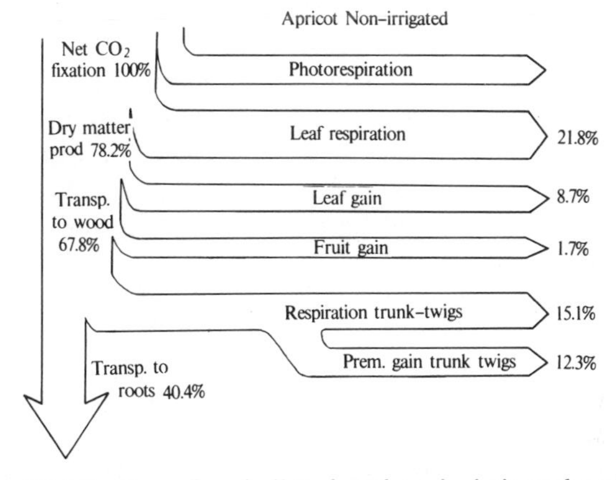

Figure 1.24 Distribution of assimilated carbon in irrigated and nonirrigated apricot. (Reproduced by permission from Schulze et al., 1977.)

(P_F), which equals 1. According to this model, any increase in cropping must reduce vegetative growth correspondingly. There are data advanced later that support such a simple relationship. However, the actual process is more complex and can only be interpreted using a more complex model (Jackson, 1983). As has been discussed previously, cropping often increases Pn efficiency. Although such increase does not compensate for the increased cropping, it complicates the preceding relationship. Heavy cropping often has a memory effect in the tree, and growth is often limited during the year that follows a heavy crop even though in this year the crop itself may be light. This especially occurs in young trees. Such trees are called 'runts'. Growers are careful not to prematurely fruit trees too heavily.

TABLE 1.8 Carbohydrate Distribution in Fruiting and Nonfruiting Apples[a]

Organ	Carbohydrate Distribution (%)		Fruiting in Previous Year Also (%)
	Nonfruiting	Fruiting	
Leaf	26.8	10.0	7.2
Fruit	0.0	62.5	67.2
Wood	53.7	22.5	20.8
Root	19.5	5.0	3.6
Leaf area, m^2	4.90	3.42	2.48
Biomass production, kg	2.82	4.65	4.14

[a] Recalculated after Lenz, 1986.

Fruit growth is generally increased at the expense of root growth. In apricots, when irrigation increased the distribution of Pn into the fruit, the distribution into the roots decreased from 40 to 10% of the total (Schulze et al., 1977) (Figure 1.24). However, irrigation may have a direct effect on decreasing root growth in addition to the fruit effect. In apple heavy cropping effectively stopped root growth of a dwarfing rootstock (Avery, 1970). Hansen (1980), reviewing his own work, summarized that cropping reduced the dry weight of the vegetative part of the apple tree by 30% over nonfruiting trees, but it virtually stopped root growth.

Fruiting not only reduces root growth, but if Pn is limiting, it can also reduce shoot growth. It has been established that fruiting decreases vegetative growth in apple (Avery, 1970; Maggs, 1963; Rogers and Booth, 1964) and in peach (Chalmers and van den Ende, 1975; and Miller, 1986; Proebsting, 1958).

Lenz (1986b) has clearly shown that fruiting in one year decreases carbohydrate transport to the vegetative part of the tree, and this is further increased by sustained fruiting on the more-than-one-year basis in Golden Delicious apples on M9 rootstock (Table 1.8).

Although it is clear from the table that the total biomass production is higher in fruiting trees with lower leaf area, Barlow (1964, 1975) estimated that the increased Pn due to the crop could only compensate for the carbohydrate needs of a very light crop and at higher levels of cropping there was a linear negative relationship between cropping and vegetative growth. Miller (1986), working with thinned and unthinned peaches, found that unthinned trees produced more total dry matter and the increase over the thinned trees was in the fruit itself. Total carbohydrates in the vegetative part of the tree was about the same in both thinned and unthinned trees. On a percentage basis thinned trees allocated 20% more to vegetative growth than did unthinned trees. It is clear that the number of fruit and not fruit size (thinned trees had larger fruit) made the difference in allocating carbohydrates.

Although a certain leaf area seems to be needed to transport carbohydrates to the wood, other experiments indicate that maintaining reserves in the bark is an important activity for the tree. It is generally assumed that the nonphotosynthetic tissues of the frame of the tree are dependent upon the availability of 'surplus' resources. Among the nonphotosynthetic tissues, bark and wood seem to have higher sink power than roots. Roots are particularly influenced by the availability of carbohydrates. Roots are affected the most by fruiting, shading, or bark ringing. Nitrogen applications first increase shoot growth, and this reduces root growth. However, enhanced shoot growth may bring about greater feedback of carbohydrate resources to encourage root growth also (Priestley, 1970). Roots appear to be the weakest sink. Root growth occurs only when shoot growth ceases (see section on root growth), and heavily cropped trees usually have very weakly developed root system. Thus in apple the relative sink strength is fruit, shoot, and root. This is not necessarily so in all fruit tree species. In cherries fruit does not seem to have a high sink strength (Flore, 1986), and consequently a high rate of cropping does not slow the growth of cherry trees. Fruit is thinned to produce a large leaf–fruit ratio and assure large fruit size among other advantages. Hansen (1982) surveyed real orchard situations and compared several sets of apple orchards that were well pruned, had good light penetration, and were in good general vigor with orchards comprised of old, dense trees in low vigor. Thus he compared orchards with high source and sink activity with orchards of low source and sink activity. Fruit size drastically decreased in orchards of high source and sink activity as production increased from 20 to 30 tons/ha. In comparison only in the low-source-, low-sink-activity orchards was a slight decrease in fruit size observed with a yield increase from 10 to 30 tons/ha. However, the overall fruit size in the latter orchards was much smaller than in the former type (Hansen, 1982).

Large variations exist within the tree, especially in fruit size. Within the tree local differences in the overall leaf–fruit ratio is not a major factor. Source activity, as determined by light exposure, may be a reason for the within-the-tree variation in fruit size only in larger trees, where fruits from well-exposed parts of the tree attain larger sizes (Hamm and Lenz, 1980; Jackson, 1967). When growth rate (fruit size) and composition of individual fruit are examined, the relationship tends to be negative, and larger fruit has a lower concentration of percentage of total dry matter (Hansen, 1982). Recalculation of Hansen's data revealed that although the concentration of total dry matter in larger fruit is lower, the overall total dry matter per fruit in large fruit is much higher than in small fruit. The overall increase in total dry matter almost keeps up with the increase in fruit size measured as weight. This indicates that proportionally more water moved into larger fruit than dry matter, which further indicates that the fruit-to-fruit variation in growth rate may be caused by differing water relations the tree establishes for individual fruits rather than variation in source–sink relationships within the tree.

The fact that fruit growth greatly depends on water movement into the fruit is well illustrated by comparing the fresh- and dry-weight growth curves of peaches. The fruit serves as a sink for dry-weight accumulation but during the two fast-growth periods in stages I and III the fresh-weight increase is much faster than the dry-weight accumulation (Lott, 1942). Miller's (1986) work clearly indicated that thinned peach trees with larger fruit size produced less carbohydrates than unthinned peach trees with smaller fruit. This emphasizes the importance of water relations as the determining factor in fruit size. Thus large fruit size does not necessarily depend only on ample Pn.

Effect of Leaf Injury on Photosynthesis

Injury to the leaf caused by diseases or insects has the potential to decrease Pn rate and cause economic damage. Ferree (1978) and Ferree et al. (1986) summarized the effect of various pests on Pn of apples. Among the most important diseases powdery mildew (*Podosphera leucotricha*) caused a 75% decrease in Pn at 35 days after infection. The change was gradual as infection developed. Leaves treated with fungicide recovered somewhat 16 days after infection but still had only 50% Pn rate of the healthy leaves 35 days after infection. The effect of apple scab (*Venturia inaequalis*) is much milder on Pn rates. The decrease is only about 20% for the same time interval. However, apple scab is a defoliating disease, and loss of foliage can very seriously impair the photosynthetic capacity of the tree itself, especially during the summer when Pn are needed for fruit growth as well as flower bud differentiation.

Simulated removal of part of the leaf blade does not affect Pn until more than 7.5% of the leaf is removed, and more than 15% has to be removed before the photosynthetic capacity on the remaining tissue is reduced. Reduction in Pn caused by physical damage may recover in about 7 days. Any injury that severs the main vein causes significant reduction in Pn.

Mites could cause considerable damage by decreasing Pn. As the population of two spotted spider mites (*Tetranychus urticae*) increases, Pn correspondingly decrease. A population of 60 mites per leaf caused a significant reduction in apple Pn 3 days after placement on the leaf. At 9 days 15 mites per leaf reduced Pn by 26%, 30 mites per leaf by 30%, and 60 mites per leaf by 43% below the uninfested levels. Mite damage to Pn is important regardless of when it occurs, but it has a special importance on spur leaves early in the season when carbohydrates are needed for fruit set and any decrease in Pn could severely reduce the number of fruit. Apple cultivars react to mite infestation differently. Reduction ranged from 0 to 20% among the eight cultivars tested. The reduction in Pn due to mite infestation does not correlate with the anatomical future of the leaf.

As early as 1933, Hoffman was able to show that lime S reduced the photosynthesis of individual leaves. Heinicke (1937) determined the Pn of an entire apple tree. He found that for about 5 days after the lime S spray the Pn were reduced by 50%. After this period a slow recovery occurred, and 15 days after the original spray the photosynthesis was similar to the original rate. After a second spraying a lesser reduction occurred for 5 days followed by a similar recovery period. Among the classical pesticides, Bordeaux mixture and oil reduced the Pn of fruit tree leaves significantly. The reduction caused by Bordeaux mixture was less than oil or lime S (Hyre, 1939). Ferree (1978) summarized the effect of the newer organic fungicides and insecticides on Pn. Since the effects of these compounds vary and some of the chemicals are discontinued, interested persons should consult the review.

Wax-based antitranspirants reduce Pn in peaches (Davenport et al., 1971) and in apples (Weller and Ferree, 1978). Certain nutrient sprays affect Pn. Potassium (K^+) sprays limited Pn severely soon after the spray was applied with a recovery in 7 days. Urea sprays had a mild effect with a rapid recovery, and $Ca(NO_3)_2$ sprays had no effect at all (Faust et al., 1986). The nature of the original shock, resulting in reduced Pn, is not known. It is clear, however, that when recovery is possible, it usually occurs in about 7 days.

Effect of Tree Size on Production Efficiency

In general, large trees have a lower production efficiency than small trees. Forshey and McKay (1970) compared carbohydrate distribution in a dwarf apple tree (on M7 rootstock with 3.1 m height) with a large apple tree (on seedling rootstock with 8.4 m height). The small tree distributed its carbohydrates to the fruit, leaves, and wood 76.0, 8.4, and 14.4%, respectively, whereas in the large tree the percentages were 45.0, 13.9, and 40.9. The large difference is in the production of wood. Similar data were obtained for dwarf (short-internode) peaches when their productivity was compared to standard peach trees (DeJong and Doyle, 1984). The dry-matter distributions for fruit, leaves, and wood for the dwarf tree were 27.8, 16.7, and 55.3%, whereas the distributions for the standard tree were 21.1, 13.5, and 65.3%, respectively. Here, again, the largest difference in the above-ground distribution of carbohydrates was in the wood versus fruit. In neither case was the root considered.

The effect of fruit on shoot growth and/or root growth is undisputable. Alternatively, if the tree is overly vigorous, its fruit production is limited.

TABLE 1.9 Distribution of Carbohydrates in Peach Trees

Location	Age of Tree (yr)	Total Carbohydrates (kg)	Fruit (%)	Leaves (%)	Wood (%)
California (De Jong 1984)	6	160.0	21.1	13.5	65.3
Maryland (Miller, 1986)	8	37.0	17.1	11.5	71.3

However, there are cases when the tree grows at a different rate yet the carbohydrate distribution is about the same. Peaches in California grow at a much faster rate than they do on the East Coast of the United States, yet the carbohydrate distribution is about the same when percentages are considered (Table 1.9). Trees of similar age in California can be four to seven times larger than on the East Coast of the United States. Respiration may account for a large part of this difference.

Relationship between Photosynthesis and Yield

It is difficult to establish a direct relationship between Pn and yield. Circumstantial evidence detailed above would tend to support that there is a direct relationship between yield and Pn. However, Flore and Sams (1986) eloquently argue, based on a sour cherry model, that when considering whether photosynthesis is limiting or not, a distinction should be made between photosynthetic rate (CO_2 fixed per unit area) and total carbon fixed, which also takes into account the leaf area and the leaf area duration. According to them, photosynthesis could limit growth of the crop during the critical growth stage of the fruit (stage 3 for stone fruits), if there is severe defoliation due to insect or diseases, or if environmental conditions are not conducive for optimum photosynthesis. They concluded that in most cases photosynthetic capacity is large enough in cherries to provide carbohydrates even for relatively large crops, and photosynthesis limits yield in only a few cases when crop loads are high and/or when stress occurs during stage 3 of fruit growth.

The relative importance of photosynthesis compared with other components of yield is illustrated in Figure 1.25. Chang et al. (1987) compiled these data. The figure illustrates that both fruit number and fruit weight are important. The involvement of Pn in fruit weight is obvious. Fruit number is usually determined by the previous-year Pn. During the previous year carbohydrates are needed to be high enough that in addition to supporting fruit and tree growth enough carbon is available for the tree to develop sufficient number of flower buds and reserves in the wood. During the spring reserves are needed for a high fruit set, and a high fruit number would assure this component for the high yield. Naturally there

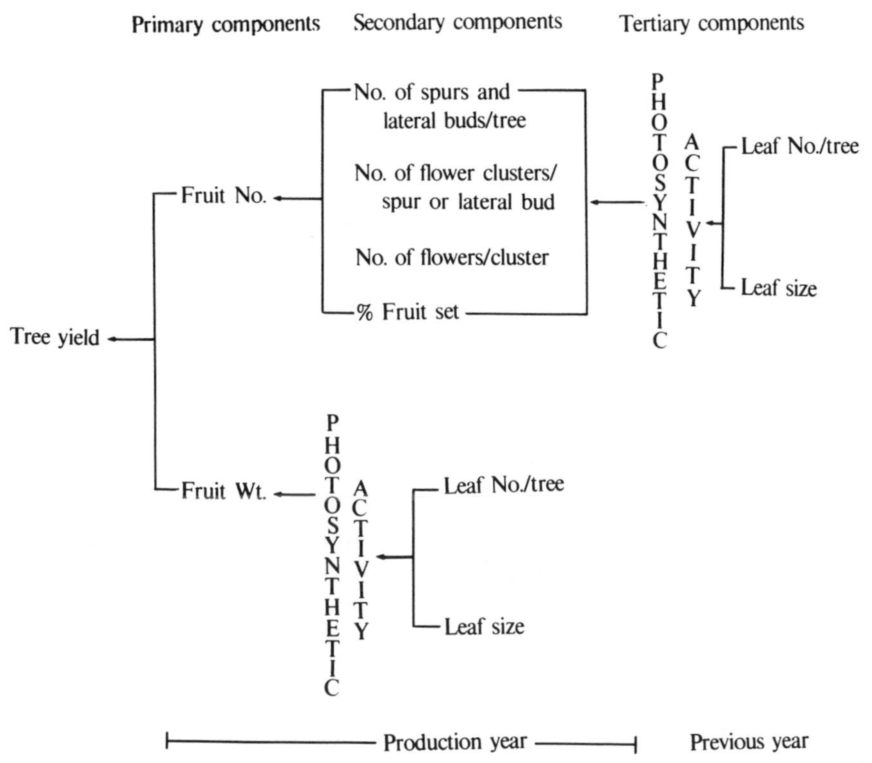

Figure 1.25 Components of yield in cherry. (Modified after Chang et al., 1987.)

are many other factors involved in determining the fruit set. At this point we only attempt to point out the involvement of Pn in yield that indirectly involves fruit number.

Cherry is a small fruit and Pn are limited only when the leaf–fruit ratio is less than 1.5 (Flore and Sams, 1986). In fruit crops such as apples where the fruit is much larger and contains a higher portion of carbohydrates produced by the leaves, carbohydrate shortages develop if the tree has 15 leaves or less per fruit. Therefore it is easier to get into stress situations in a fruit crop for Pn when the crop uses a large percentage of its carbohydrate supply for fruit growth.

For generalization Flore (1986) stated that the Pn potential in fruit crops is under two forms of control: (1) the environment, which influences the immediate physical and biochemical reactions directly and through light exposure the morphological development of the leaf indirectly, and (2) through sink demand and some type of feedback signal

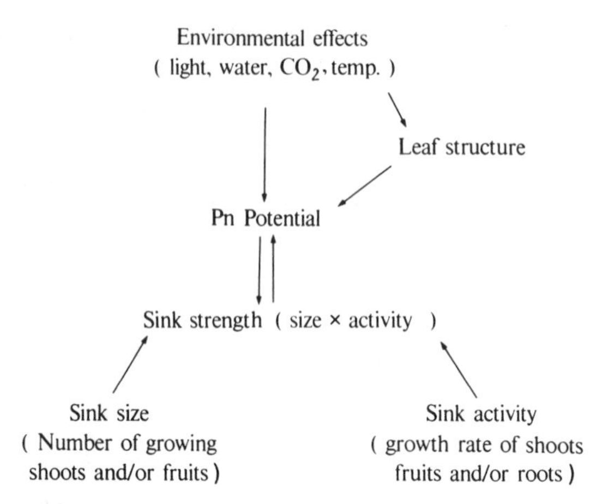

Figure 1.26 Schematic illustration of factors affecting net photosynthesis.

from the sink itself (Figure 1.26). He further emphasized that the Pn potential is seldom reached in fruit trees. Thus, when improvement of crop is considered, photosynthesis may be only one, and perhaps not the most important, of the factors considered.

REFERENCES

Anderson, J. D., P. Andrews, and L. Hough. 1959. J. Biochem. 72:9–10.

Anderson, J. D., P. Andrews, and L. Hough. 1961. Biochem. J. 81:149–154.

Avery, D. J. 1969. New Phyt. 68:323–336.

Avery, D. J. 1970. New Phyt. 69:19–30.

Barden, J. A. 1977. J. Amer. Soc. Hort. Sci. 102:391–394.

Barden, J. A. 1978. Hort. Sci. 13:644–646.

Barden, J. A., M. C. Marini, and R. E. Byers. 1986. Regulation of photosynthesis in fruit trees. Int. workshop, Davis, CA, pp. 45–48.

Barlow, H. W. B. 1964. Rpt. East Malling Res. Sta. 1963:84–93.

Barlow, H. W. B. 1969. Rpt. East Malling Res. Sta. 1968:117.

Barlow, H. W. B. 1975. Climate and the orchard. Slough, England: Farnham Royal, pp. 98–102.

Bieleski, R. L. 1969. Austr. J. Biol. Sci. 22:611–620.

Bieleski, R. L. 1976. *In* Transport and transfer processes in plant, I. T. Wardlaw and J. B. Passioura (eds.). New York: Academic Press, pp. 185–190.

Bieleski, R. L. 1977. Austr. J. Plant Phys. 4:11–24.

Bieleski, R. L. 1982. *In* Encyclopedia of plant physiology. New Series, Vol. 13A, F. A. Loewus, W. Tanner (eds.). New York: Springer Verlag, pp. 158–192.

Bieleski, R. L., and R. J. Redgewell. 1977. Austr. J. Plant Phys. 4:1–10.

Blanke, M. 1985. Ph.D. Dissertation University of Bonn, Germany.

Brown, C. S., E. Young, and D. M. Phatt. 1985. J. Amer. Soc. Hort. Sci. 110:696–701.

Cain, J. C. 1973. J. Amer. Soc. Hort. Sci. 98:357.

Chalmers, D. J., R. L. Canterford, P. H. Jerie, T. R. Jones, and T. D. Ugalde. 1975. Austr. J. Plant. Phys. 2:635–645.

Chalmers, D. J., K. A. Olsson, and T. R. Jones. 1983. *In* Water deficit and plant growth, Vol. 7, T. T. Kozlowski (ed.). New York: Academic Press.

Chalmers, D. J. and B. van den Ende. 1975. Ann Bot. 39:423–432.

Chandler, W. H., and A. J. Heinicke. 1926. Proc. Amer. Soc. Hort. Sci. 23: 36–46.

Chang, L. S., A. F. Iezzoni, and J. A. Flore. 1987. J. Amer. Soc. Hort. Sci. 112:247–251.

Chong, C. 1971. Canad. J. Plant Sci. 51:519–525.

Cowart, F. F. 1936. Proc. Amer. Soc. Hort. Sci. 33:145–148.

Crews, C. E., S. L. Williams and H. M. Vines. 1975. Planta 126:97–104.

Davenport, D. C., M. A. Fisher, and R. M. Hagan. 1971. Plant Physiol. 47:38.

DeJong, T. M. 1982. J. Amer. Soc. Hort. Sci. 107:955–959.

DeJong, T. N. 1983. J. Amer. Soc. Hort. Sci. 108:303–307.

DeJong, T. M. 1986. Regulation of photosynthesis in fruit crops. Int. workshop, Davis, CA, pp. 59–64.

DeJong, T. M., and J. F. Doyle. 1984. Acta Hort. 146:89–95.

Dozier, W. A., and J. A. Barden. 1971. J. Amer. Soc. Hort. Sci. 96:786–788.

Farquhar, G. D. and T. D. Sharkey. 1982. Ann. Rev. Plant. Physiol. 33:317–345.

Faust, M., D. Swietlik, and R. F. Korcak. 1986. The regulation of photosynthesis in fruit trees. A. N. Lakso and F. Lenz (eds.). Symp. Proc. Publ. N.Y. State Agr. Exp. Sta., Geneva, NY, pp. 50–52.

Ferree, D. C. 1978. Hort. Science 13:650–652.

Ferree, D. C., F. R. Hall, and M. A. Ellis. 1986. The regulation of photosynthesis in fruit trees. A. N. Lakso and F. Lenz (eds.). Symp. Proc. Publ. N.Y. State Agr. Exp. Sta., Geneva, NY, pp. 56–61.

Ferree, D. C., and J. W. Palmer. 1982. J. Amer. Soc. Hort. Sci. 107:1182–1186.

Ferree, M. E., and J. A. Barden. 1971. J. Amer. Soc. Hort. Sci. 96:453–457.

Flore, J. A. 1986. Regulation of photosynthesis in fruit crops. Int. workshop, Davis, CA, pp. 30–32.

Flore, J. A., and C. E. Sams. 1986. The regulation of photosynthesis in fruit trees. A. N. Lakso and F. Lenz (eds.). Symp. Proc. Publ. N.Y. State Agr. Exp. Sta., Geneva, N.Y, pp. 105–110.

Forshey, C. G., and M. W. McKay. 1970. Hort. Sci. 5:164–165.

Forshey, C. G., R. W. Weirs, B. H. Stanley, and R. C. Seem. 1983. J. Amer. Soc. Hort. Sci. 108:149–154.

Fox, T. C., R. A. Kennedy, and W. H. Loescher. 1986. Plant Phys. 82:307–311.

Grant, C. R., and T. ApRees. 1981. Physochemistry 20:1505–1511.

Gyuro, F. 1974. A gyömölestermeles alapjai (The basis of fruit production), Budapest: Mezogazdasagi Kiado.

Hamm, L., and I. Lenz. 1980. Erwerbsobstbau 22:126–129.

Hansen, P. 1967a. Phys. Plant. 20:373–391.

Hansen, P. 1967b. Phys. Plant. 20:1103–1111.

Hansen, P. 1970. Phys. Plant. 23:564–573.

Hansen, P. 1971. Physiol. Plant. 25:469–473.

Hansen, P. 1973. Acta Agric. Scand. 23:87–92.

Hansen, P. 1975. Tidsskr. Planteavl. 79:133–170.

Hansen, P. 1980. Mineral nutrition of fruit trees. D. Atkinson, J. E. Jackson, R. O. Sharples, and W. M. Waller (eds.). London: Butterworths, pp. 201–212.

Hansen, P. 1982. Proc. Int. Hort. Cong. 1:257–268.

Hansen, P., and J. Graslund. 1978. Phys. Plant. 42:129–133.

Hansen, P., and S. Stoyanor. 1972. Tidsskr. Planteval 76:646–652.

Heinicke, A. J. 1934. Proc. Amer. Soc. Hort. Sci. 32:77–79.

Heinicke, A. J. 1937. Proc. Amer. Soc. Hort. Sci. 35:256–259.

Heinicke, A. J., and N. F. Childers. 1937. Cornell Univ. Exp. Sta. Mem. 201.

Heinicke, D. R. 1963. Proc. Amer. Soc. Hort. Sci., 83:1.

Heinicke, D. R. 1964. Proc. Amer. Soc. Hort. Sci., 85:33.

Heinicke, D. R. 1966. Proc. Amer. Soc. Hort. Sci., 89:10.

Hoffman, M. B. 1933. Proc. Amer. Soc. Hort. Sci. 29:389–398.

Holland, D. A. 1968. Rpt. East Malling Res. Sta. 1967:101.

Hyre, R. A. 1939. N.Y. State Agr. Exp. Sta. Memoir 222.

Jackson, J. E. 1967. Rpt. East Malling Res. Sta. 1966:110.

Jackson, J. E. 1970. J. Appl. Ecol. 7:207.

Jackson, J. E. 1978. Acta Hort., 65:61.

Jackson, J. E. 1980a. Hort. Rev. 2:208.

Jackson, J. E. 1980b. Acta Hort. 114:69.

Jackson, J. E. 1983. Acta Hort. 146:83–88.

Johnson, R. S., and A. N. Lakso. 1986. J. Amer. Soc. Hort. Sci. 111:160–164.

Jones, H. G. 1986. Regulation of photosynthesis in fruit crops. Int. workshop, Davis, CA, pp. 4–6.

Kappel, F., and J. A. Flore. 1983. J. Amer. Soc. Hort. Sci. 108:541–544.

Kelly, G. J., and M. Gibbs. 1973. Plant Phys. 52:674–676.

Kennedy, R. A., and JoAnn Fujii. 1986. The regulation of photosynthesis in fruit trees. A. N. Lakso and F. Lenz (eds.). Symp. Proc. Publ. N.Y. State Agr. Exp. Sta., Geneva, NY, PP. 27–29.

Lakso, A. N. 1975. New York's Food Life Sciences 8:5–8.

Lakso, A. N. 1984. J. Amer. Soc. Hort. Sci., 109:861.

Lakso, A. N. 1986. The regulation of photosynthesis in fruit trees. A. N. Lakso and F. Lenz (eds.). Symp. Proc. Publ. N.Y. State Agr. Exp. Sta., Geneva, NY, pp. 6–13.

Lakso, A. N., and J. E. Barnes. 1978. HortScience 13:473–474.

Lakso, A. N., and F. Lenz. 1986. The regulation of photosynthesis in fruit trees. A. N. Lakso and F. Lenz (eds.). Symp. Proc. Publ. N.Y. State Agr. Exp. Sta., Geneva, NY, pp. 34–37.

Lakso, A. N. and R. C. Musselman, 1976. J. Amer. Soc. Hort. Sci. 101:642.

Lakso, A. N., and E. J. Seeley. 1978. Hort. Sci. 13:646–650.

Landsberg, J. J. 1980. Opportunities for increasing crop yields. R. G. Hard, F. V. Biscoe, and C. Dennis (eds.). London: Pitman, pp. 161–180.

Landsberg, J. J., C. L. Beadle, P. V. Biscoe, D. R. Butler, B. Davidson, L. D. Incoll, G. B. James, P. G. Jarvis, P. J. Martin, R. E. Neilson, D. B. B. Powell, E. M. Slack, M. R. Thorpe, N. C. Turner, B. Warrit, and W. R. Watts. 1975. J. Appl. Ecol. 12:659–684.

Landsberg, J. J., and H. G. Jones. 1981. *In* Water deficit and plant growth, Vol. 6, T. T. Kozlowski (ed.). New York: Academic Press, p. 419.

Landsberg, J. J., D. B. B. Powell, and D. R. Butler. 1973. J. App. Ecol. 10:881.

Lenz, F. 1986a. Regulation of photosynthesis in fruit crops. Int. workshop, Davis, CA, pp. 92–95.

Lenz, F. 1986b. The regulation of photosynthesis in fruit trees. A. N. Lakso and F. Lenz (eds.). Symp. Proc. Publ. N.Y. State Agr. Exp. Sta., Geneva, NY, pp. 101–104.

Loescher, W., G. C. Marlow, and R. A. Kennedy. 1982. Plant Phys. 70:335–339.

Lott, R. V. 1942. Bull. Illinois Agr. Exp. Sta. 493:323–324.

Looney, N. E. 1968. Proc. Amer. Soc. Hort. Sci. 92:34–36.

Maggs, D. H. 1963, J. Hort. Sci. 38:119–128.

Manolow, P., N. Borichenko, and B. Rangelow. 1977. Gradinarska Lozarska Nauka. 14:45–51; Hort. Abstr. 49:280–1979.

Marini, R. P., and M. C. Marini. 1983. J. Amer. Soc. Hort. Sci. 108:609–613.

Marlow, G. C. 1982. M. S. Thesis. Washington State University, Pullman, WA.

Marlow, G. C., and W. Loescher. 1980. Hort. Sci. 15:426.

Marlow, G. C., and W. H. Loescher. 1982. HortReviews 6:188–251.

Miller, A. N. 1986. Ph.D. Dissertation. University of Maryland, College Park, MD.

Monteith, J. L. 1973. Principles of environmental physics. New York: Elsevier.

Negm, F. B., and W. Loescher. 1979. Plant Phys. 64:69–73.

Negm, F. B., and W. H. Loescher. 1981. Plant Phys. 67:139–142.

Oikawa, T. and T. Saeki. 1977. Bot. Mag. Tokyo 90:1.

Palmer, J. W. 1977. J. Appl. Ecol. 14:505.

Palmer, J. W. 1981. Acta Hort. 114:80.

Palmer, J. W. 1986. The regulation of photosynthesis in fruit trees. A. N. Lakso and F. Lenz (eds.) Symp. Proc. Publ. N.Y. State Agr. Exp. Sta., Geneva, NY, pp. 30–33.

Priestley, C. A. 1970. The physiology of tree crops. L. C. Luckvill and C. V. Cutting (eds.). London: Academic Press, pp. 113–127.

Priestley, C. A. 1980. Ann. Bot. 46:77–87.

Priestley, C. A. 1986. The regulation of photosynthesis in fruit trees. A. N. Lakso and F. Lenz (eds.). Symp. Proc. Publ. N.Y. State Agr. Exp. Sta., Geneva, NY, PP. 84–87.

Priestley, C. A., and H. E. Murphy. 1980. Ann. Rep. E. Malling Res. Stu. 1979:180.

Proctor, J. T. A. 1978. Hort. Sci. 13:641.

Proctor, J. T. A., W. J. Kyle, and J. A. Davies. 1972. Canad. J. Bot. 50:1731.

Proctor, J. T. A., R. L. Watson, and J. J. Landsberg. 1976. J. Amer. Soc. Hort. Sci. 101:579–582.

Proebsting, E. L., Jr. 1958. Proc. Amer. Soc. Hort Sci. 71:103–108.

Reid, M. S., and R. L. Bieleski. 1974. *In* Mechanism of regulation of plant growth. Bull. 12. R. L. Bieleski, A. R. Ferguson, and M. M. Cresswell (eds.). Wellington, N.Z.: Roy. Soc. New Zealand, pp. 823–830.

Rogers, W. S., and G. A. Booth. 1964. J. Hort. Sci. 39:61–65.

Rom, C. R. 1985. Ph.D. Dissertation. The Ohio State University, Columbus, OH.

Rom, C. R., and D. C. Ferree. 1985. J. Amer. Soc. Hort. Sci. 110:455–461.

Rom, C. R., and D. C. Ferree. 1986. J. Amer. Soc. Hort. Sci. 111:352–356.

Roper, T. R., and R. A. Kennedy. 1986. J. Amer. Soc. Hort. Sci. 111:938–941.

Roper, T. R., W. H. Loesher and C. R. Rom. 1986. Regulation of photosynthesis in fruit crops. Int. workshop, Davis, CA, pp. 24–29.

Rumpho, M. E., G. E. Edwards, and W. H. Loescher. 1983. Plant Phys. 73: 869–873.

Sams, C. E., and J. A. Flore. 1982. J. Amer. Soc. Hort. Sci. 107:339–344.

Sansavini, S. 1982. Proc. XXI. Int. Hort. Cong., Intern. Horticultural Society, Wageningen, 1:182.

Schulze, E. D., O. L. Lange, L. Kappen, and M. Evenari. 1977. Flora, 166: 383–414.

Schulze, E. D., O. L. Lange, U. Bushbom, L. Kappen, and M. Evenari. 1972. Planta 108:259–270.

Scorza, R. 1984. J. Amer. Soc. Hort. Sci. 109:455.

Seeley, E. J., and R. Kammereck. 1977. J. Amer. Soc. Hort. Sci. 102:731–733.

Slack, E. M. 1974. J. Hort. Sci. 49:95–105.

Skene, D. S. 1974. Proc. Roy. Soc. London, B 186:75–78.

Standhill, G., G. J. Hofstede, and J. D. Kalma. 1966. Q.J.R. Meteor Soc. 92:128.

Suckling, P. W., J. A. Davies, and J. T. A. Proctor. 1975. Canad. J. Bot. 53:1428.

Tan, C. S., and B. R. Buttery, 1982. Hortsci. 17:222–223.

Thormann, H. 1978. Agroplantae 10:51–55.

Thorpe, M. R. 1978. Agric. Meteorol. 19:41.

Verheij, E. M. V. 1972. Meded. Landbourohogeschool Wageningen, p. 72.

Verheij, E. M. V., and F. L. J. A. W. Verver. 1973. Sci. Hort. 1:25.

Villiers, O. T. 1978. South Afr. J. Sci. 74:342–343.

Villiers, O. T., J. T. de Meynhart, and J. A. de Bruyn. 1974. Agroplantar. 6:33–36.

Watson, R. L., J. J. Landsberg, and M. R. Thorpe. 1978. Plant. Cell. Env. 1:51–58.

Weller, S. C., and D. C. Ferree. 1978. J. Amer. Soc. Hort. Sci. 103:17–19.

West, D. W., and D. F. Graff. 1976. Phys. Plant. 38:98–104.

Williams, M. W. 1966. Proc. Amer. Soc. Hort. Sci. 88:67–75.

Yamaki, S. 1980. J. Japan Soc. Hort. Sci. 49:429–434.

2

NUTRITION OF FRUIT TREES

Orchard trees grow and produce in one place for 15–50 years. By the time they are of bearing age, they have extensive root systems and are able to exploit nutrients from the soil. The perennial nature of the tree and the fact that it bears heavy loads of fruit regularly imposes conditions not encountered in nutrition of herbaceous plants or forest trees.

Most nutrients are applied yearly to the surface of the soil. Some nutrients do not penetrate the soil easily and therefore must be applied before the tree is planted in a quantity that may supply the tree for most of its life.

During a large portion of the year it is not possible for the tree to take up nutrients from the soil. Yet during such periods it has certain nutrient needs in some of its developing organs, such as buds or young unfolding leaves. For these periods nutrient reserves need to be established within the tree.

The tree has changing nutrient needs during the year when it sets its fruit, it raises its fruits, its shoots are growing, and finally it fills its reserves. The occasional high nutrient needs may not be met by uptake through the roots and need to be supplied through foliar application.

Fruit tree nutrition often must be concerned with the nutrition of not only the whole tree but also individual organs. For example, fruits should have a certain nutrient content for maximum storability. This requires sufficiently different techniques than nutrition for maintaining the tree. First, nutrients should be applied to the tree and then the translocation and accumulation of such nutrients into the fruit must be understood and horticultural techniques applied to direct such nutrients into the fruit in quantities to attain the goal. Translocation of B and N into the young developing fruit for fruit set and Ca for good storage quality are prime examples of this.

Thus, in addition to supplying nutrients, the nutrition of fruit trees is concerned with all the principles of plant physiology and the application of a series of special techniques to attain sustained high yield and quality fruit.

DEVELOPMENT AND FUNCTION OF FRUIT TREE ROOTS

The root is a less visible but equally important part of the tree compared with the above-ground structures. Roots anchor the tree, absorb, transport, and occasionally store nutrients and water and synthesize compounds essential for regulation of above-ground activities of the tree. Because the root is not readily visible, it has received less attention in the past. Nevertheless it is a very important structure of the tree, and considerations involving the root need to be integrated with considerations deciding horticultural manipulations affecting the above-ground parts of the tree.

All fruit trees, when planted from seed, develop a taproot system having a central axis as the main downward extending center from which the various primary and secondary roots branch off. When trees are removed from the nursery, the taproot is usually cut by the digging process. Thus, orchard trees planted on seedling rootstocks depend on lateral roots. A considerable percentage of rootstocks for fruit trees are vegetatively propagated. These roots do not have a strong central structure as the main axis of the root. The primary side roots are about equal in size, and the entire root system is more horizontal. Root systems of fruit trees are extensively reviewed by Kolesnikov (1971), Papp and Tamasi (1979), Atkinson (1980), and Tamasi (1986).

Figure 2.1 Root distribution of 'Jonathan' apple trees planted in sandy soil at distances 7 × 4 m on sandy soil. Dotted line designates the edge of the canopy. (Reproduced by permission from Papp and Tamasi, 1979..)

TABLE 2.1 Size of Root System of Fruit Trees (m^2)[a]

Age of Tree	Apple	Pear	Plum	Apricot	Cherry	Sour Cherry
1	3.2	1.1	0.9	1.7	1.9	2.4
2	5.7	3.4	1.6	4.1	3.0	3.1
3	10.5	8.1	2.7	5.0	4.3	3.9
4	15.5	19.7	4.6	10.9	9.7	11.4
5	18.2	21.3	7.9	15.8	11.8	13.4

[a] Rootstocks were as follows: 'Jonathan, M4; pear, seedling; plum "Myrabolon"; apricot, seedling of apricot; cherry and sour cherry, Mahaleb.' From data of Papp and Tamasi, 1979.

Distribution of Roots of Fruit Trees

Roots of fruit trees are basically shallow scaffolds occupying the soil volume between 20 and 80 cm in depth with deep vertical roots descending to about 3 m. The spread of roots is generally much greater than that of the branches (Figure 2.1). Obviously the size of the root system greatly depends on the age and type of tree. Atkinson (1980) listed available data on the size of fruit tree root systems of many different trees. Generalizations are difficult because age, rootstock, soil type, and water supply all greatly influence the size of the root. Nevertheless, overall size of roots of various species of fruit trees are summarized in Table 2.1.

When a tree is removed from the nursery, its root system is usually cut back. During planting operations, the root is further pruned and requires time to regain its original proportion as far as the entire tree is concerned.

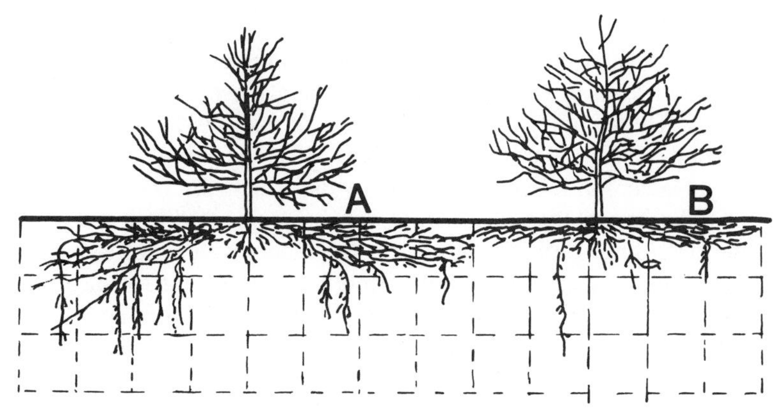

Figure 2.2 Root penetration of apple trees on sandy and clay soils: A, sandy; B, clay. (Reproduced by permission from Gyuro, 1974.)

Consequently, the root–top ratio of trees based on the weight of the root and branch system is different in young compared with older trees. For example, the root–shoot ratio of a Jonathan apple on a seedling rootstock is 1:2.5 at the age of 1 year; it changes to 1:1.9 to 1:1.4 between 2 to 5 years, and finally around 6 years of age a balance is reached between the roots and the above-ground portion by establishing a more or less permanent ratio of 1:1.2 (Papp and Tamasi, 1979). Even though the root system is quite large, the majority of roots are concentrated near the center. Atkinson reviewed cases where 36–82% of the root systems for apples and pears were found within the center 1 m^2 area of the trunk. Distribution of roots differed according to the size of the root. A higher proportion of roots less than 1 mm in diameter have been found at greater distances from the trunk.

Although roots have the potential to spread widely, they are often restricted by the presence of roots of neighboring trees. Intermixing of roots rarely occurs, even in high-density apple plantings (Atkinson, 1976) and at densities of 2000 trees/ha (Schultz, 1972). The depth to which apple roots penetrate varies. Atkinson (1980) gives the range of 0.4–8.6 m with 1–2 m as the most common range. The majority of roots occur in the range of 0–80 cm with about 70% at 0–30 cm depth. Soil aeration is often the determining factor in how deep the majority of roots penetrate (Figure 2.2). In general, while root penetration of fruit trees can be quite deep, the large portion of the roots are located close to the surface.

Root Density of Fruit Trees

The root density of fruit trees has important implications for utilization of water and nutrients. Root density can be expressed relative to the surface area (length of root per surface area of leaves, L_A = cm cm^{-2}) or relative to soil volume (length of root per volume of soil, L_V = cm cm^{-3}). Values for L_A range from 0.8 to 23.8 with 2–6 being the most common (Atkinson, 1980). These values are several orders of magnitude lower than those in Gramineae (100–4000) or herbs (52–310). Occasional L_A values for peach and pear have been recorded in the range for conifers, which is 68–126. The importance of low root density, expressed in low L_A values, is in the size of the root–soil interface. Replacing water that transpired by the plant initially comes from soil immediately adjacent to the root, and water in this zone is being replenished from the bulk of the soil. If the rate of withdrawal from the immediate root zone exceeds the rate of movement through the soil, the soil adjacent to the roots becomes dryer than the bulk of the soil, and the rate of water flow into the root decreases, creating a water stress. If root density is low, one can expect a high rate of water withdrawal from the immediate root zone in order to satisfy the transpirational demand of the tree. This creates high gradients at the root surface. Newman (1969) calculated that when L_A is less than 10, soil resistance at the root surface is larger than 0.2 MPa. As mentioned, fruit trees have low L_A values that influence the rate of nutrient uptake per unit root length, which must be high in order to satisfy the nutrient need of the tree. Atkinson and Wilson (1980) measured nutrient inflow into roots in water cultures and suggested an inflow of 0.56 pmol cm^{-1} s^{-1} for P and 8.5 pmol cm^{-1} s^{-1} for N for apple. These values are five to eight times larger than those Brewster and Tinker (1972) suggested for a number of species of plants. Thus, fruit tree roots can be characterized by relatively limited soil contact because of low L_A values and high inflow rates for nutrients. These require high nutrient availability in the soil; otherwise the trees may enter not only water stress but also nutrient stress very easily.

The relatively limited root system of fruit trees in comparison to other plants is not altered if the size and presence of root hairs is considered. Kolesnikov (1971) mentions that a 1-year-old apple tree may have 17 million root hairs with a total length of 3 km. This must be contrasted with a rye plant that may have 15 billion root hairs with a total length of 600 km. In general, Kolesnikov estimated root hairs to be between 300 and 500 mm^{-2} of primary apple root surface. The thickness of root hairs is also variable. Seedling roots of apples may have root hairs of 328 μm thick, whereas the thickness of root hairs of M9 rootstock ('Paradise IX') may only be 61 μm. The root hairs of apples are shorter than those of most plants. Root hairs of *Prunus* species are longer but tend to be irregular (Head, 1968: Rogers, 1939).

The contact between the root and the surrounding soil varies considerably. A large portion of apple roots visible in a root laboratory were either not in contact or only partially in contact with the soil. The reasons for this are varied. Soil fauna, particularly nonparasitic nematodes adjacent to the root surface, can greatly reduce the root–soil contact (Atkinson, 1980). Water stress can cause substantial root shrinkage, greatly reducing root–soil contact. Rogers (1968) indicated that a reduction of 50% in root diameter can occur, which leaves the root suspended in a root cavity. In contrast the root–soil contact can be improved if roots undergo secondary thickening (Head, 1968).

The absorption of water and mineral nutrients by plants is often assumed to occur through the younger parts of the root system through root tips and through areas with root hairs. In cherries, which seem to be typical of fruit trees, absorption of ^{32}P and water are equal by white and woody roots on the basis of surface area but higher in white roots on the basis of volume. Absorption into woody roots through injury points or through lenticels was discounted. Woody roots also absorbed ^{45}Ca and ^{86}Rb (Atkinson and Wilson, 1979). Since thicker, woody roots have better soil contact, they may have special importance under stress conditions in fruit trees.

Growth of Fruit Tree Roots

Young roots of a fruit tree are initially white and succulent with short root hairs. After 1–4 weeks the root begins to turn brown and the root hairs shrivel (Rogers, 1939). Browning occurs during a 2–3-week period during the summer, but it may take as long as 12 weeks during the winter (Head, 1966). The browning may spread in waves from older regions toward the tip. The browning of the cortex is followed by decay and disintegration. After loss of the cortex, thickening occurs in some roots, which become the perennial root system.

The maximum rate of root growth for apple (Rogers, 1939) and for cherry (Head, 1968) is about 1 cm day^{-1}. Roots grow mostly during the night (Head, 1965; Hilton and Khatamian, 1973).

Unlike the top of the tree, roots do not have rest periods. The beginning of root growth of fruit trees depends on temperature. Kolesnikov (1971) reported that in controlled studies root growth of apples started at 4–5°C, those of pears at 6–7°C, and those of apricots and peaches at 12°C. This corresponded to orchard observations that root growth started at approximately the time when the soil reached the above temperatures. Occasionally, slightly higher temperatures were reported for the beginning of growth. The root growth of apple has been reported to start at 6.2°C (Rogers, 1939). After its initiation, root growth follows an irregular course with periods of active growth alternating with less active periods. Some cultivars exhibit a growth peak in May–June, June–July, and September (Rogers, 1939).

Even though roots begin to grow at a given temperature and appear to have no rest period, apple roots have a reduced regeneration potential that corresponds to shoot dormancy but appears somewhat independent from it. This reduced root regeneration potential seems to be overcome by chilling applied to the root. Thus roots may begin to grow whenever temperature is right, but there is a large difference in the rate and amount of growth (both shoot and root) depending on the stage of dormancy (Young and Werner, 1984, 1985). While Young and Werner's (1984) data indicate that for maximum budbreak and new shoot and root growth in apple the rootstock must be chilled, in peach only the scion chilling is required, and chilling the rootstock delays budbreak and shoot growth.

Chilling of apple roots may enhance the production of some hormone. Application of 6-benzyladenine (BA) substituted for root chilling in apple trees that received only shoot chilling and resulted in budbreak equal to trees that had both the shoot and root chilled (Young and Werner, 1986). Because of the high budbreak potential of peach, this species may be higher in root-produced cytokinins and, therefore, does not require root chilling for maximum cytokinin production. There is no data available backing up this supposition.

The periodicity in root growth greatly depends on shoot growth and fruit load of the tree. The end of the initial peak in root growth usually corresponds with the beginning of active shoot growth, and the second peak starts after shoot growth ceases (Head, 1967; Rogers and Head, 1969). Quinlan (1965) suggested that the bimodal periodicity for root growth is due to competition between the shoots and roots for carbohydrate reserves. Several independent studies appear to corroborate the hypothesis that photosynthates are translocated to the roots only after the main period of shoot growth (Katzfuss, 1973; Priestley et al., 1976). Cropping of the tree further reduces root growth (Atkinson, 1977, 1980; Head, 1969).

In species that have weaker shoot growth, the period of root growth may be longer. Head (1967) reported only one long peak of root growth in plums extending from May to late July. Atkinson and Wilson (1980) observed even longer periods of root growth for plum and cherry rootstocks, Iglanov (1976) for apricot, and Head (1968) for pears. In all these species the longer period of root growth has been attributed to weaker shoot growth.

Growth regulators that decrease shoot growth generally increase root growth. Paclobutrazol, a gibberellin biosynthesis inhibitor, greatly reduces shoot growth of apple but induces the growth of roots. The newly formed roots are short and thick. When trees are resupplied with gibberellic acid by sprays, roots elongate again and grow in a normal fashion (Steffens, personal communication). In contrast, the root system under dwarf trees is considerably smaller than under trees of normal growth habit. This produces a shoot–root ratio near unity (Faust unpublished information).

Effect of Management Factors on Root Growth and Distribution

Irrigation Many reports are reviewed by Atkinson (1980) on the positive influence of irrigation on root growth. Irrigation (or the lack of it) has a twofold effect on root growth. First, as Richards and Cockroft (1975) suggest, soil drying determines the root growth of peaches in the surface soil. The implication is that this is a direct effect of dry soil on root growth. Second, low soil water content has an indirect effect: Trees do not receive sufficient water and thus a water stress develops and photosynthetic efficiency decreases. Under such conditions the roots do not receive sufficient carbohydrates and do not grow. It is notable that root growth depends essentially on a daily supply of carbohydrates for its activities. Lucic (1967) reported that reducing light intensity by 50% reduced root growth within 2–3 days in 1-year-old pear seedlings. Faust (1980) reported a similar decrease in Ca uptake within a few days after limiting carbohydrate supply to the roots. Uptake could be restored by feeding sucrose to the phloem. These studies indicate that despite the relatively extensive structure of the root, the root reserves cannot support root activity, including growth. In this respect photosynthesis and irrigation, which maintains a high rate of photosynthesis, are very important.

When trickle irrigation is used, only a portion of the root area is irrigated. Tree roots adapt to situations when only one part of the root is wetted. Black and Mitchell (1974) observed proliferation of roots of mature pear trees in the wetted zone when irrigation was converted to trickle irrigation. Similar results were reported for peach (Willoughby and Cockroft, 1974). The root system of a tree is restricted to the irrigated area under arid conditions. In humid regions, where trickle irrigation is used, some root growth would be anticipated in the unirrigated soil. Willoughby and Cockroft (1974) found four to five times the root concentrations under peach and apple trees in the wetted zone of trickle irrigation compared with areas of the root system receiving only natural precipitation. Direct evidence from apple root distribution studies suggests that the depth of the root system of the tree decreases under trickle irrigation (Goode et al., 1978).

Soil Management Management of soil has a definite effect on root development. Cocker (1959) studied the effect of cultivation and sod on root systems of apples. Under cultivation, roots growing above 12 cm are pruned annually. The sod produces a better soil structure and a more favorable soil–root interface, but the sod presents competition to the tree root for water and nutrients. The commercial solution is often a compromise: sod between rows and a herbicide strip in the row. The herbicide strip greatly increases the number of roots less than 1 mm in diameter (Atkinson and White, 1980; White and Holloway, 1967), but it has little effect on the distribution and depth of the roots (Atkinson and

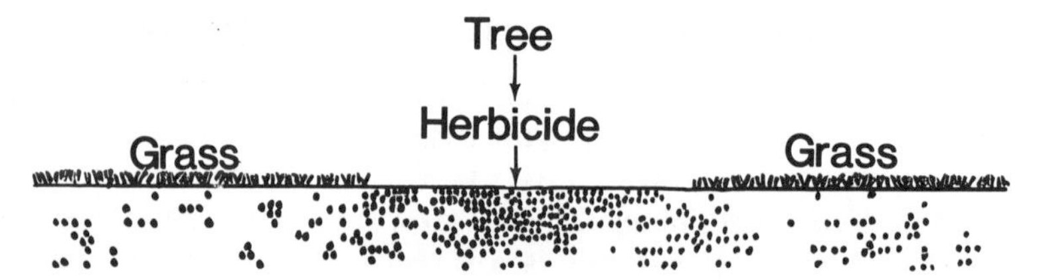

Figure 2.3 Apple root distribution under herbicide strip and grass between the rows. (Reproduced by permission from Atkinson and White, 1976.)

TABLE 2.2 Development of Root Systems of Fruit Trees in Relation to Compaction.[a]

Densitometer Reading (kg/cm^2)	Soil State	Behavior of Root System of Fruit Trees
0–10	Loose	Roots spread freely
10–20	Loosely packed	Roots spread freely
20–30	Slightly compacted	Roots penetrate soil
30–60	Compact	Roots encounter obstacles; amount of roots reduced almost by half
60–100	Heavily compacted	Roots penetrate only along cracks

[a] Reproduced by permission from Kolesnikov, 1971.

White, 1980). In young apple trees, root growth is higher under the herbicide strip than under the sod alley, and root growth begins earlier in the year (Figure 2.3). As a result, the majority of roots are within the herbicide strip (Atkinson and White, 1976). As the tree ages, more use is made of the sod strip, but even after 12 years apple trees develop relatively sparse root systems into this area. Orchard machinery tends to move through the orchard over the same path all the time. Fritzcshe and Nyfeler (1974) could not find evidence that root growth was reduced under the wheel track marks in the orchard.

Permanent sod with herbicide strips creates a favorable soil condition that is not compacted easily. Tree roots are sensitive to soil compaction and do not grow into the compacted area. Kolesnikov (1971) adapted data of Vashchenko on the effect of soil density on root growth of fruit tree roots. Soil densities up to $30\,kg/cm^2$ had no effect on the ability of roots to penetrate the soil in all directions. Above this value, roots encounter obstacles and eventually only penetrate along cracks (Table 2.2). Mechanical impedance is correlated with a threefold increase in the IAA content of plants in general. Elongation of roots is severely inhibited by high indole acetic acid (IAA) content (Butcher and Pilet, 1983), and

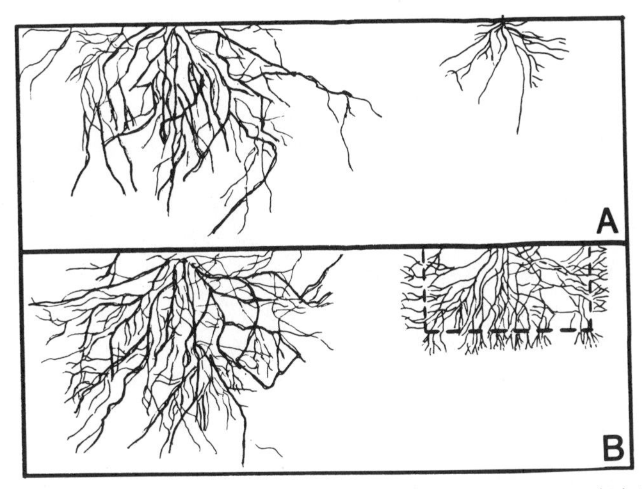

Figure 2.4 Effect of root pruning on regrowth of roots of apples. (A) Entire half of root system of 15-year-old 'Jonathan' apple trees has been removed; illustration shows regrowth 4 years after root removal. (B) Similar tree before part of root system is removed at dotted line and subsequent regrowth after 4 years. Left: roots without pruning. (Reproduced by permission from Gyuro, 1974.)

concomitantly the lateral root formation is increased (Lachmo et al., 1982). It is very likely that similar hormonal control exists in fruit trees as well.

Mulch Root growth at the surface is increased by mulching. Root growth of young apple trees was higher in all diameters, particularly at 0–8 cm depth (White and Holloway, 1967) under mulch. Mulch increased L_V values of roots in pears from 0.12 to 0.14 compared with sod (Reckruhm, 1974). Mulching also increased the surface rooting in peaches (Hill, 1966).

Pruning of Roots For at least 100 years root pruning has been used to reduce the size of fruit trees. Following root pruning, both root activity and root growth rate increase in apple and peach (Maggs, 1964; Richards and Rowe, 1977) but only at the depth that the cut occurred. The overall growth of the tree is redistributed in favor of roots, and shoot growth is retarded. The effect of root pruning on shoot growth is immediate. In peaches, 25 days after root pruning, a redistribution of growth was evidenced by a 20% increase in root dry weight and a 23% reduction in the top dry weight (Richards and Rowe, 1977) (Figure 2.4). Root regeneration requires the same conditions as normal root growth. Regeneration requires a certain amount of carbohydrate input from the top portion of the tree. If this is prevented, regeneration cannot occur. Shading of

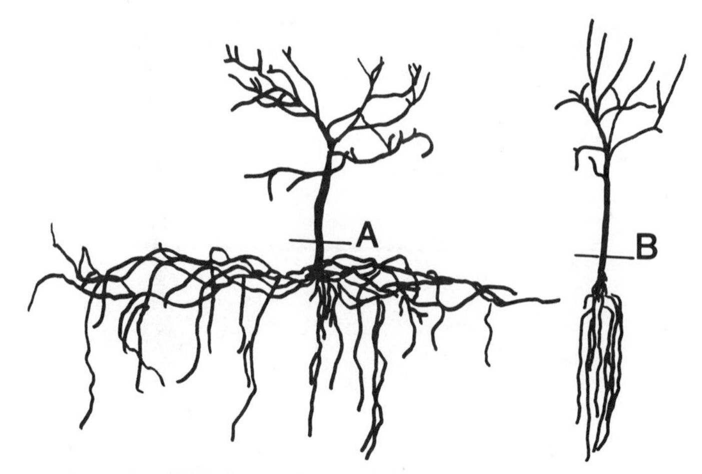

Figure 2.5 Penetration of young apple roots in wide and in close plantings: (A) wide planting, 2.4 m^2 per tree; (B) close planting, 0.3 m^2 per tree. (Reproduced by permission from Atkinson et al., 1980.)

root-pruned apple seedlings retarded regeneration of roots (Faust, personal communication), and large fruit load on root-pruned apple trees prevented root regeneration (Hoad and Abbott, 1983).

There are two prevailing theories as to why root pruning decreases shoot growth. First, according to Kramer and Kozlowski (1979), the water deficit created by root pruning could account for decreased shoot growth because a minimum water level, and thus a certain turgor pressure, is necessary for cell expansion. Second, Richards and Rowe (1977) have suggested that decreased shoot growth in root-pruned trees is caused by limited supplies of growth substances, especially cytokinins from the roots. Cytokinins exogenously applied to leaves decreased root–shoot ratios, whereas application to the roots increased them (Richards, 1980; Richards and Rowe, 1977). They concluded that one of the functions of cytokinin is to draw assimilates to the site of application. Other aspects of root pruning have been reviewed by Geisler and Ferree (1984).

Root Restriction The root system cannot spread beyond the available soil volume when the root is restricted. Root restriction could result if trees are planted in a container, in ridges over heavy clay, which restricts downward penetration, or too close in a high-density orchard. At wider spacings, the root system of apples is mainly composed of horizontal roots with relatively few vertical sinkers, but at higher tree densities the root system consists mainly of vertical sinkers (Atkinson et al., 1976) (Figure 2.5). In high-density orchards the weight, length, volume, and surface area of roots of individual trees decreased in comparison to similar trees planted at wider within-row distances. Atkinson (1980) calculated the leaf area [cm^2/root length (cm)] for a range of tree densities of apples. He obtained values of 1.3–1.7 for LAIs less than 1.3 (low-density orchard) and comparable values of 2.1–3.4 for LAIs between 5.8 and 9.7 (high-density orchard). He concluded that in a high-density orchard twice the length of the root is needed to supply a given evaporative surface.

Effect of Tree Nutrition on Root Growth Root growth is greatly influenced by the general nutritional status of the tree. The effect is most likely indirect and acts through influencing the above-ground part of the tree first. Nitrogen stimulates primary growth of the absorbing roots. In comparison to N deficiency, the weight of the roots may increase by 50% when N is adequately supplied the tree, creating a 200% increase in absorbing surface (Kolesnikov, 1971). In contrast, excessive N suppresses root growth, probably because high N promotes a high rate of shoot growth, which in turn opposes root growth. Phosphorus and K both promote branching of the root system. In addition K increases the weight of roots more efficiently than the above-ground portion of the tree (Kolesnikov, 1971). Calcium is essential for the growth of shoot tip. In Ca-deficient or even in Ca-insufficient conditions, when deficiency cannot be noticed on the above-ground portion of the tree, root tips often die.

Cyclic Renewal of Roots

Death of the roots is part of the natural process. Roots are dying off and renewed systematically. Dying off occurs at different times depending on the root length. Short lateral roots die off during the first or second week of spring growth. As growth continues, roots of the next door of branching begin to die off (Atkinson, 1980). The percentage of roots with dead tips can be quite high. As a rule, the longer the root is, the more dead root tips it has. Kolesnikov (1971) determined dead root tips in relation to root length for apple and pear (Table 2.3).

In cooler climates root shedding takes place between April and November; in warmer climates it apparently occurs year around. The number of dead roots apparently increases with age. In apple from 5 to 9 years of age, the percentage of dead root tips was 0, 17, 24, 59, and 53%, respectively (Kolesnikov, 1971).

The reason for dying off is not known. An unusually high percentage of dead root tips has been observed in connection with nutrient deficiencies, especially Ca. However, the systematic die-off cannot be associated with the irregular occurrence of nutrient deficiencies.

TABLE 2.3 Percentage of Dead Root Tips in Apple and Pear by Length of Root[a]

	Overall	0.1–5 mm	5–10 mm	11–15 mm	26–30 mm	36–40 mm	66–70 mm	101–105 mm
Apple	2.0	0.3	2.6	5.2	9.7	16.7	20.5	43.0
Pear	4.5	1.4	3.9	7.4	11.1	14.3	23.5	33.9

[a] Reproduced by permission from Kolesnikov, 1971.

DETERMINING NUTRITIONAL STATUS OF FRUIT TREES

Nutrition is one of the most effective ways to influence the productivity of fruit trees. In order to apply the right amount of nutrients to achieve the desired results, one has to know the nutritional status of the tree.

In agronomic crops, soil analysis is used to determine the nutrient status of the soil. Before the yearly planting of such crops, the soil nutrient status can be adjusted. In most cases leaf or tissue analysis correlate well with soil tests (Hipp and Thomas, 1968) in agronomic crops and vice versa. In fruit trees, the condition of the soil and the condition of the tree together determine the uptake rate of nutrients, which could be very different from what is available in the soil. Therefore, in fruit trees, leaf analysis is the most convenient and the most accurate guide in determining the nutritional status of this perennial plant.

In contrast to soil analysis, leaf or tissue analysis is the integration of all circumstances that influence the availability of nutrients (Lundegardh, 1945) and reflects nutrient uptake conditions rather than the nutrient status of the soil (Mengel and Kirkby, 1982). Low nutrient uptake can result from poor soil aeration, low moisture conditions, or low metabolic activity of the roots. High nutrient uptake can reflect optimum moisture conditions, large, well-developed root systems that can explore the soil well, or high photosynthetic rates that supply the root with sufficient carbohydrates for optimum root metabolism.

A high content of one nutrient in the tree may reflect an inadequate supply of another nutrient. When growth is depressed for any reason, certain nutrients may accumulate more than if growth had continued at a higher rate. If growth rate is too high, certain nutrients may be diluted more than one would expect at a normal growth rate.

Synergistic or antagonistic relationships between nutrients must also be taken into account when leaf analysis data is interpreted. Shear et al. (1946) advocated that tree growth is a function of two variables of nutrition, intensity and balance. When the nutrient status of the leaves is determined at any given level of nutritional intensity, a multiplicity of ratios may exist between the various elements. Maximum growth and productivity occur only upon the coincidence of optimum intensity and balance. This concept has become especially important when it is applied to the storability of fruit. At N–Ca ratios of 10 on a dry-weight basis in the flesh of apples, the fruit can be stored for a long time in good condition. In contrast, if N–Ca ratios increase to 30, the fruit certainly will suffer breakdown and cannot be stored (Shear, 1974). The importance of the proper ratios between Ca and B and between K and B has been pointed out by several authors. A high Ca+K–Mg ratio consistently resulted in B toxicity, but a high K+Mg–Ca ratio had little effect on the appearance of B toxicity at the same B concentrations (Shear et al., 1946).

Despite the important considerations advocated by Shear et al. (1946), the majority of leaf analysis data is still reported as absolute percentages. This method of reporting has the advantage of ease of computerization and simplicity of interpretation.

Nutrient Composition of Leaves

Leaf nutrient concentration changes with age of the leaves. Some nutrients, such as N, P, and K, decrease with age of the leaf; others, such as Ca and Mg, increase. There is a certain plateau in this change between 110 and 125 days after bloom (Figure 2.6). Because the nutrient concentrations during this period are relatively stable, this is the period when leaves are used to measure the nutritional status of the tree. For application of leaf analysis to tree nutrition and fertilizer practice, it is essential that various ranges of the nutritional status of the tree be established:

(a) leaf nutrient concentrations associated with deficiency symptoms,
(b) range of nutrient concentrations in the absence of symptoms over which the tree responds to applications of the insufficient nutrient,
(c) optimal nutrient concentration associated with maximum yield and fruit quality, and
(d) range of concentrations at which nutrients cause toxicity or produce undesirable effects.

Such ranges are illustrated for apple in Table 2.4.

Concentration of nutrient requirements varies with fruit species. Peaches require much higher N concentration for optimum growth and cropping than apples. Nutrient requirements also change with geographic location. Trees grown in southern locations require higher levels of N than those grown in the north.

It is generally recognized that leaf nutrient concentrations change with leaf position, which perhaps also reflects the age of the leaf. There is a general increase in the concentration of Mg from the base to the tip of the shoot. If trees are sufficiently supplied with K, there is no concentration gradient along the shoot. However, if K is present in insufficient quantity, a gradient with increasing concentration toward the tip is present (Kidson, 1947). The same trend is observable for N. In contrast, Ca, Fe, and Mn are highest in the basal leaves. This is important because, for uniformity, midshoot leaves should be used for leaf analysis.

The presence of the crop usually decreases leaf concentrations of P and K and increases Ca and Mg. However, if the tree is low in Mg, cropping

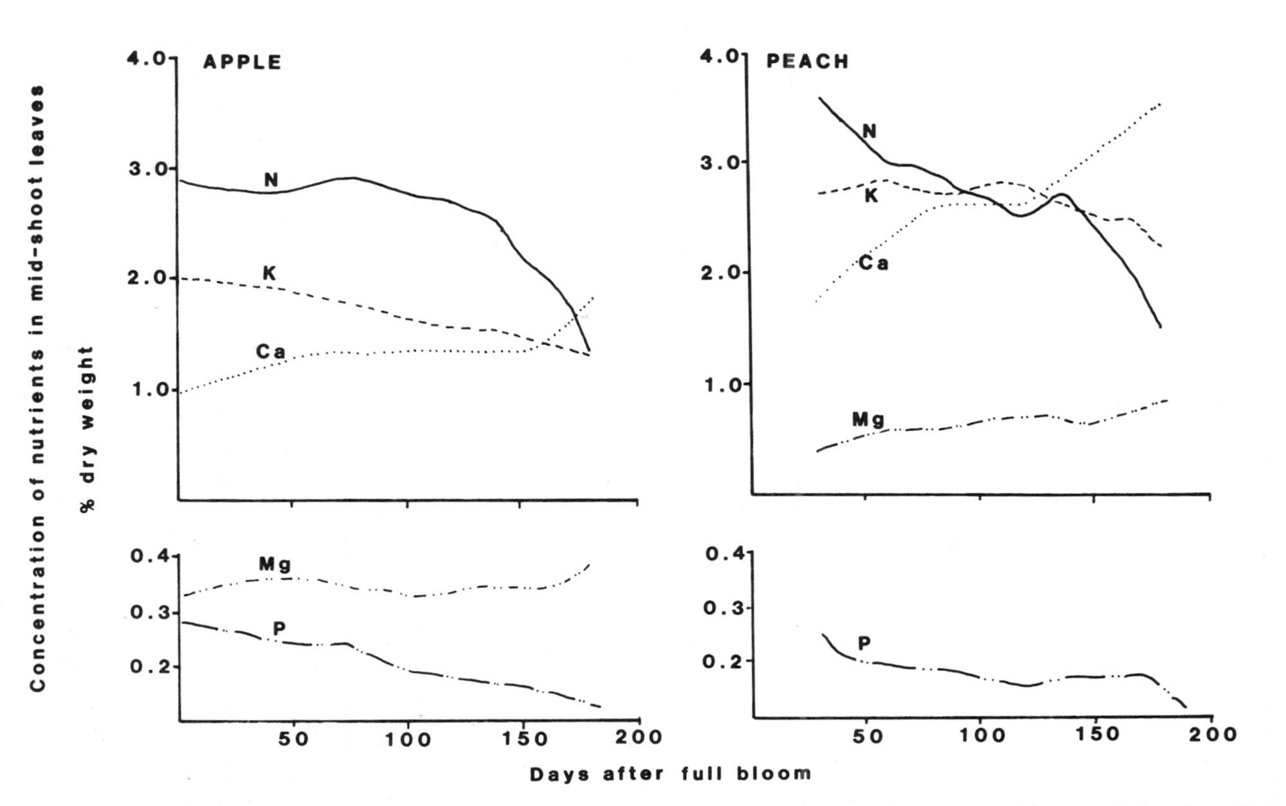

Figure 2.6 Nutrient concentration in leaves of apple and peach during the growing season. (Reproduced by permission from Rogers et al., 1953.)

TABLE 2.4 Nutrient Ranges in Apple (% dw)

Element	Deficient	Insufficient	Optimum	High
N	1.7	1.8–2.0	2.1–2.2	2.3
K	0.8	0.9–1.5	1.5–2.0	2.1
Mg	0.15	0.25	0.35	0.40
P	0.11	0.15	0.22	—
Ca	1.0	1.0–1.4	1.5–1.8	—

TABLE 2.5 Optimum Nutrient Concentrations in Leaves of Fruit Trees

	Percentage of Dry Weight					ppm				
	N	P	K	Ca	Mg	Fe	B	Zn	Cu	Mn
Apple	2.0	0.2	1.5	1.8	0.4	85	35	25	6	25
Apricot	2.0	0.1	2.8	1.5	0.4	100	45	35	30	30
Sour cherry	3.0	0.3	2.5	1.5	0.4	180	45	30	10	30
Sweet cherry	2.5	0.3	1.5	1.5	0.4	100	45	30	10	30
Peach	3.2	0.3	2.3	2.0	0.6	120	45	30	10	80
Pear	2.5	0.2	2.0	1.5	0.4	120	45	30	10	60
Plum	2.5	0.2	2.5	2.5	0.4	120	35	30	10	50

trees will show Mg deficiency in their old leaves while noncropping trees will not show deficiency symptoms at the same Mg concentrations.

For diagnostic purposes it is most convenient to use absolute percentages to determine the nutrient status of the tree. This is opposed to relative comparisons of ion pairs or ratios between various nutrients present in the leaf. Kenworthy and Martin (1966) and Shear and Faust (1980) list a wide range of absolute concentrations obtained by leaf analysis. Average values for midshoot leaf nutrient concentrations obtained for productive trees about 110 days after bloom are listed in Table 2.5.

Fruits also have their internal nutrient gradient. In general, the skin and the core in pome fruit are high in nutrients. This is in contrast to that in the pulp, which is relatively low. Usually the calix end is lower in nutrients than the stem end.

In apple there are two distinct types of gradients. One gradient is represented by K, Ca, Mg, and Mn. The skin and core tissues contain higher amounts of these elements than the flesh, where the concentration is much lower. The other gradient is represented by P, Fe, Zn, B, and Cu. The skin contains the highest concentration, and the concentration decreases toward the inside of the fruit (Faust et al., 1967). Such gra-

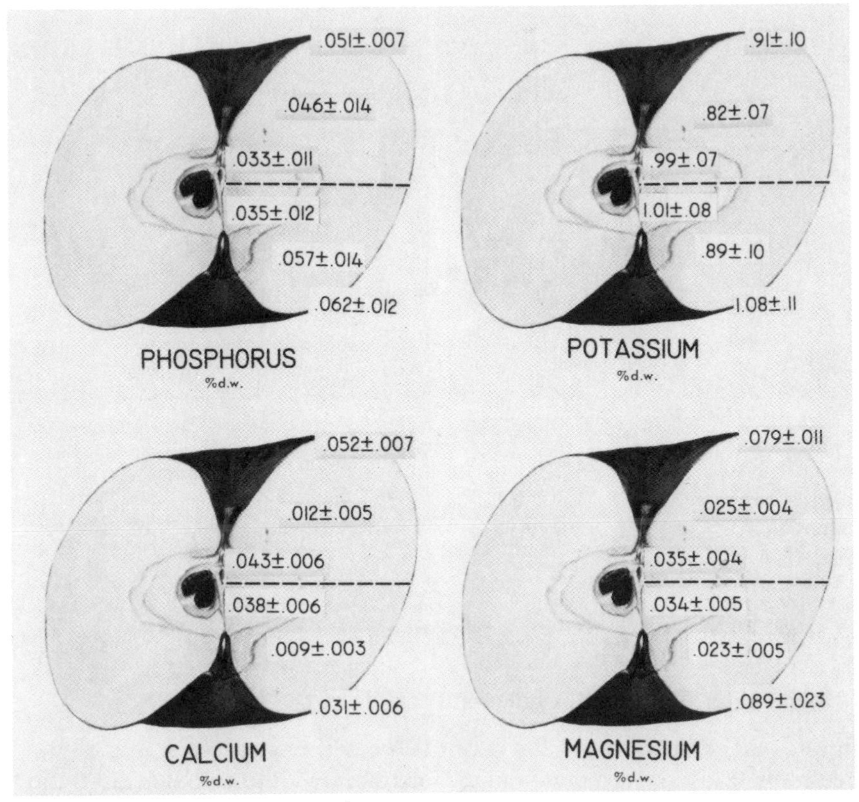

Figure 2.7 Concentration gradients of Ca, P, K, and Mg in apple. (Reproduced by permission from Faust et al., 1967.)

dients for P and Ca are illustrated in Figure 2.7. Concentrations in the proximal half of fruit are higher than near the calix end. The fruit-to-fruit variation is large. Thus, for fruit analysis, a blend of 10 fruits should be used.

Most nutrients migrate to the site of high metabolic activity. If there is a location within a given tissue where for various reasons the metabolic activity is high, nutrients will migrate to this location. In apple, bitter pit, cork spot, and other metabolic disorders are localized and develop in low-Ca fruit or at locations within the fruit where the Ca concentration is the lowest. However, the spot itself could be high in Ca because the high metabolic activity associated with the disorder attracts nutrients (Ca, Mg, and other ions) into the affected tissue. Internal bark necrosis of apples is very similar in this respect. The necrotic brown spots accumulate Mn in high quantities, usually after the disorder has started and the metabolic activity increased.

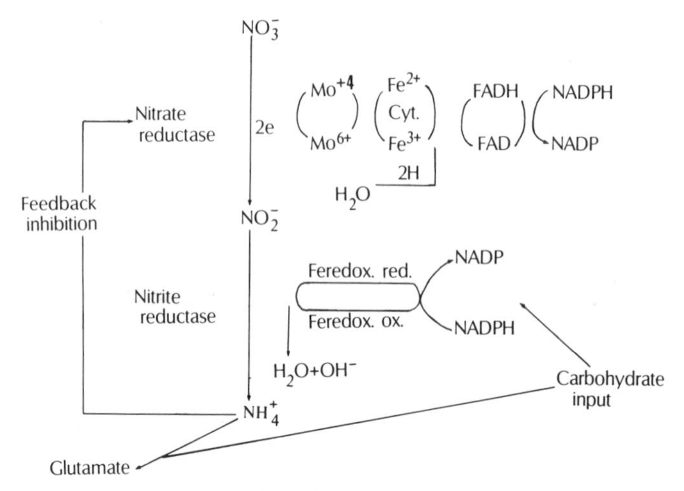

Figure 2.8 Schematic illustration of reactions involving reduction of nitrate.

UTILIZATION OF NITROGEN BY FRUIT TREES

Absorption of Nitrate and Ammonium Ions

Nitrate and ammonium are the major sources of nitrogen taken up by the tree roots. Most of the ammonium is incorporated into amino acids in the roots, whereas the nitrate may be mobile in the xylem under certain circumstances, and it may be utilized by either the roots or the leaves.

The reduction of nitrate to ammonia is mediated by two enzymes: nitrate reductase and nitrite reductase (Beevers and Hageman, 1980). Nitrate reductase is a complex enzyme containing several prosthetic groups including flavin adenire dinucleotide (FAD), cytochrome, and molybdenum, and it requires NADH or NADPH as an electron donor (Figure 2.8). Nitrate reductase is an inducible enzyme with a short half-life of only a few hours (Beevers and Hageman, 1983). It is present in low levels in plants that do not receive NO_3^-, but it can be induced within a few hours upon NO_3^- fertilization (Oaks et al., 1972). The two reducing enzymes together produce ammonium ions that are processed by the plant into amino acids.

The ammonium ion is toxic to the plant (Givan, 1979) depending on the pH of the cell. Therefore, the plant cannot tolerate the presence of NH_4^+, and when NH_4^+ accumulates, nitrate reductase is inhibited by a feedback inhibition.

Early reports of Thomas (1927a) and Eckerson (1931) placed nitrate reduction into the roots of fruit trees. Their work revealed no activity in the areal parts of the trees. Several decades later Titus and Ozerol (1966),

Leece et al. (1972), and Klepper and Hageman (1969) demonstrated nitrate reductase in leaves of apple, apricot, sour and sweet cherry, and plum but not in peach. Perez and Kliever (1978) demonstrated that leaves of plum, pear, and cherry all contained nitrate reductase, and nitrate reductase in these trees is an inducible enzyme not different from other plants.

Nitrate reductase in the leaves of trees was usually demonstrated under conditions of high nitrate in the medium or after application of large amounts of nitrate in the orchard. Titus and Kang (1982) interpreted this as due to the saturation of nitrate reductase in the roots, which allows NO_3^- to be translocated to aerial tissues. They considered the occurrence of nitrate and nitrate reductase in the leaves of fruit trees as an exception rather than the rule. Ammonium ions are known to be inhibitory to nitrate reductase (Frith and Nichols, 1975b). Thus, it is also possible that NO_3^- can bypass the inhibited enzyme when ammonium ions are supplied concurrently with nitrate from NH_4NO_3. This inhibition also explains reported results that NO_3^- is utilized to a lesser extent if the two ions are present together (Grasmanis and Nicholas, 1966, 1971; Shear and Faust, 1971).

Uptake of N and the resulting growth responses have been measured after supplying trees with either nitrate or ammonium alone or in combination. The uptake of either form is continuous throughout the active growing period with a relatively high peak during the summer (Grasmanis and Nicholas, 1971). The growth response to nitrate and ammonium nutrition has been controversial in apple. Nitrate induced more growth on M7 rootstock (Buban et al., 1978) in the orchard and in sand-cultured York Imperial apple trees (Shear and Faust, 1971). Apples, however, initiated more flowers if exposed briefly to ammonium than when grown on nitrate N continuously (Grasmanis and Edwards, 1974). Thus certain benefits can be obtained from either form of the N, but the benefits are different depending on the N source.

Trees utilize N from the soil depending on the metabolic processes of the above-ground part of the tree. Nitrogen, regardless of the form absorbed by the root, is largely utilized in the root, which requires considerable amounts of carbohydrates. Thus, it is not surprising that N uptake efficiency is high only when the tree produces photosynthates. Weinbaum et al. (1978) studied uptake and measured *nitrogen utilization efficiency* (NUE) in plum during nine 10-day periods.

They defined NUE as

$$\text{NUE} = \frac{\text{total fertilizer N absorbed/tree/10 days}}{\text{total fertilizer N applied/tree/10 days}}$$

It is clear from their data that N is not taken up before rapid shoot

TABLE 2.6 Nitrogen Uptake by Plum Trees[a]

Stage	KNO_3–^{15}N Application Period	NUE[b] (%)
Dormant	January 16–26	4.75
Bud swell	March 5–15	4.34
Rapid shoot growth	April 2–12	30.52
Shoot growth ceased	May 14–24	39.02
	July 9–19	32.73
	August 6–16	35.91
	September 10–20	32.73
Mid–leaf fall	October 22–November 1	16.14
Dormant	December 3–13	3.66

[a] After Weinbaum et al., 1978.
[b] NUE = nitrogen utilization uptake.

growth begins, decreases when leaves are senescing, and ceases when leaves fall (Table 2.6).

Utilization of Reserve Nitrogen by Fruit Trees

Since N is not taken up before shoot growth commences, early spring activities within the tree draw on N reserves. The importance of N reserves for early spring growth and fruit set has been recognized since the early 1920s in apple (Harley et al., 1958; Murneek, 1930; Roberts, 1921; Thomas, 1927b), peach (Taylor, 1967; Taylor and May, 1967), and pear (Taylor et al., 1975). Taylor et al. (1975) indicated a highly positive correlation between the level of storage N and the extent of new extension growth during spring. Data on application of N indicate that when N fertilizers were applied late in the season, the N concentration of spurs was greater during the following spring before growth commenced (Hooker, 1922). Late application of N is an ideal way to increase the reserves of the tree for early spring activities. However, there are practical reasons why late applications are limited. Late N applications can delay coloring of fruit, especially apples, and reduce hardiness of the shoots because shoots grow too long and have insufficient time to harden. If N is applied very late in the growing season after shoot growth has ceased, the storage N needs of the tree could be satisfied without the undesirable effects of the late N application (Heinicke, 1934). Since N can be applied in the form of sprays, late season N sprays are effective in increasing the N reserves of the tree (Swietlik and Faust, 1984; Titus and Kang, 1982).

Nitrogen is stored in the form of storage proteins. The rate of hydrolysis of storage proteins depends on the temperature at which regrowth occurs. O'Kennedy and Titus (1979) placed dormant apple trees into growth chambers at 20, 25, and 30°C. The temperature of the chamber had a direct effect on rate of protein hydrolysis in the bark. At 25°C, 83% of bark proteins were hydrolyzed in three weeks. Kang et al. (1981) approached the decomposition of proteins in apple bark in the orchard by examining the low-molecular-weight peptides of the bark. He could not account for the overall decline of 60% in proteins by the loss of low-molecular-weight peptides. Kang and Titus (1980) suggested that as growth proceeds, early spring N needs are met by mobilization of amino acids present in dormant tissues, which in turn are produced by hydrolysis of proteins. Kang and Titus (1980) isolated an endopeptidase from dormant apple shoot bark. They postulated that the activation of this endopeptidase is responsible for the rapid metabolism of storage proteins that accompanies budbreak. Kang et al. (1981) further modified this by suggesting that proteins in apple bark undergo modifications prior to hydrolysis, which requires a multienzyme system.

O'Kennedy and Titus (1979) emphasized the role of buds in the proteolysis of apple bark. Titus and Kang (1982) reviewed and interpreted the effect of buds on proteolysis. The implication is that the buds may control proteolysis by some unknown hormonal or other signal. In contrast, Tromp and Ovaa (1971) could not demonstrate the effect of debudding on protein breakdown in apple. Tromp (1970) speculated that the commencement of protein hydrolysis may mark the end of winter dormancy, and the signal that induces protein hydrolysis may be connected with factors influencing dormancy rather than an active call from the buds.

Assimilation of Ammonium

Regardless of whether NH_4^+ is produced as an end product of nitrate reduction or absorbed by the root from fertilizer applications, the tree must metabolize it. Titus et al. (1968) demonstrated that both apple and peach roots are capable of combining oxoglutarate and NH_4^+ into amino acids. Titus and Splittstoesser (1969) reported metabolism of fumarate to glutamate and aspartate in both of the above-mentioned species. Recently, Titus and Kang (1982) reviewed the incorporation of the ammonium ion into amino acids in apple roots.

The important aspect of N metabolism is that it is intimately linked to carbohydrate metabolism. Glycolysis, the tricarboxylic acid cycle, and the pentose pathway are all closely linked to amino acid biosynthesis with regard to the supply of carbon skeletons for the amino acids and energy for conversion. During the incorporation of NH_4^+ into amino acids the demand for carbohydrates in the root is very high. The tree must satisfy this demand to utilize N efficiently. In shaded trees with LAIs larger than

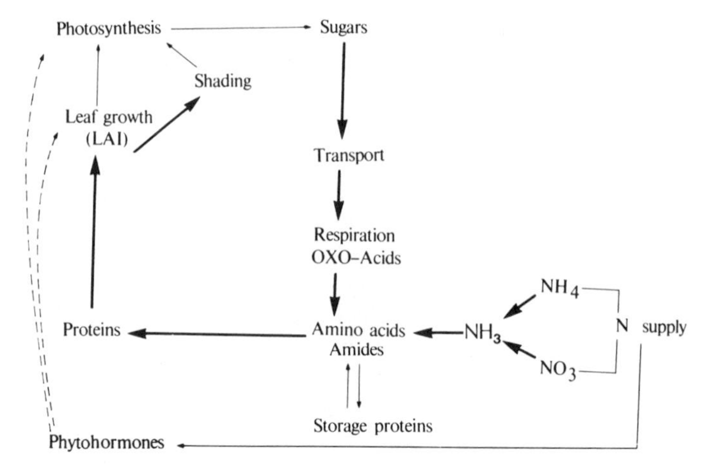

Figure 2.9 Schematic representation of physiological processes involved in absorption and assimilation of N by fruit trees.

3 for apple and 8 for peach, the application of N fertilizers should take this into account. The need for carbohydrates appears especially high soon after the N is applied in NH_4^+ form to the soil or as a spray to the leaves. Soil nitrification bacteria need time to convert NH_4^+ to nitrate, and during this time NH_4^+ is absorbed. Naturally, NH_4^+ is absorbed immediately if applied as a spray. In both cases photosynthesis must occur at sufficient levels or the reserves of the tree should be available to immediately utilize NH_4^+ for amino acids. A schematic diagram of the involvement of photosynthesis in N metabolism is presented in Figure 2.9.

While the need for carbohydrates is clear for the utilization of N, the total amount of carbohydrate utilized for this purpose is not known. Priestley (1972) and Catlin and Priestley (1976) could not detect the short-term changes in the carbohydrate status of young apple trees soon after NH_4^+ application, although the soluble N of the tree increased. They could not associate the increase in soluble N with the carbohydrate status of the tree, tree size, root mass, or active or ceased status of growth. They concluded that the tree produced sufficient carbohydrates for the utilization of N and carbohydrate availability has never limited nitrogen uptake. Since arginine is abundant in the soluble N fraction of the tree, it is possible that at times when carbohydrates are limiting, the tree links four N molecules to a carbohydrate carrier instead of one N, as when producing glutamate. That arginine is abundant in fruit trees argues for the fact that this may be the case.

The synthesis of glutamate from 2-oxoglutarate and NH_4^+ is generally catalyzed by the glutamate dehydrogenase (GDH) enzyme. The presence

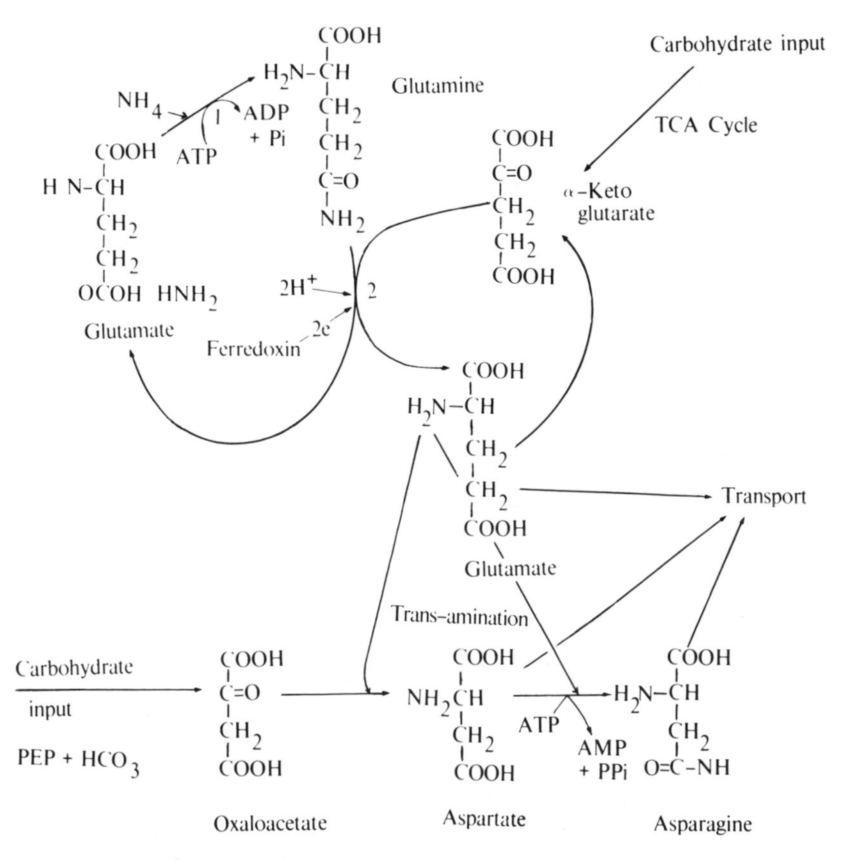

Figure 2.10 Schematic representation of physiological processes involved in absorption and assimilation of N by fruit trees.

of this enzyme has been demonstrated in apple trees, including the roots (Cooper and Hill Cottingham, 1974; Kang and Titus, 1980; Spencer and Titus, 1972). The synthesis of glutamate by GDH may not function as the main system of N incorporation into amino acids. Under steady-state conditions glutamine is the first product of NH_4^+ assimilation catalyzed by glutamine synthetase (GS) (Titus and Kang, 1982). Once glutamine is synthesized, the formation of glutamate proceeds from glutamine in the presence of glutamate synthase (GOGAT) utilizing 2-oxoglutarate. Evidence has been presented by Kang and Titus (1980, 1981) that in apple bark and leaf tissues the GS–GOGAT system is the major route for NH_4^+ assimilation. The occurrence of these enzymes in roots has been expected but not demonstrated as yet (Figure 2.10).

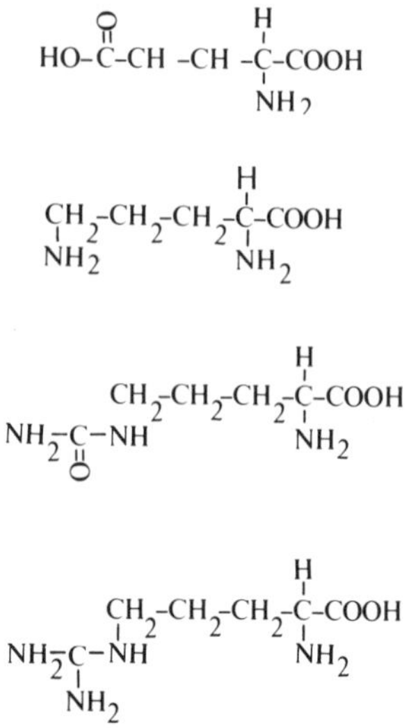

$HO-C(=O)-CH_2-CH_2-CH(NH_2)-COOH$ Glutamate

$CH_2(NH_2)-CH_2-CH_2-CH(NH_2)-COOH$ Ornithine

$NH_2-C(=O)-NH-CH_2-CH_2-CH_2-CH(NH_2)-COOH$ Citrulline

$NH_2-C(NH_2)-NH-CH_2-CH_2-CH_2-CH(NH_2)-COOH$ Arginine

Figure 2.11 Structure of amino acids with various numbers of amino groups.

Not only glutamate but also aspartate is synthetized in roots, bark, and leaves of apple. The major route for synthesis of aspartate is through the glutamate–oxaloacetate transaminase system (Titus and Kang, 1982). The presence of this transaminase enzyme has been demonstrated in the above-mentioned tissues (Cooper and Hill Cottingham, 1974). Aspartate can serve as an additional NH_4^+ acceptor. In this reaction the amide asparagine is synthetized. Recent evidence suggests (Titus and Kang, 1982) that the glutamine-dependent asparagine synthetase may be the major role of asparagine synthesis in plants in general, including fruit trees. However, the presence of this enzyme in fruit trees has not been demonstrated.

Hill Cottingham and Lloyd-Jones (1973) investigated the catabolism of arginine as a storage form of N, but the synthesis of arginine received little attention. The N molecule is incorporated from glutamate (one N) to ornithine (two N), from ornithine to citrulline (three N), and from citrulline to arginine (four N) (Figure 2.11). The only step that is characterized in apple is the ornithine-to-citrulline reaction catalyzed by ornithinecarbamoyltransferase (Spencer and Titus, 1974). Nevertheless, arginine was found in young apple roots as a result of N application (Hill Cottingham and Cooper, 1970; Tromp and Ovaa, 1976, 1979). Thus it appears that at least in this fruit species, the synthesis of arginine is possible in the roots.

Since the same enzyme system exists for nitrate reduction in both roots and leaves of fruit trees (Titus and Kang, 1982), the same conditions exist in the leaves when nitrogen is applied through sprays in the form of NO_3^-

or NH_4^+ from a variety of sources independently or combined (Swietlik and Faust, 1984). This should be taken into account when nutrient spray formulations are made.

Response of Fruit Trees to Nitrogen Application

Most fruit tree responses to nitrogen nutrition may be classified into three groups: (1) vegetative responses, (2) responses involving fruiting, and (3) responses involving fruit characteristics. Studies on responses of N were among the first experiments on fruit trees. Lyon et al. (1923) reported that the growth of apple trees was directly proportional to the amount of N applied up to a point that appeared to be the maximum growth rate. Restriction of growth due to N deficiency was expressed by narrowing the top–root ratio, decreasing trunk circumference, and decreasing spread and height of the tree. Similar results were observed on cherry (Tukey, 1927) and peach (Bell and Childers, 1954). Not only the quantity but also the formulation of N applied makes a difference in the growth of the trees. Shear and Faust (1971) reported that growth was greatly limited in apple as a result of NH_4^+ nutrition in comparison to when the same amount of N was applied as NO_3^- (Figure 2.12). In orchards the NH_4^+ effect is difficult to observe. Shear and Faust (1971) used sand cultures and applied the nutrient solution every 5 days. As a result, nitrification did not occur, and the effect of the form of N on growth was dramatic.

Bradford (1924) made the first discovery that N fertilization increased bloom of apples. Nitrogen fertilization stimulated blossom formation in spur buds and in terminal and lateral shoot buds and reduced the tendency toward alteration of bloom. Harley et al. (1942) interpreted this phenomenon by observing early leaf surface development, especially the large size of spur leaves, which has particular importance in determining flower initiation. Nitrogen application apparently increased spur leaf size and early rate of photosynthesis, which apparently enhanced flower bud initiation.

In apples, flower initiation is affected by the time of application and the form of applied N in addition to the total level of N. Nitrogen applied in the form of ammonium during the flower bud development, even for a short period (24 hr), greatly increased flower bud formation (Grasmanis and Edwards, 1974) (Table 2.7).

Since the trees had ample supply of N throughout the season, it is unlikely that the N played a nutritional role. Ammonium application could play a role in developing growth hormones, as Marschner (1986) interpreted it, or it could slow growth temporarily, which could enhance flower bud development.

The possibility that ammonium caused hormonal changes is supported by Buban et al. (1978), who measured the cytokinin content of apples in

Figure 2.12 Effect of nitrate and ammonium on growth of apple trees: (A) low N in the form of NH_4^+; (B) low N in the form of NO_3^-; (C) high N in the form of NH_4^+; (D) high N in the form of NO_3^-. Trees are in sand culture, York Imperial on M26. (Reproduced by permission from Shear, 1971.)

the xylem sap and found that it was considerably higher with ammonium nutrition (Table 2.8).

In the same experiment, growth was also affected similar to that reported by Shear and Faust (1971). Trees receiving NH_4^+ grew less than

TABLE 2.7 Effect of Ammonium versus Nitrate Supply of Flower Bud Differentiation in Apples[a]

Treatment	Number of Buds per tree		
	Total	Flower	% Flower
Continuous NO_3 supply	35.5	12.8	38.7
Nitrate N supply interrupted by 8 weeks of ammonium supply	33.3	21.2	63.7
Nitrate N supply interrupted by 24 hr of ammonium supply	38.8	30.3	77.5

[a] Reproduced by permission from Grasmanis and Edwards, 1974.

TABLE 2.8 Effect of Nitrogen Nutrition on Cytokinin Concentration in Xylem Sap of Apple[a]

Treatment	Zeatin Concentration 1 day after Start of N Supply (μg/ml)
Nitrogen supply withheld	0.05
Nitrate N	0.82
Ammonium N	1.95

[a] Reproduced by permission from Buban et al., 1978.

trees receiving NO_3^-. For these reasons, it is not possible to separate the growth effect from the hormonal effect due to ammonium supply on enhancement of flower bud induction.

High N affects young trees in the opposite direction. High N tends to delay bloom, and within limits low N tends to cause earlier blossoming. Young trees are pruned to induce the maximum number of shoots that would develop into future branches of the tree. Consequently, pruned young trees also have the maximum number of young leaves near the growing shoot tips. The main sites of gibberellin synthesis are the shoot apex and the expanding leaves. In plants in general (Marschner, 1986) N is the element that influences gibberellin (GA) synthesis the most. The effect of N is presumably indirect on GA synthesis. Independently high N supply influences the cytokinin production in the root, and the export of cytokinins from the root maintains a high rate of growth of the shoot apex and the young leaves. In young trees there is a combined effect of N, presumably acting through the root in maintaining growth rate and pruning, which maximizes the number of shoots by removing apical dominance and inducing budbreak. The result of the combined effect is that the tree produces a large number of GA-producing sites and maintains the GA production for an extended time because of the N effect. As described in Chapter 4, flower bud formation is prevented by high GA concentrations. Therefore, it is not surprising that in young apple and pear trees, the high N level synergistic with pruning delays flowering.

In peach the effect of N on flowering is different. As illustrated in Chapter 6, peach trees must have a certain growth rate to develop flower buds. Trees must be pruned and fertilized with N to attain the desired growth rate. The effect of N combined with pruning on flowering of young and/or old peach trees cannot be easily explained with possible hormonal changes in the tree.

The effect of N on flower bud development also has been demonstrated in plums. October sprays of urea gave the highest yields compared with spring soil applications (Mishra, 1984). Calcium cyanamide (2.5–7.5%) was applied to 5-year-old plum trees during the dormant period to stimulate flowering and cropping. Calcium cyanamide, in addition to breaking dormancy, also supplies considerable amounts of N to the trees. The 2.5% Ca cyanamide spray gave the best results, the 7.5% concentration producing excessive fruit set requiring thinning (Pereira et al., 1982, 1984). Trees with optimal leaf N (2.68%) showed high (39.43%) flower fertility, whereas those with minimal leaf N (1.67%) had low (11.16%) fertility. Too high N could cause poor fruit set. The pistils of plums with low leaf N content were higher in cystine, methionine, lysine, and arginine and were unproductive, with poor ovary development and low fertilization and cropping (Peikic and Boskovic, 1985). The fall or dormant application of N seems very effective in increasing fruit set, indicating the involvement of storage N in the fruit set.

The rate of translocation of urea from the leaves of apples to other parts of the tree has caused some controversy. Boynton et al. (1953) reported that about 50% of the urea N absorbed by apple leaves attached to actively growing or nongrowing shoots was translocated out of the absorbing leaves within 24 h. Forshey (1963) reported rather limited translocation of foliar-absorbed urea N. His study showed poorer translocation of N to permanent structures of apple trees when this element was supplied via the leaves as compared to soil application. Forshey (1963) observed that supplying N exclusively via sprays resulted in low levels of N in the bark but maintained adequate levels of N in the leaves. He suggested that this distribution explained the low vigor and productivity of trees exclusively supplied with N via the foliage. When ^{15}N was supplied via foliage or soil to sour cherry (Swietlik and Slowik, 1981) and apple trees (Hill-Cottingham and Lloyd-Jones, 1975), the percentage of absorbed N found in leaves in August was only 7–10% greater in leaf- than in soil-supplied trees. In these experiments the amount of N absorbed constituted only a modest portion of total tree N, whereas in Forshey's (1963) experiment it constituted almost 50% of total tree N.

Weinbaum and Neumann (1977) suggested that metabolism and translocation of foliar-absorbed N from ^{15}N-labeled potassium nitrate spray applied to prune trees were enhanced by surfactants.

The amount of foliar-absorbed N subsequently exported to other parts of woody plants may differ according to the physiological stage of growth. During senescence, 23–70% of initial apple leaf N is reabsorbed by the tree (O'Kennedy et al., 1975; Oland 1963; Shim et al., 1972, 1973). Shim et al. (1973) established that N absorbed by senescing apple leaves is exported to the tree as amino acids and urea. However, only 5% of urea C was found in permanent tissues at the end of leaf abscission, indicating that urea sprays are rather poor suppliers of C. More recently, Swietlik and Slowik (1981) reported that in sour cherry trees as much as 80–87% of leaf-absorbed N from autumn application of ^{15}N urea migrated to the rest of the tree before abscission. Depending on the year of study, 49–64% of fertilizer N reabsorbed from the leaves was found in the roots during dormancy. Using ^{15}N techniques, Hill-Cottingham and Lloyd-Jones (1975) found that about 62% of leaf-absorbed N from autumn application of foliar urea to apple trees was recovered in permanent tissues during dormancy and that this N was evenly distributed among root and stem tissues of the stock and scion. The chemical form of stored N and the storage sites for N derived from senescing leaves were recently discussed in detail by Titus and Kang (1982).

These data indicate that N absorbed by leaves during senescence is highly mobile in fruit trees. The translocation of N from the leaves in fall may depend on temperature. If the temperature is such that leaves senesce slowly, most of the N is translocated. However, N application in late fall also increases the rate of photosynthesis of leaves, and they stay green longer. This increases the possibility that they may be killed by frost before the N is translocated to the tree.

UTILIZATION OF MAJOR NUTRIENTS BY FRUIT TREES

Calcium

Calcium is perhaps the most important mineral element determining the quality of fruit. It is especially important in apples and pears because these fruits are stored for a long period of time, and the effect of Ca on storage quality cannot be substituted by other factors. Calcium is also important in other fruits because of its general effect in delaying fruit ripening. High-Ca fruits can be transported better and remain in good condition longer. The tissue Ca concentration at which these desirable effects are achieved is usually higher than concentrations the fruit normally accumulates. Therefore, the desire to produce 'high'-Ca fruit is somewhat unnatural, and to achieve it requires mastering all the factors that influence Ca uptake and transport.

Calcium nutrition is complicated by the fact that Ca is needed most in the fruit. Consequently, Ca not only needs to be taken up by the tree but also needs to be transported to the fruit. This often requires methods other than those used by traditional nutrition.

Calcium Requirement of Fruit Trees Removel of Ca from the soil by trees requires more attention than removal of other elements. Nitrogen, P and K all can be supplied by broadcasting them to the soil surface annually and are expected to reach the root zone within a short period of time. In contrast, Ca applied to the soil surface penetrates the soil profile very slowly and for all practical purposes must be applied before planting when it can be mixed into depths useful for root uptake. Calcium applied before planting must last for the entire lifetime of the tree.

Atkinson (1986) assembled available data on the Ca need of apple trees. The estimates vary considerably on the use of Ca. A western U.S. 'Delicious' apple orchard used 168 kg $ha^{-1} y^{-1}$; an Australian orchard including only the crop and prunings removed 58 kg $ha^{-1} y^{-1}$. The various parts of the tree required varying amounts. The fruit, pruning, framework, and roots and leaves of the 'Delicious' orchard accounted for 8, 28, 46, and 86 kg $ha^{-1} y^{-1}$, respectively. The growth of the roots in an English orchard required 31.3 kg Ca $ha^{-1} y^{-1}$. Varietal differences may account for some of the differences in estimations. Leaves of 'Golden Delicious' contained 45 kg ha^{-1} while leaves of 'Granny Smith' had only 34 kg ha^{-1} in comparable conditions. The Delicious fruit contained 4 kg ha^{-1} while Cox fruit only had 2 kg ha^{-1}. The fact that 'Cox' fruit contains the minimal amount of Ca is well known, which is why this apple cultivar cannot be grown in warm areas where higher Ca concentrations are needed for assuring good fruit quality. The uptake of low Ca by 'Cox' is a genetic characteristic that can be transmitted to the progeny (Faust et al., 1971). The amount of Ca used per hectare also varies with tree density (Atkinson, 1986). In summary, assuming that on average fruit trees use 100–150 kg Ca $ha^{-1} y^{-1}$, the soil must contain 2500–3750 kg Ca to satisfy the lifetime (25 years) Ca requirement of the orchard. On acid soils it is important to make sure that the specified Ca is available before planting.

Function of Calcium in Physiology of Fruit Trees Hanson (1984) remarked that it is difficult to define the role of Ca in plant nutrition. This is even more difficult if fruit nutrition is concerned. Calcium is a macronutrient as far as its supply is concerned. Its concentration in the leaves of fruit trees is at the percentage level on a dry-weight basis. Yet at the cellular level it functions as a micronutrient. Calcium exerts its physiological and biochemical roles in the apoplast and in the cytoplasm. The difference in effective concentration between the two sites is large. Effective concentrations of Ca^{2+} are 1–5 mM in the apoplast and 0.1–1 μM in the cytosol (Poovaiah and Reddy, 1987). Only a few of the discovered effects of Ca are demonstrated in fruit trees or in the fruits themselves. Here we discuss only the effects in fruits or fruit trees. For other roles of Ca the review by Hanson (1984) should be consulted.

Calcium in the apoplast plays a binding role in the complex of polysaccharides and proteins forming the cell wall. Rossignol et al. (1977) estimated that at least 60% of the total Ca^{2+} in plants is associated with the cell wall. In apple fruits the cell wall also acts as the major site for Ca deposition. Although there is 'enough' Ca^{2+} in the cell wall of the apple, the concentration can be greatly increased with artificial infiltration (Conway and Sams, 1987; Poovaiah, 1986). In contrast, treatment with chelating agents reduces the Ca concentration of the cell wall (Letham, 1962). Pectic acids are the major sites for Ca^{2+} binding in the apple fruit (Knee, 1987). Cell wall proteins and hemicellulosic polysaccharides are also likely binding sites (see review by Hanson, 1984). As pectic compounds degrade during the course of ripening or by invasion of pathogens, the fruit softens. Increasing the Ca^{2+} levels of the cell wall by infiltration preserves the firmness of the fruit (Poovaiah and Reddy, 1987) and protects to a large degree against microorganisms that gain entry into the fruit by downgrading pectins (Conway and Sams, 1987).

In contrast, there is no evidence that in the leaves or in the branches additional Ca^{2+} would make any difference. Calcium deficiency symptoms in the leaves can only be seen if deficiency is induced in experimental conditions in sand cultures. Even in deficiency conditions symptoms of Ca deficiency can only be observed if the trees are in rapid growth (Shear, 1971). In orchards Ca deficiency of leaves has not been reported. It appears that in most conditions fruit trees have enough Ca^{2+} to satisfy the formation of the cell wall in leaves and in the wood.

Reaction of Ca^{2+} with the outer surface of the plasma membrane is the other documented effect in the apoplast. In low-Ca tissues or if Ca^{2+} ions are removed from the tissues by chelation, the cells become leaky (Van Steveninck, 1965). Leakage can be prevented by millimolar Ca concentrations. The immediate importance of leakiness of the plasma membranes is not apparent. The outer membranes are the ones that become leaky. Leakiness of tonoplast could cause compartmentalization problems within the cell. However, millimolar concentrations of Ca^{2+} that prevent leakiness would cause severe problems in the cytosol and thus the tonoplast cannot be the leaky membrane (Hanson, 1984).

Calcium in the medium also promotes K^+ uptake over Na^+. In apples Na toxicity is often observed when in the nutrient solutions Ca^{2+} is omitted and the Na forms of other salts are used. Sodium toxicity is expressed as leaf edge 'burning' in apples, cherries, and other tree leaves (Shear, 1971). Epstein (1961), studying the relative Na^+ and K^+ fluxes into roots, concluded that the Ca^{2+} promotion of K^+ uptake over Na^+ was primarily due to cation discrimination. The protective effect of Ca^{2+} against the influx of Na^+ is very important in irrigated orchards where the irrigation water has slightly increased levels of Na^+.

Calcium in the cytoplasm may regulate several enzyme activities none of which have been examined in detail in fruit trees. There is evidence that Ca also serves as a secondary messenger in fruits through its association with calmodulin (Poovaiah and Reddy, 1987). The most important effect of Ca in the cytosol is the regulation of respirational activity. Low-Ca fruit has a very high respirational rate, and the respiration is inversely related to Ca concentrations (Faust and Shear, 1972). The high respiration rate decreases the storability of apples where this effect is most important since apples are stored the longest among all fruits. In addition to the effect on CO_2 production, ethylene production also inversely relates to Ca concentrations (Conway and Sams, 1987). 'Ripening' is expressed by CO_2, C_2H_4, softening, and cellular disorganization. The effect of Ca^{2+} in preserving cellular organization in addition to preserving cell wall integrity is clearly visible in electron micrographs (Figure 2.13). Thus Ca exerts its effect upon ripening, a complex phenomenon, at each level where it has a controlling activity.

Calcium ions also regulate microtubule assembly and microfilament formation and through this enter into the cycle of cell division (Hanson, 1984). In low-Ca apples about 1 month after normal cell division is completed, the healthy enlarging cells undergo a redifferentiation via direct nuclear or amitotic divisions. This results in a promiscuous intracellular proliferation of daughter cells. The daughter cells remain in the confines of the original mother cell wall until it ruptures (Figure 2.14). The disorder is called 'cork spot' for its brown, hard, and dry-looking appearance. Once the proliferation is initiated, it is continuous throughout the season (Miller, 1980). Increasing the Ca content of the fruit prevents the disorder, but there is no evidence that Ca deficiency initiates it.

Uptake and Transport of Calcium Generally, it is believed that Ca^{2+} absorption takes place near the young root tips through the nonsuberized portion of the root. The uptake of Ca^{2+} has been extensively studied by Russel and Clarkson (1976). Their investigations revealed that in contrast to potassium and phosphate the uptake of Ca and Mg is restricted to an area just behind the root tips. The difference in uptake has been explained in terms of root structure and particularly the development of the Casparian strip. This suberized zone of the endodermis cell walls represents a strong barrier to water and solute movement, and therefore

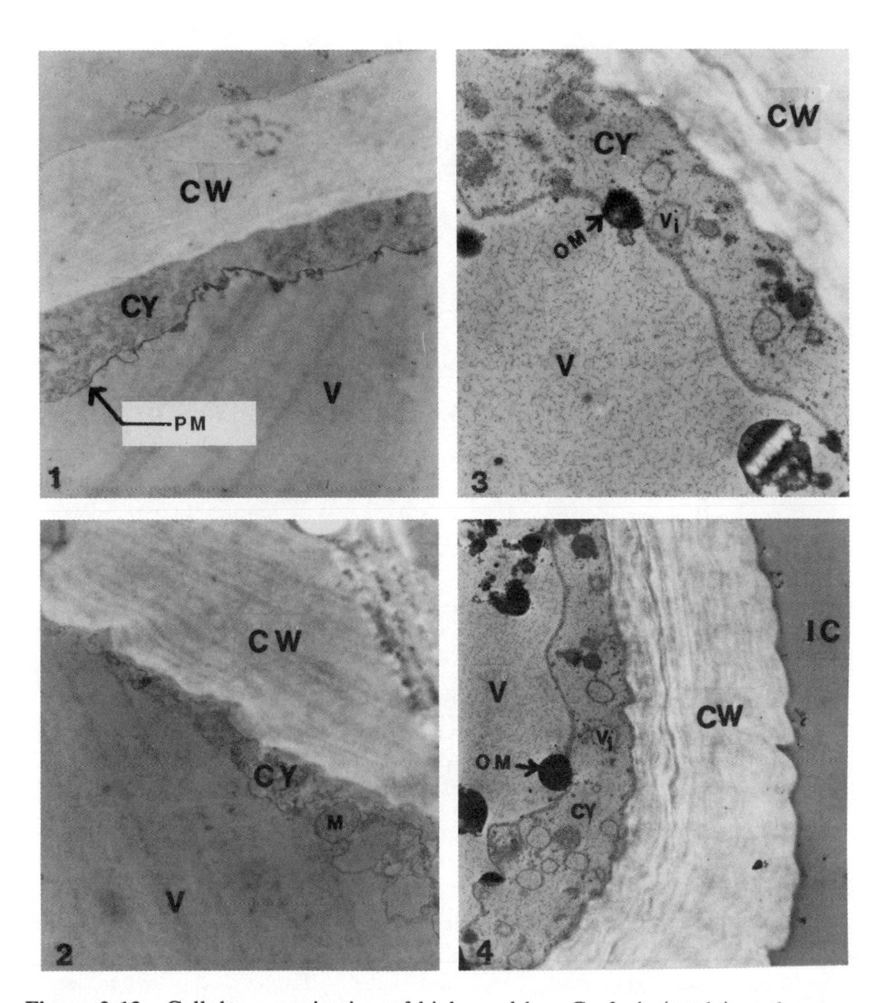

Figure 2.13 Cellular organization of high- and low-Ca fruit (apple) at the same time. Note the disorganization will also occur in high-Ca fruit upon ripening but at a later time: 1, 2, high-Ca fruit; 3, 4, low-Ca fruit. Abbreviation: CW, cell wall; CY, cytoplasm; V, Vacuole; OM, osmophilic body; PM, plasma membrane; M, mitochondrion; IC, intercellular space; Vi, vesicle (enlargement 10,640×).

continuous flow of water from the soil through the apoplast to the central cylinder is prevented. Water flow in roots with developed Casparian strips across the endodermis to the central cylinder has to follow the symplastic pathway. Since the endodermal cell walls are rich in plasmodesmata, water transfer between the cortex tissue and the central cylinder is not affected. However, Ca^{2+} is not transported effectively by the symplast. The movement of Ca^{2+} from the cortex to the central cylinder is there-

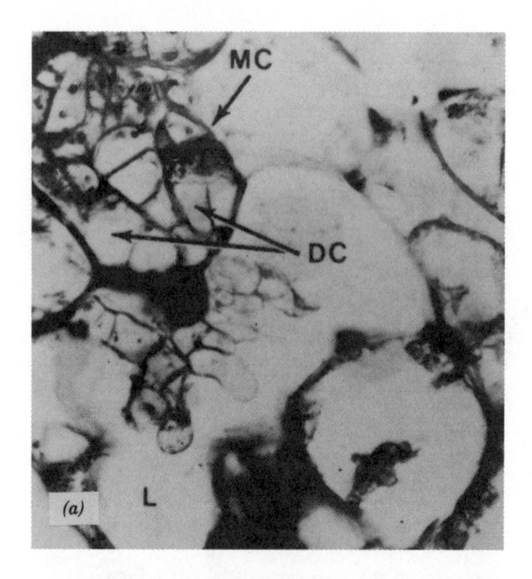

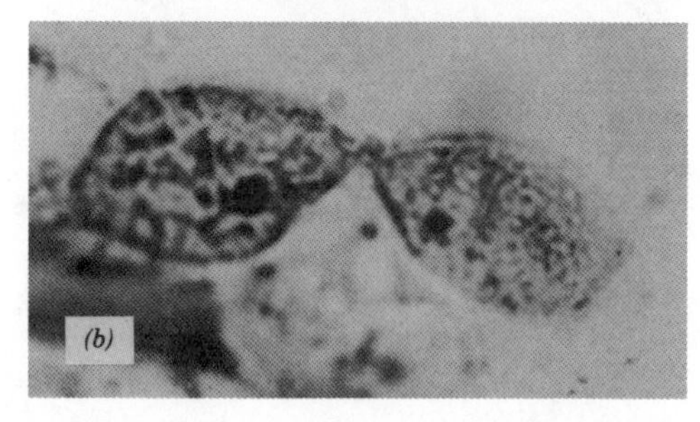

Figure 2.14 Amitotic cell division in low-Ca apples occurring after the normal cell division. The disorder called 'cork spot'. (*a*) Daughter cells, DC, formed within the mother cell, MC. The cell walls of the daughter cells braces the mother cell and the tissue is very firm. (*b*) Dividing nucleus. (Reproduced by permission from Miller, 1980.)

fore restricted to the apoplastic pathway, which is only accessible in nonsuberized roots (Russel and Clarkson, 1976). Although in general this may be the case, Atkinson (1986) reported that Ca has also been absorbed by the suberized portion of the cherry roots.

Although the apoplastic movement of ions is considered a passive process, in apples Ca uptake requires considerable quantities of photosynthates as a daily supply. Calcium uptake in apple seedlings in solution culture was greatly impaired when photosynthesis had been partially inhibited by herbicides or carbohydrate transport to the root had been prevented by removing a ring of bark. Calcium uptake could be restored in these experiments when sucrose was fed to the plant below the ring

(Faust, 1980). It is possible that the carbohydrate supply was needed in the preceding experiment to assure root growth for entry of Ca^{2+} through the nonsuberized new roots rather than to provide energy for uptake of Ca^{2+} per se. Nevertheless, Ca uptake in fruit trees must be considered as an energy-input-requiring process when the overall process is considered.

Translocation of Ca^{2+} has been also studied in plants in general and in fruit trees in particular (Faust and Shear, 1973). Similar to uptake, Ca^{2+} translocation is a passive process. Calcium in the xylem sap is translocated upward with the transpiration stream. Thus, to a large extent, transpiration controls upward translocation rate of Ca^{2+}. Shear (1980) could explain almost all metabolic disorders of fruit connected with Ca on the basis of water use by the trees. However, the movement of Ca^{2+} in the xylem vessels cannot be explained simply in terms of mass flow. Calcium ions are absorbed on the cation exchange capacity of the xylem, which is largely made up of carboxylic groups of pectins, phosphorylic and phenolic hydroxyl groups, and components of lignin. The xylem cylinder of apples operates as an iron exchange column, and the absorbed Ca^{2+} can be exchanged by other double-charged cation species (Shear and Faust, 1970). Distilled water cannot move Ca^{2+} upward, but Mg^{2+}, Zn^{2+}, Mn^{2+}, or Ba^{2+} solutions can effectively do so (Faust and Shear 1973; Shear and Faust, 1970). The common nonmetabolic use of the ion exchange system in the xylem vessels by a number of double-charged cations can easily explain the two-sided effect of Mg^{2+} on Ca^{2+} accumulation. When both elements are in low concentrations in the soil, the addition of Mg^{2+} enhances the accumulation of Ca^{2+} in the leaves of apples. In contrast, when Ca^{2+} is in sufficiently high concentration in the soil, addition of Mg^{2+} depresses Ca^{2+} accumulation. It appears that when Ca ions are insufficient in quantity to saturate the ion exchange complex, the addition of Mg ions would help in the ascent of Ca ions. In contrast, when Ca alone is able to saturate the system, any interference by Mg ions would decrease Ca movement.

There are other important aspects of Ca^{2+} translocation. In the growing plants Ca is translocated preferentially toward the shoot apex even though the transpiration rate of young leaves is much lower than of older leaves. This preferential movement in apples has been demonstrated by Shear and Faust (1970). The transport to the growing point is induced by the auxin indole acetic acid (IAA), which is synthesized in the shoot apex. It is believed that during growth, an IAA-stimulated proton efflux pump in the elongation zones of the shoot apex increases the formation of new cation exchange sites so that the growing tip becomes a center for Ca accumulation (Marschner and Ossenberg-Neuhaus, 1977). Bangerth (1979) suggested that basipetal IAA transport forces Ca^{2+} to be translocated acropetally. This is important in apple fruit because Ca^{2+} transport into the fruit greatly decreases about the time IAA synthesis in the fruit subsides. Studies with 2,3,5-triiodobenzoic acid (TIBA) with apple fruit have shown that treatment of the petiole with TIBA decreased fruit Ca concentration and increased metabolic disorders connected with Ca insufficiency (Oberly, 1973; Stahly and Benson, 1970).

An important aspect of the preferential transport of Ca to the shoot tip is that when shoot growth is strong during the summer, Ca bypasses the fruit and is transported to the growing shoots. This limits the Ca accumulation in the fruit only to the period before rapid shoot growth. Therefore shoot growth needs to be controlled to improve Ca accumulation into the fruit.

The rate of downward movement of Ca^{2+} is very low due to the fact that Ca^{2+} is transported only to a very small extent in the phloem in plants in general (Wiersum, 1979). This indicates that Ca supplied to fruit trees as a spray to increase fruit Ca must be applied to the fruit itself because Ca applied to the leaf will not translocate to the fruit or to any other part of the tree. Wienecke and Führ (1973) claimed that ^{45}Ca applied in one year to the leaves of the tree was eventually deposited to the wood and remobilized the next spring when the tree needed Ca. This indicates that in late fall, before leaf fall, translocation of Ca had to occur. Faust (unpublished) fed high concentrations of ^{45}Ca to the cut midrib of apple leaves. Under these conditions Ca translocated to other parts of the leaves. Apparently if the concentration gradient from the leaf to the tree is higher than from the tree to the leaf, translocation will occur probably through the ion exchange system of the xylem. Such an occurrence is unusual and only occurs in experimental conditions or if cell structures of the senescing leaves disintegrate and Ca is no longer bound to the cellular components, creating relatively high concentration in the leaf. This explains Wienicke's finding that Ca of leaves can be reused.

Calcium Accumulation in Leaves and Fruit Calcium accumulation in leaves of fruit trees continues throughout the entire season. Since Ca is translocated with the transpiration stream and the leaves receive water throughout the season, they also accumulate Ca. In old leaves Ca usually remains in the major veins of the leaf. If Ca concentration exceed 0.8–1.0% of the dry weight of the leaf, Ca is sufficient for growth-related activities. However, leaf concentrations closer to 2% are required to assure that the fruit is high enough in Ca. Leaf Ca concentration is not a good predictor for fruit Ca levels. The only statement one can make on this subject is that when leaf Ca concentration is above 1.8% of the dry weight, Ca influx into the tree had to be continuous throughout the entire season. In such a case it is highly likely that the fruit also received sufficient levels of Ca and fruit Ca concentrations are also high.

Calcium accumulation into apple and pear has been studied more than into any other fruit. In general, accumulation of Ca occurs only during

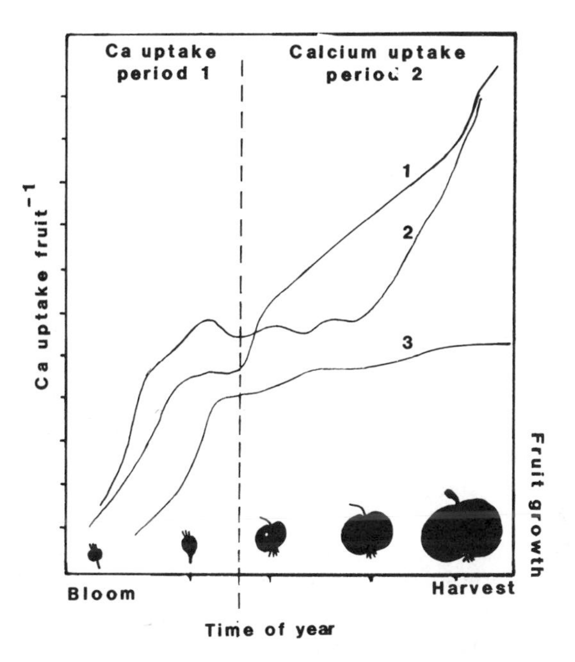

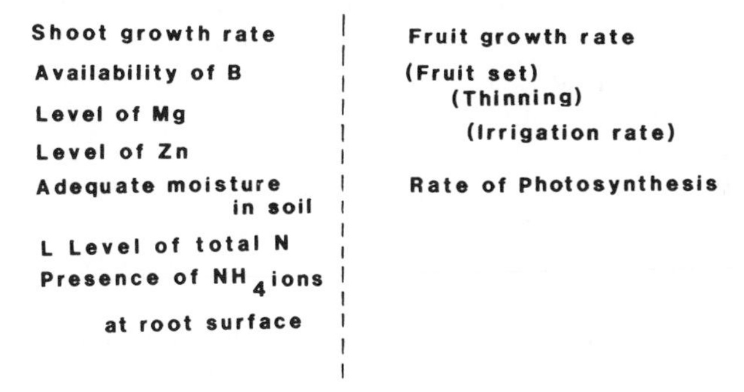

Figure 2.15 Calcium accumulation in fruit. Curve (1) Ca uptake in Washington state; (2) Ca uptake in occasional years in New York state; and (3) Ca uptake in most years in New York state.

the first portion of fruit growth. Thereafter the total Ca remains constant (Figure 2.15). However, because fruit growth continues, Ca concentration decreases depending on the rate of final swell of the fruit. This is important in determining storage quality. Large fruit, in which Ca dilution rate has been high, usually has very poor storage quality. Small fruit, in contrast, has high Ca concentration and stores well. There are reports that Ca uptake into the fruit may continue later into the season at certain locations. Accumulation of Ca into apple fruit in Washington state has been reported (Rogers and Batjer, 1954), to continue throughout the summer. The reason for this is not apparent. It is possible that the rate of photosynthesis is better at this location than other locations when Ca uptake into the fruit is restricted to the first period of growth. Higher rates of photosynthesis may assure root growth and Ca uptake at levels that can assure supply to the fruit after satisfying the need of the growing shoots.

Factors affecting accumulation of Ca are illustrated schematically for apples in Figure 2.15. In influencing Ca uptake, the uptake curve should be divided into two major sections. The first section is the rapid uptake, which usually lasts 4–6 weeks after bloom. During the second section, Ca uptake is limited and fruit growth is rapid, which dilutes the Ca that has already accumulated in the fruit. Thus during the first period, all the prerequisites for uptake need to be assured, while during the second period the excessive fruit growth and dilution need to be prevented. Factors that are important in the uptake of Ca are the presence of NH_4^+ in the root zone, which decrease Ca absorption (Shear, 1971); Mg ions, which compete with Ca transport (Shear and Faust, 1971); rapid shoot growth, which presents competition to Ca transport into the fruit; B ions, whose role is not clear, but they may function by facilitating higher IAA levels in the stem or in the roots; and sufficient moisture content in the root zone, which assures continuous Ca movement into the root and up the xylem to the fruits (Shear, 1982). Strong winter pruning induces a high rate of shoot growth and produces relatively few and consequently large fruit. Strong pruning is known to influence all the Ca-related problems in the fruit. High rates of N fertilization may have the same results as far as promoting shoot growth is concerned. Effects of strong pruning and high rates of N applications are additive for inducing Ca-related problems. Influencing Ca concentrations of the fruit during the second period requires avoiding all action that make the fruit overly large. These include overthinning, strong pruning, and excessive irrigation during the period before harvest.

Calcium Deficiency Disorders There is ample evidence that deficient Ca levels cause various disorders. In case of severe deficiency the symptoms can be observed as deformed and chlorotic leaves. Chlorotic leaves most often occur when shoots are expanding at a fast rate (Shear, 1971). The affected tissues may become soft because of insufficient development of cell walls. Brown substances develop in the intercellular spaces and also in the vascular system. Leaf margin burning is often described in fruit trees as part of the symptomatology of Ca deficiency. Ca deficiency symptoms are usually induced in sand cultures or in other experimental

TABLE 2.9 Calcium-related Disorders of Apple Fruit

Disorder	Cause	Reference
Cork spot	Abnormal cell division	Miller, 1980
Watercore	Sorbitol accumulation	Marlow and Loesher, 1984
Bitter pit	Unknown	—
Jonathan spot	Unknown	—
Deep cracking	Water influx	Shear, 1971
Lenticel spot	Unknown	—
Gleosporium rot	Fungus	Sharples, 1980
Raised lenticels	Unknown	—

systems. However, in the nutrient solutions used in studies inducing Ca deficiency, Ca salts were most often substituted with Na salts. The consequence of this is that the burning of leaf margins is caused by sodium toxicity rather than Ca deficiency. Thus burning of leaf margins cannot be accepted as a Ca deficiency symptom (Shear, 1971). Insufficient Ca levels are also associated with various disorders. Disorders develop most often in fruit where Ca concentrations are low. In apples alone a considerable number of Ca-related disorders can be listed (Table 2.9). Although a high concentration of Ca will prevent all of these disorders, they do not always develop when Ca is low, and they cannot be labeled Ca deficiency disorders. Each of these disorders has a set of causing circumstances, some of which are known and others are not. If Ca concentrations are low and the causal circumstances are present, the disorder will develop; if the causes are not present, fruit will stay free of the disorder. Even though high Ca concentrations prevent these disorders, many of them can be prevented by methods other than Ca.

On rare occasions in peaches, a disorder has been observed that is quite similar to the blossom end rot of tomatoes, and the affected peach fruit is lower in Ca than the fruit free from the disorder. Cork spot in pears (Woodbridge, 1971) is also a Ca-related disorder that affects deciduous fruits other than apples. Many tropical or subtropical fruits are also subject to disorders preventable by Ca.

Calcium-related disorders develop in low-Ca fruit. Yet analysis of the affected tissues reveals high Ca concentrations in the lesion itself. Faust and Shear (1968) claim that the high rate of metabolic activity that accompanies the development of most disorders attracts Ca into the affected tissue. The Ca level of tissue in the incipient stages of lesion development is low, and it becomes high by the time the lesion is fully developed. The distance from which Ca is attracted into the affected tissue is not known.

The tissue N concentration in apple has an overriding effect on the development of Ca-related metabolic disorders. Shear (1974) reported

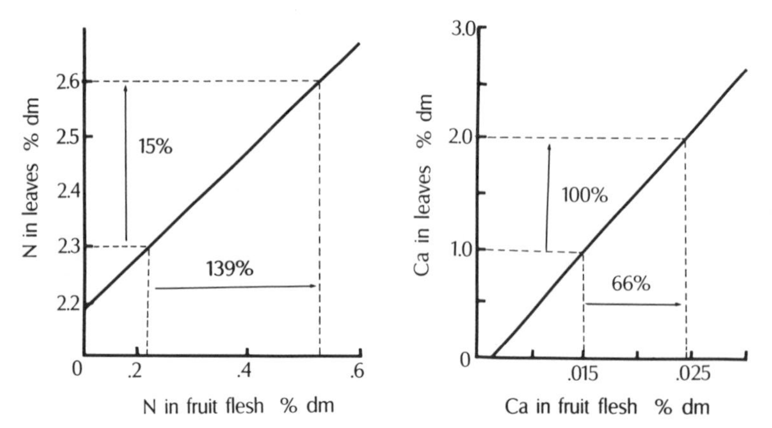

Figure 2.16 Relationship between leaf and fruit N and Ca. (Reproduced by permission from Shear, 1974.)

that N–Ca ratios, determined as percentage of dry weight in the flesh of the apple, varied between 10 and 30. When the ratio was 10, almost certainly no metabolic disorder develops. In contrast as a N–Ca ratio of 30 the occurrence of the disorder was very high even at the same level of Ca concentrations. Shear worked with two disorders, bitter pit and cork spot. Thus in many cases it is simpler to prevent Ca disorders by lowering the N concentration of the tree rather than by increasing the Ca concentration of the fruit. Most horticulturists do not realize that high-N nutrition increases the N concentration in the fruit unproportionally higher than N concentration in the leaves. Figure 2.16 illustrates the relationship between leaf and fruit flesh N and Ca. A relatively modest increase in leaf N concentration (15%) in apple is associated with more than doubling of N concentration in the fruit flesh (135% increase). In contrast, doubling the leaf Ca concentration only increases fruit Ca concentrations by 66%. Therefore, avoiding Ca-related disorders, the leaf N levels in apple must not be higher than 2%.

High concentration of Mg, especially if used in the form of sprays, aggravate the occurrence of bitter pit of apples. We have discussed the fact that Mg interferes with Ca transport and could lower the fruit Ca concentrations. However, when Mg is supplied as a spray, the interference with Ca transport in the xylem during the upward movement of Ca must be discounted. Thus Mg is likely to interfere with one or more Ca function at the cellular level in producing Mg-induced, Ca-related metabolic disorders.

Several workers have considered antagonistic relationships between Ca^{2+}, Mg^{2+}, and K^{+} in connection to the severity of bitter pit. They developed ratios of Mg to Ca and Mg+K to Ca, none of which explains meaningfully the relationship between these nutrients and the development of bitter pit. By considering the major effect of both K and Mg and substituting them in the preceding relationships, interesting speculations can be developed. The major role of K involves protein synthesis, essentially the same as the N effect. Magnesium interferes with the function of Ca. Thus the relationship can be rewritten using the roles of K and Mg as follows: promotion of protein synthesis plus interfering with Ca function over the existing Ca concentration. If the relationship is rephrased as stated above, it is easy to see that both Mg and K have major roles in the development of bitter pit.

Potassium

Potassium is a major nutrient that needs to be supplied in relatively large quantities to crop plants in general and to fruit trees in particular. Plant nutritionists have been intrigued for decades by this high requirement of K^+. The concentrations required by K^+-activated enzymes do not explain this high demand (Hsiao and Lauchli, 1986). However, the role of K^+ as a major determinant for protein synthesis and its role as an osmoticum maintaining the water status of the cells can account for the quantities needed. It is likely that these two functions are predominant in the fruit trees as well. For basic aspects of K^+ nutrition, reviews by Marschner (1986) and Hsiao and Lauchli (1986) should be consulted.

Potassium Requirement of Fruit Trees The potassium requirement of fruit trees is nearly equal to the requirement for N and Ca. In general 1.5–1.8% of the dry matter is considered adequate for most trees in the midshoot leaves sampled at the end of July. Deficiency symptoms are visible only when leaf dry-weight concentrations are below 0.7%. The main visible symptom of the deficiency in all fruit tree species is leaf scorch. First the leaf loses its normal green color; then it displays a water-soaked appearance and finally develops necrosis. In general scorch proceeds inward from the edge of the leaf (Ballinger et al., 1966; Boynton and Oberly, 1966). At the 1% leaf K level the leaves are free of visible symptoms, but the fruit does not color normally (Cain, 1948). The effect of K on fruit color is probably indirect. Color development occurs in apple when considerable quantities of sugars are present in the fruit (Faust, 1965). It is likely that insufficient K concentration decreases the photosynthesis of leaves, which in turn lowers sugar concentrations.

Growth of the trees is also affected when the level of K is below 1% in the leaves. Batjer and Degman (1940) presented data obtained in greenhouse experiments that with such low levels growth decreased, as reflected in total height of trees, linear length of branches, total linear growth, and trunk diameter. The K effect could be attributed to lower net

CO_2 assimilation. Although it was relatively easy to demonstrate the association between K nutrition at suboptimal level and growth in apples in greenhouse studies, such association could not be demonstrated in orchards (Boynton and Oberly, 1966). In orchard conditions growth is only affected when deficiency is present, as evidenced by leaf scorch.

In spite of marked differences obtained in shoot dry weight in a series of experiments in apple (Tromp, 1980), the uptake of K per shoot dry weight was unaffected by the relative humidity of the air and light intensity. This indicates a linear relationship between growth and K^+ uptake. Tromp (1980) argued that K^+ is taken up in proportion to the demand exerted by the growing shoot. In determining the total demand of a fruit tree for K^+, the requirement of the leaves as well as the fruit needs to be considered. As discussed later, the demand of the fruit for K^+ is high, and fruiting trees must be supplied with K to the point that the total 'demand' can be satisfied.

Role of Potassium in the Physiology of Fruit Trees Potassium is the most abundant cation in the cytoplasm. It has important roles in pH stabilization, osmoregulation, enzyme activation, protein synthesis, stomatal movement and photosynthesis, and cell extension. The various functions of K^+ in plants have been reviewed by Lauchli and Pfluger (1979).

It has been known for a half century that K^+ deficiency can reduce transpiration in apples (Childers and Cowart, 1935) similar to other plants. Stomatal opening requires K^+ accumulation in the guard cells, and K^+ deficiency could lead to stomatal closure, which results in reduced transpiration. For a review on the stomatal role of K^+ see Hsiao and Lauchli (1986).

Potassium is an important solute in expanding cells, and expansive growth is very sensitive to K^+ deficiency. Cell extension involves the formation of a large central vacuole that especially in the fruit can occupy 90% or more of the cell volume. It is well established that cell extension is the consequence of the accumulation of K^+ in the cells and in the vacuole (Marschner, 1986). Potassium is needed for the enlargement of fruit (Fisher et al., 1959). In pears a response in fruit size was obtained between 0.7 and 1% K in leaf dry matter. No effect was observed if the leaf K concentration was above 1% prior to the application of K (Fisher et al., 1959). Potassium and reducing sugars can act in a complementary manner to produce the osmotic potential needed for cell enlargement in some plants (Marschner, 1986). Whether K^+ and sorbitol, the major osmotically active compound in fruits, can act in a same complementary manner in the enlargement of fruits is not known.

Potassium plays an important role in counterbalancing immobile anions in the cytoplasm and mobile anions in the vacuoles as well as in the xylem and phloem (Marschner, 1986). The accumulation of organic acids is often the consequence of K^+ transport without accompanying anions into the cytoplasm. The unbalanced K^+ requires stoichiometric synthesis of organic acids for charge balance. In peaches K fertilization resulted in an increase in titrable acidity in the fruit (Kwong and Fisher, 1962). The acid produced in fruit trees is often malate, which may be retranslocated to the roots or accumulates in the tissue. Malate is synthesized in large quantities in Fe-deficient apples where protein synthesis is impaired; thus K^+ is in excess, requiring the synthesis of acids for charge balance (Sun et al., 1987). Organic acid and K^+ also accumulate in the affected spots of the apple disorder 'Jonathan spot'. Investigators looked upon the potassium–acid accumulation complex as the cause of the disorder. It is more likely that other synthetic processes (perhaps proteins) were impaired, leading to the accumulation of free K^+, which in turn triggered malate synthesis.

Potassium Accumulation in Leaves and Fruit In heavily cropping apple and prune trees both dry matter and K^+ accumulate in the fruits in comparison to defruited trees, which accumulate the dry matter and K^+ in their leaves (Hansen, 1980). The total dry matter and K^+ content of both types of trees is the same. Thus it is clear that the sink power of the fruit is the same for carbohydrates and K^+. In contrast, very little Ca moves into the fruit and the distributions of P, N, and Mg are intermediate (Hansen, 1980).

The association between carbohydrate content and K^+ in fruits is evident from several different types of data. In apple various fruit–leaf ratios cm^2 from 200 to 900 changed the potassium percentage of dry matter very little, indicating that the K^+ content of tissues is directly related to their carbohydrate content (Hansen, 1980). A positive correlation between K and the soluble dry matter or acid content of fruits is well demonstrated (Perring and Preston, 1974; Wilkinson, 1958).

The sink power of fruits over other parts of the tree for K^+ is illustrated by additional observations. In prunes K deficiency symptoms usually develop when the tree carries excessive fruit load. The fruit apparently has such high requirements for K that it competes to the extent that the vegetative parts of the tree suffer and the leaves show scorch (Lindner and Benson, 1954). Similar results have also been reported for peaches. Heavy crop lowers the K level of leaves to the magnitude that this should be taken into account when interpreting leaf analysis data (Popenoe and Scott, 1956).

Phosphorus

In plant cells, phosphorus (P) either remains as inorganic phosphate (P_i), esterified to a carbon chain, or is attached to another P by an energy-rich pyrophosphate bond. Phosphorus also functions as a structural element in the macromolecules of DNA and RNA and serves as the bridging molecule in phospholipid biomembranes. One of the most known roles of P involves the energy transfer mechanism including the generation of ATP and the formation of sugar and alcohol esters. In addition to these roles, P also has a regulatory function in many enzymic processes where P_i controls the rate of the reaction.

Uptake of Phosphorus by Fruit Trees Roots of apple trees take up P when the soil solution contains concentrations between 0.5 and 10 μM. The roots are unable to take up P if the solution concentration is in the range 0.25–0.50 μM (Bhat, 1983). There are large differences in calculated nutrient inflow rates into the roots. Calculations range from the low 0.1–0.2 pmol $cm^{-1} s^{-1}$ for apple (Bhat, 1983) to 1.3–2.8 pmol $cm^{-1} s^{-1}$ for cherry (Atkinson and Wilson, 1979). Nutrient uptake in August is apparently higher than earlier in the season. Asamoah (1984) measured 0.2 pmol $cm^{-1} s^{-1}$ uptake rates early in the season in contrast to rates of 2.1 pmol $cm^{-1} s^{-1}$ in August. During the early season, when uptake rates are low, the initial P requirement of the tree can be met by redistribution. Atkinson (1986) estimated that about 14% of the P in nursery trees was available for retranslocation during resumption of growth in early spring. The conditions in orchards are usually not optimal for nutrient uptake, and consequently a lower uptake rate can be expected where orchard trees are concerned. Atkinson (1986) recalculated the data obtained by Batjer et al. (1952) on the balance of P in various tissues and found a probable uptake rate of 0.19 pmol $cm^{-1} s^{-1}$ in orchards. Atkinson (1986) also suggested that the rate of inflow needed to supply the demand of both leaves and fruit is around 0.12 pmol $cm^{-1} s^{-1}$. Thus at least in orchards in the western United States, where Batjer et al. (1952) gathered their data, the inflow rate of P satisfies the P demand of the tree.

Accumulation of Phosphorus in Leaves and Fruit Accumulation of P in leaves reaches the maximum in July. The uptake of P into fruit of apple closely follows the weight increase of the fruit, and uptake continues until harvest (Wilkinson and Perring, 1964). There is a correlation between leaf and fruit P content in apple (Atkinson, 1986). When leaf P concentration on a dry-weight basis is 0.26–0.28%, the fruit concentration is about 3 times less, 155–165 $\mu g\, g^{-1}$ fw. When leaf P concentration is in the range of 0.18–0.20%, fruit concentration decreases to 107–112 $\mu g\, g^{-1}$ fw. Sharples (1980) estimated that for acceptable apple quality a minimum level of 110 $\mu g\, g^{-1}$ fw of the fruit is required. Phosphorus deficiency symptoms on leaves or other parts of the tree have not been observed in orchards. The concern is the fruit P concentration. Phosphorus levels in the apple have been positively correlated with fruit firmness and negatively with low-temperature breakdown (Sharples, 1980); therefore it is imperative to have high enough P levels in the tree. The difference in apple leaf concentrations between satisfactory and insufficient P concentrations is very little. Based on fruit quality analysis, Sharples (1980) suggested a satisfactory range of 0.20–0.25% in leaves on a dry-weight basis and a minimum of 0.24% for good fruit quality. The range of P concentration in orchard leaves of apple varies greatly. Kenworthy and Martin (1966) assembled an impressive list of reported P concentrations in apple that ranged from 0.08 to 0.39% of leaf dry weight. Phosphorus concentrations of the wood and bark are much less, usually in the range of 0.04–0.09%.

Leaf concentration of P in other fruit species (peaches, plums, and cherries) is essentially the same as in apples with pears containing somewhat less.

Effect of Soil Management System on Phosphorus Uptake It has been noted that low-temperature breakdown in apple fruits, associated with low fruit P levels, most often develops in fruit grown in clean-cultivated orchards. Such reports have been generated in England, where clean cultivation was experimentally tried (Atkinson, 1986), and from Hungary, where apples are traditionally grown with clean cultivation (Kallay and Szucs, 1982). In contrast, no concern has been expressed about low fruit P levels in the United States, where apples are grown in sod with herbicide strips in the row.

Adoption of total-herbicide clean cultivation reduced the average apple P concentration from 116 to 104 $\mu g\, g^{-1}$ fw of fruit (Johnson et al., 1983). Atkinson (1986) attributed the diminished P concentrations to limited root growth during the period P is taken up to the fruit. In early June P uptake decreased sharply regardless of soil management treatments, with the exception of irrigated sod (Atkinson and White, 1980). At the same time (June), root length increased under irrigated sod, and it was maintained throughout the fruiting season. In contrast, root length was not increased but was maintained in nonirrigated sod treatments, and it was not maintained and it has never increased in total herbicide treatment or in the use of herbicide strips. The irrigated sod treatment had five times the root length visible in the observation windows of the laboratory compared to total herbicide (Atkinson and White, 1980). From this data Atkinson (1986) concluded that in the herbicide-treated clean plots, root length was insufficient to absorb the P necessary for the fruit.

The fruit itself apparently absorbs sufficient quantities of P from the xylem sap so that xylem sap concentrations of P are greatly lowered. Hansen (1980) found much higher quantities of P in the xylem sap of defruited trees than fruiting trees. At the same time the leaf concentrations of P were not different from normal levels (0.23–0.24% dw).

The Importance of Mycorrhizae on Phosphorus Absorption Endomycorrhizae grow both inter- and intracellularly in the root cortex of fruit trees. Although there are several groups, the most common is the vesicular–arbuscular mycorrhiza (VAM). The fungus grows in the cortex and develops lipid-rich bodies (vesicles) and highly branched structures (arbuscules) within the host cells. The VAM are obligate symbiotic fungi, and they are not very host specific. Mycorrhizae often increase the growth of fruit trees. In most cases the increased phosphorus uptake is the primary cause of observed growth or performance improvement of mycorrhizal plants. In general growth responses due to mycorrhiza are not observed when sufficient quantities of P are supplied to the soil. Mosse (1957) found that apple seedlings or cuttings infected with *Endogone* sp. produced larger plants. Doud-Miller (1983) investigated several sites in the United States with a wide range of available soil P concentrations. There were three to eight species of fungi per site, the most common genera being *Glomus* and *Gighaspora*. Granger et al. (1983) inoculated M7 and MM111 apple rootstocks with *Glomus epigaeum*, but only M7 responded to the inoculation with increased P concentrations. On low-P soils, with 13 ppm exchangeable P, *Glomus fasciculatum, Glomus mosseae*, and *Gighaspora margarita* all increase the growth of apple seedlings (Geddeda et al., 1984). In another set of experiments in the absence of P fertilizer, *G. mosseae* increased growth in three of five soils, but when P fertilizer was applied, only one of five soils gave positive results (Hoepfner et al., 1983).

In general, when P is applied, the effect of mycorrhizae in increasing growth is minimal. In high-P soils even mycorrhiza infection is limited on apples. The use of P fertilization decreased infection, although infection was found up to 400 mg P kg^{-1} soil on all soil types used but only on some soil types when the rate of application was increased to 600 mg P kg^{-1} soil (Hoepfner et al., 1983). The rate at which infection occurred was lower in another set of conditions. Infection with 1 of 14 mycorrhizal fungi occurred in soils amended with 30–200 mg P kg^{-1} soil, but no growth effect was observable at 200 mg P kg^{-1} soil (Doud-Miller, 1983). Atkinson (1983) assessed the degree of infection on the roots of Cox MM106 apples under grass and under clean, herbicided cultivation. Early in the season infection was higher under irrigated sod in comparison to the clean orchard floor obtained by herbicides. He suggested that the early season difference was due to increased root production under irrigated sod (Atkinson, 1986). The length of infected roots is likely to be influenced by the total length available for infection.

Seedling growth of peaches and growth and survival of tissue-culture-propagated nursery material is often unsatisfactory in fumigated nursery soils. Inoculation of such soils with soil containing mycorrhizal spores or growing strawberry plants on such soil for 1 year usually restores the growth of the planted tree material to the normal level. Among the horticultural plants, strawberries are known to be the best in supporting mycorrhizal growth.

Responses of Trees to Phosphorus Nutrition Cherries usually do not respond to P fertilization. Westwood and Wann (1966) summarized the available data in this respect. Anne et al. (1952) reported that trees with low N content were unusually high in P, and such trees were unhealthy. In general, abundant P supply tend to accentuate N deficiency, and in contrast, many tree species grown with high N tend to respond to P applications. Peach trees are more sensitive to P deficiency than apple trees. However, peach trees do not show much growth response under heavy P treatment (Ballinger et al., 1966).

A few reports in the literature indicate a definite response of apricots, prunes, and plums to P fertilization. In California, where annual crops show a marked increase in growth following P applications, stone fruits on the same P-deficient soils show no indication of insufficient P levels and do not respond to P fertilization except when newly planted (Lilleland and McCollam, 1961). Seedlings of stone fruits have a higher requirement to P and respond to application of P more like annual crops. In some cases seedlings only respond to P fertilization when N is also applied (Bohomaz, 1960). In nutrient solution plum trees started to show deficiency symptoms at the time when P levels were 0.14% in the dry matter of the leaves. Pears could tolerate lower levels of P than other fruit crop. Bould (1966) summarized the available data, which indicate that 0.14–0.16% P in the dry weight of leaves is satisfactory for pears.

Magnesium

Magnesium is needed and taken up by fruit trees in lower quantities than Ca. The content of Mg in the leaves of fruit trees is on the order of 0.3–0.5% of dry matter, which compares to 1.2–1.8% of Ca. The rate of Mg uptake can be strongly depressed by competing cations such as K^+, NH_4^+, Ca^{2+}, and Mn^{2+} as well as H^+, that is, low pH (Marschner, 1986). In fruit trees, among the ions listed above, the influence of K in repressing Mg uptake is the most important because orchards are fertilized relatively heavily with K. Early studies (Boynton and Burell, 1944; Wallace, 1939) Kidson et al., 1940–1941; Southwick, 1943; all reported that K fertilization to some extent aggravated or induced Mg deficiency symptoms. It is noteworthy that in orchards, when apple leaves show Mg deficiency symptoms, they are usually low in Mg but high in K even on soil that has little available K (Boyton and Oberly, 1966). While K fertilization depresses leaf Mg concentration in peach similar to that described above for apple, the Mg concentration in other tissues of the tree (fruit, branches, and trunk) is not affected (Cummings, 1973).

In plant tissues a high proportion of Mg, often over 70%, is diffusible and is associated with inorganic anions and organic acid anions such as malate and citrate (Mengel and Kirkby, 1982). About 10–20% Mg is localized in the chloroplast, less than half bound to chlorophyll; the other half serves as activator of ribulose biphosphate carboxylase. Magnesium also has an essential role as a bridging element for the aggregation of ribosome subunits (Marschner, 1986). When Mg is insufficient or K is excessive, the subunits dissociate and protein synthesis ceases. In leaf cells about 25% of the total protein is in the chloroplast. This explains why Mg deficiency particularly affects the size, structure, and function of the chloroplast, including electron transfer in photosystem II (Marschner, 1986). Among the other roles of Mg, its role in the transfer of high-energy phosphate in ATP metabolism is significant. The substrate for most ATPases is Mg–ATP (Marschner, 1986).

It is already mentioned that Mg deficiency causes chlorosis. The chlorosis is interveinal, and only extreme cases become necrotic. The affected leaves drop later in the season. By mid-August Mg-deficient trees are usually partially defoliated. The degree to which the symptoms develop on affected trees varies greatly. Only a few branches may be affected or the whole tree uniformly scorched and defoliated. Magnesium is a mobile ion in trees, and in case of deficiency Mg is transported from the old to the young leaves. Consequently, the old leaves are the first affected by Mg-induced chlorosis and the first to abscise.

The fruit requires considerable amount of Mg. While in the leaves of apple the Ca concentration on a dry-weight basis is about 5 times that of Mg in the fruit, the Mg concentration is twice as high as that of Ca. If trees are marginally supplied with Mg, those with fruit will show the Mg deficiency first or the Mg deficiency in the leaves of fruiting trees will be more severe. Starch usually accumulates in Mg-deficient tissues (starch decomposition, sucrose formation, and phloem loading require energy and ATPase activity, which all depend on Mg), and thus photosynthates are not partitioned into the fruit. Magnesium-deficient trees also usually produce small fruits. The overall photosynthesis of the tree is also severely affected by defoliation. Thus the small fruit size of Mg-deficient trees may be due to more than one factor affecting photosynthesis.

Similar to Ca, the rate of uptake of Mg appears to be genetically controlled (Korbam and Swiader, 1984). Fruit of certain progenies of apple contained more Mg than others. The same progenies were also susceptible to bitter pit, a disorder of apples. Fruit that is high in Mg is also low in Ca, pointing to the relationship between these elements discussed previously.

Magnesium deficiency does not affect the rate of growth of fruit trees. Edwards and Horton (1981) reported that the leaf Mg concentration of peaches was higher, stem Mg concentration was slightly higher, and root Mg concentration was not affected by increasing the Mg supply in the nutrient solution. In contrast, growth rate remained the same. Rootstocks rarely if ever influence leaf or stem Mg levels in apple (Poling and Oberly, 1979) or in peach (Knowles et al., 1984), but self-rooted peaches contain significantly more Mg when compared with leaves of the same cultivar grafted on four different rootstocks (Couvillon, 1982). The year-to-year variation in the Mg concentration of leaves is relatively large compared to yearly variation in N, P, or Ca. The variation could be as high as 20% from year to year (Knowles et al., 1984).

As the tree ages, the Ca concentration of the annual rings decreases. In one study (Blanpied and Oberly, 1978) the rate of decrease was from 2000–2200 ppm to about 600 ppm during about 35 years at two different locations, New York state and British Columbia, Canada. Magnesium has shown a very different pattern across the wood of apples with essentially no change during the 35-year period in British Columbia and only some decrease from 300 to 250 ppm in New York. It is not known why the tree in its young years can deposit Ca but cannot maintain this activity as time goes on. Similarly it is not known why the tree can maintain the uptake and deposition of Mg throughout its life. However, the practical consequence of this is that it is impossible to correct Ca insufficiency in an old tree while correcting Mg insufficiency is not a problem.

Magnesium toxicity has not been observed as such in fruit trees. However, some disorders may be accentuated or their development may be triggered by a high concentration of Mg. All these effects are known either to lower Ca concentration in the fruit or perhaps counteract the beneficial effect of Ca. The most well known of these disorders is bitter pit of apples, a collapse of cell clusters usually after storage of the fruit. As was pointed out previously, the cause of bitter pit is not known; the presence of sufficient Ca prevents it and high Mg accentuates it (Askew et al., 1960; Bunemann, 1962; Martin et al., 1962).

UTILIZATION OF MINOR NUTRIENTS BY FRUIT TREES

Iron

Iron is an essential constituent of redox systems in plants and occurs naturally in fruit trees. Iron is capable of valency changes, and this can facilitate the shifting of electrons in the redox systems of which Fe is an

TABLE 2.10 Chlorophyll Content ($\mu g\ cm^{-2}$) of Iron-deficient Apple Leaves[a]

	Chlorophyll a	Chlorophyll b
High Fe	130	33
High Fe + $CaCO_3$	95	22
Low Fe	71	16
Low Fe + $CaCO_3$	35	7
No Fe	18	3
No Fe + $CaCO_3$	4	1

[a] After Sun et al., 1987.

essential component. The iron-containing redox systems are the hemoproteins and the Fe–S proteins. Cytochromes, the most important of the hemoproteins, are located in the chloroplast and in the mitochondria. Other heme enzymes are the catalase and peroxidases, which facilitate the dismutation of H_2O_2 to water and O_2 in case of catalase and water and an organic hydrogen donor in case of peroxidases.

Among the nonheme iron-containing redox systems ferredoxin, an Fe–S–containing protein is the most important. Ferredoxin serves as an electron acceptor–donor system in fruit trees in photosynthesis and nitrite reduction. Iron is a component of aconitase, which catalyzes the isomerization of citrate to isocitrate. This is important in fruit trees when inhibition of this enzyme causes accumulation of organic acids.

Symptoms of Iron Deficiency Most of the Fe is in green leaves (over 80%) and is located in the chloroplast (Terry and Low, 1982). Naturally Fe-deficient trees show various degree of interveinal chlorosis (Korcak, 1987). Chlorophyll formation under Fe deficiency conditions is greatly decreased largely due to impaired protein synthesis (Table 2.10). Chloroplasts are small in leaves of apples suffering from Fe chlorosis; the stoma of chlorotic apple leaves is poorly developed and does not have the characteristic structure (Zhou et al., 1984a) (Figure 2.17).

Light emission data of green leaves of apple seedlings grown before the plants were subjected to Fe stress show great changes in the emission region characteristic to disturbances in photosystem 2. Thus, apparently, when chlorosis develops later in the season, leaves developed before the chlorosis that apparently green are still affected (Faust, unpublished).

The expression of Fe chlorosis may be confounded by the occurrence of simultaneous micronutrient deficiencies (Korcak, 1987). When multiple deficiencies of Mn, Zn, and Fe were induced in apple seedlings, the expression of Fe deficiency predominated (Zhou et al., 1986).

Physiological Changes in Iron-deficient Fruit Trees The most obvious result of Fe chlorosis is the decrease in photosynthesis of the affected

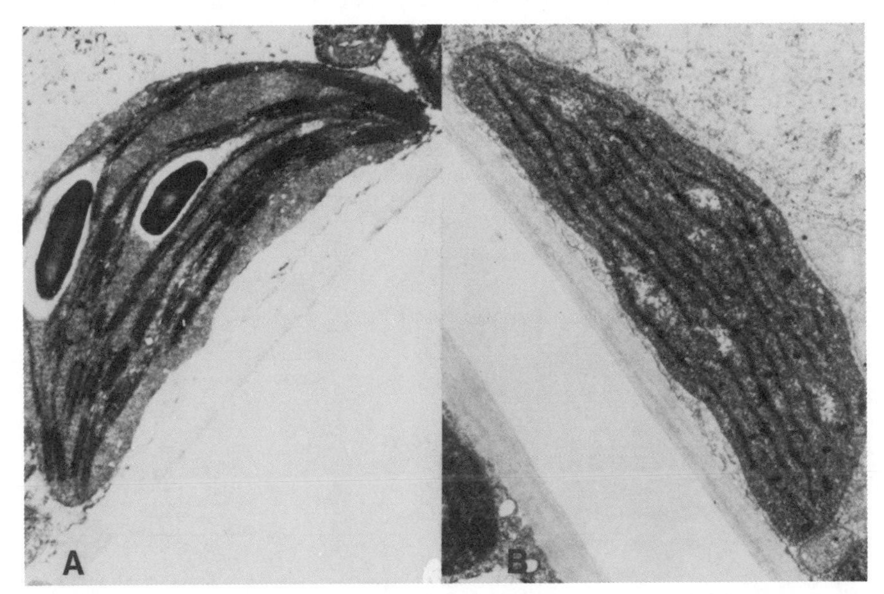

Figure 2.17 Chloroplast structure of (*A*) Fe-sufficient apple and (*B*) Fe-deficient leaves.

TABLE 2.11 Photosynthetic Rate and Root Respiration of Apple Seedlings Affected by Iron Chlorosis[a]

	Net Photosynthesis (CO_2 mg dm^{-2} s^{-1})	Respiration (CO_2 μl g^{-1} fw hr^{-1})
High Fe	34	360
High Fe + $CaCO_3$	30	416
Low Fe	25	520
Low Fe + $CaCO_3$	20	544
No Fe	6	520
No Fe + $CaCO_3$	3	544

[a] After Sun et al., 1987.

leaves. The magnitude of decrease is quite important (Table 2.11).

Acids, especially malic acid, accumulate in Fe-deficient plants (Table 2.12). Perhaps the inhibition of aconitase, an enzyme of the tricarboxilic acid cycle, or the decreased rate of protein synthesis leads to free K, which in turn may induce acid formation (discussed under Potassium). Pear leaves are unusually high in K and low in Ca in lime-induced chlorotic leaves (Lindner and Harley; 1944; Thorne and Wallace, 1944).

TABLE 2.12 Acid Content ($\mu g\ g^{-1}$ fw) of Various Parts of Apple Seedlings Affected by Iron Chlorosis[a]

	Succinic Acid	Malic Acid	Quinic Acid	Citric Acid	Total
		Young leaves			
With Fe	23	249	6421	164	6858
Without Fe	84	2670	465	4286	7506
		Old leaves			
With Fe	0	285	3982	275	4543
Without Fe	2	1792	3320	2035	7150
		Fibrous roots			
With Fe	0	760	213	1020	1984
Without Fe	83	1822	113	3593	5613

[a] After Sun et al., 1987.

While acid of the citric acid cycle increases due to Fe stress, quinic acid decreases. In 'old leaves' grown before the Fe stress was imposed on apple seedlings that remain green, the decrease of quinic acid is negligible, but the leaves still respond by increasing the acids of the citric acid cycle.

Uptake of Iron In soils the primary form of iron is Fe^{3+}. It is usually taken up by plants as Fe^{2+}, with the exception of Gramineae. Thus the trees face the double role to first solubilize the Fe as Fe^{3+} chelate and then reduce the solubilized Fe^{3+} to Fe^{2+} to be absorbed into the root (Korcak, 1987). Marschner (1986) lists several possibilities of how plants dissolve Fe^{3+} from the soil surface. The most likely possibility that takes place in the fruit tree rhizosphere is the solubilization of Fe^{3+} by malate as an organic ion. Apples are known to produce large quantities of malic acid, and this acid leaks out to the root zone (Sun et al., 1987). Forming iron–phosphate–citrate polymers of Fe–Al chelates is an unlikely means of iron uptake in fruit trees since high-P soils or P fertilization usually aggravates Fe deficiency (Thorne and Wann, 1950).

The exact area in, on, or near the root where reduction of Fe^{3+} takes place is not known (Korcak, 1987). Marschner (1986) assembled information concerning the reduction at the root surface, but he placed the emphasis on the plasma membranes of rhizodermal cells to reduce iron. When integrity of these membranes is impaired, the reduction of Fe^{3+} to Fe^{2+} is drastically reduced. However, Marschner (1986) does not rule out the possibility that phenolics excreted by the roots reduce Fe during the solubilization process. There is no available evidence that fruit trees extrude phenolic compounds at a high rate. In apples, both oxygen and photosynthates are required for reduction of Fe in the root (Tong et al., 1986). Thus the possibility is strong that reduction of Fe in fruit trees takes place at the rhizodermal membrane level.

Bicarbonate-induced Iron Chlorosis 'Lime-induced chlorosis' is a term often used for chlorosis associated with disturbed Fe metabolism on high Ca-containing soils. Juritz (1913) reported chlorosis of apple, apricot, peach, and prune along with other nondeciduous fruits associated with calcareous soils in South Africa. Other observations reported the occurrence of chlorosis from several fruit-growing areas around the world (McGeorge, 1949; Shen and Tseng, 1949; Thorne and Wann, 1950). Korcak (1987) distinguished soils having high pH (in H_2O greater than 7.0) from calcareous soils that he defined as any soil containing sufficient calcium carbonate or calcium–magnesium carbonate to effervesce visibly when treated with cold 0.1 *N* HC1. The bicarbonate ion, which forms on calcareous soils, is the most important soil factor associated with lime-induced Fe chlorosis in apples and pears (Boxma, 1972) along with many other plants (Marschner, 1986).

In dry calcareous soils the development of bicarbonate (HCO_3^-) is governed on a soil-by-soil bases by the equilibrium of formation and decomposition of HCO_3^-. For details see the review by Korcak (1987). The level of bicarbonate is rarely high enough in dry soils to cause major problems in Fe uptake. In wet calcareous soils the dynamics of bicarbonate formation is much higher and depends on high CO_2 pressure in the soil and hydrolysis of $CaCO_3$, which requires the presence of water. The overall reaction, in which water is the driving force, can be expressed as follows:

$$CaCO_3 + H_2O + CO_2 = Ca^{2+} + 2HCO_3^-$$

The percentage of $CaCO_3$ in the soil is less important than its hydrolysis to form HCO_3^-.

Despite many studies on the subject, the mechanism by which high HCO_3^- concentration induces chlorosis is still not well understood. Marschner (1986) lists eight possible effects of high bicarbonate in regard to Fe utilization, including decreased uptake, impaired reduction at the plasma membrane, reduced transport, unavailability of Fe for chlorophyll formation in the leaves, and uneven distribution of Fe within the leaf tissue. Few data are available as to which of these mechanisms may occur in fruit trees. There are occasional reports in the literature that chlorotic leaves of peach were frequently not different in Fe content from green leaves, but the 'active' Fe content was lower (McGeorge, 1949). This presumably points to Fe reduction problems. Iron is obviously not available for chloroplast formation, at least in apples, a problem discussed previously. The leaves are almost white when bicarbonate-induced chlorosis develops in apples compared to pale yellow when Fe deficiency is the cause of chlorosis (Zhou et al., 1984a).

Sensitivity of Fruit Trees to Lime-Induced Chlorosis There is an apparent difference in sensitivity among deciduous fruit trees to lime-induced chlorosis. The most general ranking in the order from most sensitive to least sensitive species is peach and pear > sweet cherry > plum > apricot > apple > sour cherry (Barney et al., 1984; Thorne and Wann, 1950). Kessler (1957) stated that plum is somewhat more sensitive than peach. Vose (1982) classified peaches, plums, and cherries as highly sensitive but apples as relatively tolerant.

Among apple species *Malus micromalus* and *M. domestica* cv. York and Golden Delicious were affected less by Fe chlorosis on calcareous soils than *M. zumi, M. hupehensis, M. honanensis, M. sieboldii*, and *M. baccata*. In the visual expression of chlorosis on a scale of 0 (no chlorosis) to 5 (severe chlorosis and necrosis), the former group was rated around 3 and the latter group between 4.5 and 5.0 (Korcak, 1978). Pear is generally more subject to iron chlorosis than apple (Harley and Lindner, 1945). The cultivars Bartlett, Winter Nelis, and Comice are more affected than Hardy and Clairgeau (Hendrickson, 1924). Trees on quince and *Prunus pyrifolia* are more affected with lime-induced chlorosis than on *P. communis* rootstock. Almond is more resistant than peach, myrabolan, and apricot when they are used as rootstock for peach on calcareous soils (Day, 1953; Kessler, 1958). Peach–almond hybrids such as GF677 consistently showed only about half the susceptibility to lime-induced chlorosis as peach seedlings when both were compared as rootstocks grafted with peach (Syrgiannidis, 1985). Mahaleb and mazzard cherry rootstocks are also susceptible to Fe chlorosis (Day, 1951). Other *Prunus* species such as the Japanese plums are relatively resistant to chlorosis (Rom, 1983).

The high lime content of the soil and moisture conditions repeatedly come up when practical observations are made. Lime-induced chlorosis develops usually on soils higher than 20% Ca content (Marschner, 1986). Chlorosis was noted in orchards in South Africa on calcareous soils around limestone outcrops or primarily in low-lying areas within a field (Beyers and Terblanche, 1971). In China iron-deficiency chlorosis of fruit trees is more prominent during the rainy season (Shen and Tseng, 1949). The author (Faust, unpublished) noted in China that the spring-developed leaves of apples and peaches grown during a dry period are green, and chlorotic leaves develop on the same shoots when the soil has plenty of moisture during the rainy season.

Correcting Iron-Deficiency Chlorosis Because of the difficulties encountered to absorb Fe from the soil, direct soil applications were bypassed and ferrous sulfate trunk injections were used beginning at about 1949. Usually about 400 ml of 1–2% ferrous sulfate injected into the trunk has at least a partial corrective effect on apples (McGeorge, 1949), cherries (Thorne and Wann, 1950), pears (Raese and Parish, 1984) and plums (Yoshikawa et al., 1982). Beginning in the early 1950s Fe chelates were used to correct Fe deficiency chlorosis. The large volume of information developed in this area was reviewed by Wallace and Wallace (1982). First Fe–EDTA was used on acid soils with reasonable results, but it could not satisfactorily correct chlorosis on calcareous soils (Korcak, 1987). At higher pH Fe–EDDHA has a greater stability, but it still looses its effectiveness with time (Hamze et al., 1985). Many tests have been made on the relative effectiveness of chelated Fe applied through soil and foliar sprays. Foliar applications are reviewed by Swietlik and Faust (1984).

In nutrient solutions increasing K tended to ameliorate the bicarbonate effect in apple by increasing both leaf Fe concentrations and growth (Zhou et al., 1984a). A higher chelator–Fe ratio (EDDHA–Fe, 10:1) in nutrient solution tended to reduce the chlorosis in apple seedlings compared to a low ratio (1:1) (Tong et al., 1986).

The other obvious control measures were discussed in the preceeding and include the use of tolerant rootstocks. Among all fruit species, only the peach–almond hybrids as peach rootstock are tolerant enough to have commercial importance. Correcting soil drainage and the use of controlled irrigation are important measures to decrease soil water content, which in turn may control bicarbonate formation. One-way irrigation can be controlled as drip irrigation. Drip irrigation with added Fe–EDDHA is apparently very useful to control lime-induced chlorosis.

Boron

At soil pH < 8 boron occurs as weak boric acid, which accepts OH^- rather than donates H^+, as illustrated by

$$B(OH)_3 + 2H_2O \rightarrow B(OH)_4^- + H_3O^+$$

Boric acid is also the most likely form of B taken up by plants. The influence of B on the physiology of plants is well described by Marschner (1986) and can be summarized as follows. Boron is strongly complexed in the roots, and the degree of complexing with cell wall constituents is the determining factor for B requirement in various species. Boron forms stable mono- and diesters with carbohydrates that have cis–diol configurations. A substantial portion of total B in higher plants exists as stable cis-borate ester complexes in the cell wall. A number of sugar derivatives

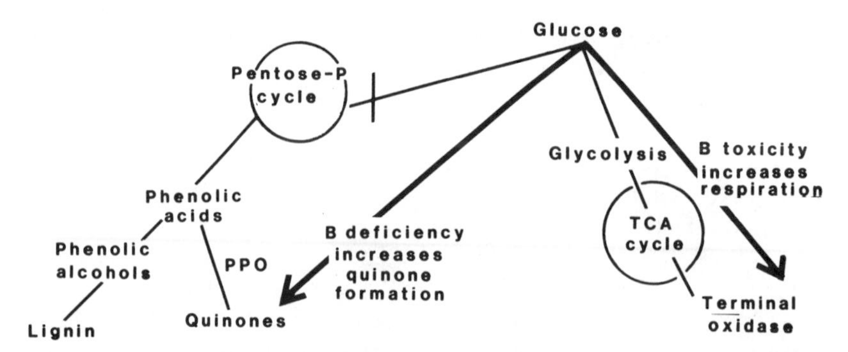

Figure 2.18 Schematic representation of B deficiency and toxicity on metabolic processes in plant cells. Boron restricts the flow of metabolites through the pentose phosphate cycle indicated by a bar before the cycle.

such as mannitol, mannan, polymannuronic acid, and *o*-diphenolic acids such as caffeic acid and ferulic acid all have a cis–diol configuration and form stable borate complexes. There is no record on borate complex formation of sorbitol, the most important polyol in fruit trees.

Boron is also able to form stable 6-P-gluconate-borate complexes and thus restricts the flow through the pentose pathway and the synthesis of phenols. This has important implications in both B deficiency and B toxicity. The damaging effect of B deficiency is interpreted by the excessive formation and accumulation of phenolics in B-deficient tissues. The accumulation of phenolics leads to the increased activity of polyphenol oxidase (PPO), leading to highly reactive phenolic intermediates in the cell wall (Marschner, 1986). The effect of phenolics on the permeability of membranes is well documented. Thus damage in the cell membranes is also expected when B deficiency conditions prevail. In case of high B in the tissues, B toxicity conditions may exist. The pentose phosphate pathway is blocked by high-B glycolysis and the synthesis of cellulose in the cell wall increases. The increase of glycolysis by B may be important because high-B fruit is known to ripen faster, an activity associated with increased respiration (Figure 2.18).

In experimental conditions when B supply is cut off, there is a dramatic increase in the activity of IAA oxidase in a variety of plants. When B is resupplied, IAA oxidase activity immediately decreases (Marschner, 1986). Although such information is not available for fruit trees, based on observations, many investigators working with fruit trees equate B application with IAA activity.

Enhanced cell division in a radial direction with a distinct proliferation of cambial cells and impaired xylem differentiation in the subapical shoot tissue are also typical in B-deficient plants (Marschner, 1986). Cell elongation usually ceases and a decrease in ribonucleic acid (RNA) synthesis rapidly follows when B is deficient. In severe B deficiency, the protein content of young leaves decreases and soluble nitrogen compound increases. When the supply of B is insufficient, both the production and export of cytokinins into the shoot decrease (Marschner, 1986). Although the preceding effects are documented in a variety of plants, relatively few effects have been shown in fruit trees. Nevertheless, these effects can explain the variety of responses fruit trees show when B is deficient or too high for optimal production.

Symptoms of Boron Deficiency There are several ranges of B deficiency at which the tree respond with increasing severity of symptoms. The normal range of B concentration in midshoot leaves of apple is 35–40 ppm. At around 25 ppm, B is insufficient. At this level only secondary problems involving Ca can be observed. (See the section on Ca for B–Ca interactions.) Deficiency develops in the fruit when the B concentration of the leaves is between 14 and 21 ppm (Oberly, 1963). At this level, the so-called drought spot is visible (Latimer, 1941). When B concentration is below 12 ppm, symptoms of the vegetative part of the tree are highly visible. Other species follow about the same order of response: in peach the corresponding ranges are 10 ppm severe deficiency, 11–17 ppm deficient, 18–30 ppm insufficient, 30–59 ppm optimal, 60–80 ppm high, and 81–155 ppm excessive (Kamali and Childers, 1970).

As mentioned, symptoms of B deficiency are often noticeable in the fruit before symptoms show in the shoot. In apple the mildest effect of B deficiency is the flattening of the fruit. If deficiency is slightly more severe, internal cork, round or irregular brown regions within the core area, is clearly visible upon cutting the fruit. The dead cell masses become dry, hard, and corky (Boynton and Oberly, 1966). In pears similar brown areas are closer to the surface, and if they develop early, the surface above the spots is depressed. In apricot internal browning develops often around the stone cavity, and plums only show malformed fruit without browning (Benson et al., 1966).

If B deficiency develops early, external cork develops on apple before the fruit is half grown. In the early stages the areas appear water soaked. Then they harden and crack. The fruit appears dwarfed and misshapen. The incidence of internal or external cork depends on the severity of B deficiency and variety and time of development of deficiency (Boynton and Oberly, 1966). In apricot external cracking, shriveling, and acute malformation is visible. In prune, fruit is normal even though vegetative parts are affected. Boron-deficient prunes color early and drop before harvest (Benson et al., 1966; Hansen, 1962).

If B deficiency is acute enough, the apple tree will show dieback of shoots in late summer, yellowing and red veins on the terminal shoots, and death of small areas of bark near the tip followed by progressive death of the inner bark and the cambium. Internodes of growing shoots are shortened and the leaves are strap shaped. The short internodes usually from a 'rosette' of leaves (Burrell, 1940). In pears shoots are short with dwarfed terminal leaves, and the basal leaves fail to develop. In cherry terminal growth is reduced, defoliation and dieback occurs at the terminals, and leaves are narrow with irregular margins and necrotic edges. In plum, prune, and apricot dieback occurs, the leaves are curled, the midrib is thickened, and a shoot-holing effect is observable (Benson et al., 1966; Woodbridge, 1954).

There is very little margin between sufficiency and toxicity when B is concerned. When normal level of B is between 35 and 40 ppm, adverse effect of toxicity can be observed at 70 ppm in apple (Faust, unpublished) and at 80 ppm in peach (Kamali and Childers, 1970). This is why only one spray of B is recommended in early spring to correct for B insufficiency. More than one spray within the same season often causes B toxicity (Swietlik and Faust, 1984). Boron toxicity in apple manifests itself in early maturation of fruit, premature fruit drop, shortened storage life, and senescence breakdown in storage (Haller and Batjer, 1946; Phillips and Johnston, 1943). In cherry B toxicity is associated with gumming of shoot, a common response of stone fruit to any kind of injury. Leaves are normal size, unlike in the case of B deficiency, but have necrotic areas on the margin. In plums, thickening of leaves, corkiness along the midrib, enlarged nodes, gumming, bark necrosis, and death at the shoot tip are all caused by excess B. Prunes and plums do not set normal crops when B is high. Apricots are quite tolerant to high B (Benson et al., 1966). Boron toxicity is often caused by B in the irrigation water. Essentially the irrigation water must be free of B. As low as 1.5 ppm B in the irrigation water can cause B toxicity (Eaton et al., 1941).

Role of Boron in Physiology of Fruit Trees The major role of B in fruit trees involves fruit set. Flowers are very high in B in apple (Crassweller et al., 1981), pear (Johnson et al., 1954) and cherry (Woodbridge et al., 1971). If the B content of flowers is insufficient, especially in pears, the flowers wilt, die, and persist on the tree. This phenomenon is called 'blossom blast' (Batjer and Rogers, 1953). The B needed in the flower is largely transported from reserves in the wood during development of the flower (Callan et al., 1978). Boron sprays applied the previous fall or before bloom during spring are effective in increasing the B content of flowers. Fall sprays are more effective than spring sprays (Table 2.13). The B remobilized and transported into the flowers is from the adjacent branches and not from the root (Callan et al., 1978). Some of the transport is through the xylem, but Hanson and Breen (1985) could only account for 26% of the remobilized B transport based on xylem exudate

TABLE 2.13 Effect of Previous Fall and Prebloom Sprays on Boron Content (ppm) of Flowers of Prune at Full Bloom[a]

Condition	Anther	Ovary	Style
Unsprayed	47	59	52
Fall spray	120	171	133
Spring spray	63	98	79

[a] Reproduced by permission from Callan et al., 1978.

analysis. Apparently some B is transported through the phloem, as indicated for apples by van Goor and van Lune (1980).

The temperature may play a role in the rate of transport. Overtree misting that reduces bud temperature also reduces B transport into the flowers of apple (Crassweller et al., 1981). Year-to-year variation in the B content of flowers may be caused by temperature differences, bloom density (the existing reserve is distributed to more flowers), and conditions influencing accumulation in the previous year (apricot fruit is a strong accumulator of B, and the bark may not accumulate sufficient reserves in a high-cropping year).

The importance of flower B content centers on the fact that B influences fruit set. Boron sprays applied to trees not deficient in B based on leaf analysis either during fall or spring are effective in increasing fruit set in a number of fruit tree species (Batjer and Rogers, 1953; Batjer and Thompson, 1949; Chaplin et al., 1977; Hanson and Breen, 1985). The effectiveness of sprays is higher during a cool spring than in warm springs when the set is generally high (Hanson and Breen, 1985). Boron increases pollen germination and pollen tube growth in vitro (Thompson and Batjer, 1950a), but high B content of buds does not enhance pollen tube growth in vivo (Hanson and Breen, 1985). There is no evidence that the increased fruit set is caused by better pollen germination or pollen tube growth (Callan et al., 1978). Boron may enhance the cell division or nucleic acid synthesis in the developing fruit, which can easily influence fruit set.

Boron transport into the leaves and to the fruit occurs as the development of these organs proceed. There is essentially a straight-line relationship between fruit weight and B per fruit (Figure 2.19), indicating a continuous transport as the fruit develops (van Goor and van Lune, 1980). Apparently accumulation into the bark is similar. Blossom blast, indicating insufficient B in the reserves, usually occurs after dry summers when low soil moisture prevents B accumulation (Johnson et al., 1954). The bark content of peach can easily be increased from 18 to 32 ppm with proper B nutrition (Kamali and Childers, 1970).

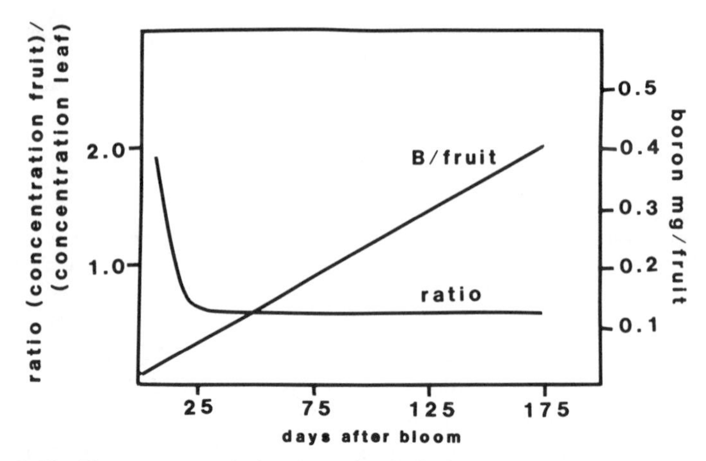

Figure 2.19 Boron accumulation into the fruit during the growing season. (Reproduced by permission from van Goor and van Lune, 1980.)

Trees low in B resume spring growth 1 week to 10 days later than trees sufficient in B. Thus incipient B deficiency can be observed in early spring. After the trees resume growth, no difference between the low and sufficient B level trees is observable. The June leaf analysis often does not show the difference in leaf B content.

Zinc

Zinc is predominantly taken up as a divalent cation, Zn^{2+} at lower pH and as a monovalent cation, $ZnOH^{+}$, at higher pH. High concentrations of other divalent cations such as Ca^{2+} inhibit Zn uptake. It is transported as a free ion or bound to organic acids. In plants, Zn is not oxidized or reduced. It functions primarily as a divalent cation (Marschner, 1986). It acts either as a metal component of enzymes or as a cofactor for a large number of enzymes. Among the most important enzymes that contain Zn alcohol dehydrogenase, superoxide dismutase, carbonic anhydrase, and RNA polymerase must be mentioned. As a cofactor for RNA polymerase, Zn is involved in protein synthesis. The rate of protein synthesis in Zn-deficient plants is drastically reduced, and amino acids and amides accumulate. The involvement of Zn in alcohol dehydrogenase may be important in fruit trees. Ripening fruit always has an overabundance of pyruvate, which 'spills over' to acetaldehyde, which is further converted to ethanol by alcohol dehydrogenase. Acetaldehyde is toxic to cells and causes browning in the fruit. Thus its conversion to ethanol is of major importance to prevent browning (Faust et al., 1969). If Zn is insufficient, this reaction may be imperiled and browning occurs.

The most important symptom of Zn deficiency, 'little leaf,' is associated with the reduced auxin content of shoot apices. Apparently Zn is required for the synthesis of trytophan, a precursor for the formation of IAA. Tryptophan levels of Zn-deficient plants are low, and the level increases when plants are resupplied with Zn (Marschner, 1986).

Chandler (1937) observed that fruit trees are more subject to Zn deficiency than many annual crops. He found that cherry is more susceptible than apple and apple is more susceptible than many other deciduous fruits.

The most characteristic symptom of Zn deficiency is rosetting of the leaves at the shoot tip. The leaves produced are small, mottled, abnormally shaped, narrow, and bunched together at the tip of the shoot, hence the name 'little leaf disease'. Rosetting is most easily recognizable during the first flush of growth. Death of the terminals may occur during the following season with a number of laterals producing nonvigorous growth. These laterals are susceptible to winter injury (Boynton and Oberly, 1966). Narrowing and dwarfing of terminal leaves is usually the best indicator of incipient Zn deficiency when rosetting is not evident. Leaf Zn levels below 25 ppm on a dry-weight basis are indicative of deficiency (Boynton and Oberly, 1966). There is little variation from the general appearance of Zn deficiency in the various fruit species. Some plums show stippled foliage due to small chlorotic areas scattered over the leaf. In apricot, foliage often is not low in Zn, but the growing points and buds are low in Zn when the tree is affected by incipient Zn deficiency (Benson et al., 1966).

Little leaf is associated with sandy soils containing limited quantities of Zn (Wann and Thorne, 1950) and with soils containing high P and Ca and little clay or organic material (Mulder, 1950). Loneragan et al. (1979) provided evidence that Zn uptake may be inhibited by cations such as Ca^{2+}. Phosphorus enhanced Zn adsorption in the soil to Fe and Al oxides and hydroxides and to $CaCO_3$. In orchards P is not a major concern since P fertilization in orchards is relatively limited. Calcium, however, is important in many orchard soils. Light intensity is a factor in Zn deficiency. Leaves affected with little leaf may attain normal green color if they are partially shaded (Benson et al., 1966).

Controlling Zn deficiency in fruit trees is best done by sprays applied during dormancy at a rate of 37 kg Zn/ha, which is 113 kg of 32% $ZnSO_4$. One kilogram of Zn chelate per tree applied to the soil also produced satisfactory recovery (Benson et al., 1957).

Manganese

Manganese activates a number of enzymes in plants in vitro, but its role in vivo is presumably as a structural element of metalloproteins, where it acts as an active binding site or serves as a redox system (Marschner,

1986). Among the few Mn-containing enzymes, superoxide dismutase is the most important. Superoxide dismutase protects tissues from the deleterious effect of free oxygen radicals (superoxide) by forming H_2O_2 and O_2.

In Mn-deficient plants, the chlorophyll content and typical chloroplast membrane constituents such as phospholipids and glycolipids are reduced. In fruit trees Mn deficiency has been reported in many species. The typical expression of the deficiency is interveinal chlorosis, somewhat like Fe chlorosis. Manganese deficiency develops on older leaves during midsummer and becomes more pronounced as the season progresses. In case of severe deficiency, leaves are lacking color, and heavy defoliation may be induced by strong wind (Thorne and Wann, 1950).

Normal leaves of apple, pear, peach, and cherry range around 70–85 ppm dw. Occasional samples could go as high as 220 ppm. In contrast, leaves that show Mn deficiency chlorosis contain usually less than 17 ppm and often as low as 5 or 6 ppm. In fruit trees on acid soils Mn toxicity is much more common than Mn deficiency. Toxicity is often associated with Fe deficiency, and the brown spots on Fe-deficient chlorotic leaves are usually high in Mn, indicating Mn toxicity.

The chlorosis develops first on old leaves and does not seem to be associated with loss of vigor (Boynton and Oberly, 1966). In cherry Mn deficiency chlorosis is quite similar to the one caused by Zn deficiency, although the size of the leaves are apparently not affected.

In acid soils, high in Mn availability, trees can take up a considerable amount of Mn. The 'Delicious' apple cultivar is apparently quite sensitive to high Mn concentration in its tissues. Manganese toxicity in apples is expressed as leaf chlorosis, early leaf abscission, reduced flower bud formation, reduced growth, and internal bark necrosis (IBN) (Cooper and Thompson, 1972).

Internal bark necrosis, commonly called 'measles,' first appears on the bark as small raised areas called pimples. They progress through a stage of cracked pimples, cracked islands of bark, and rough scaly bark. The tissue under the pimples appears dead surrounded with water-soaked tissue. Freshly cut bark or affected branches oxidize rapidly. Internal bark necrosis is usually associated with interveinal chlorosis, yellowing and early abscission of leaves, and reduced growth of the trees. If trees are affected by IBN early in their life, they never develop to be productive (Ferree and Thompson, 1970).

Lesions of IBN usually originate adjacent to the phloem fiber bundles in the bark. A single cell or a few adjoining cells of the cortical parenchyma are affected, each containing a dark amorphous mass (Figure 2.20). The advanced development of lesions usually associated with the development of meristematic activity are often at a distance of several cells from those containing deteriorating protoplasts (Eggert and Hayden, 1970).

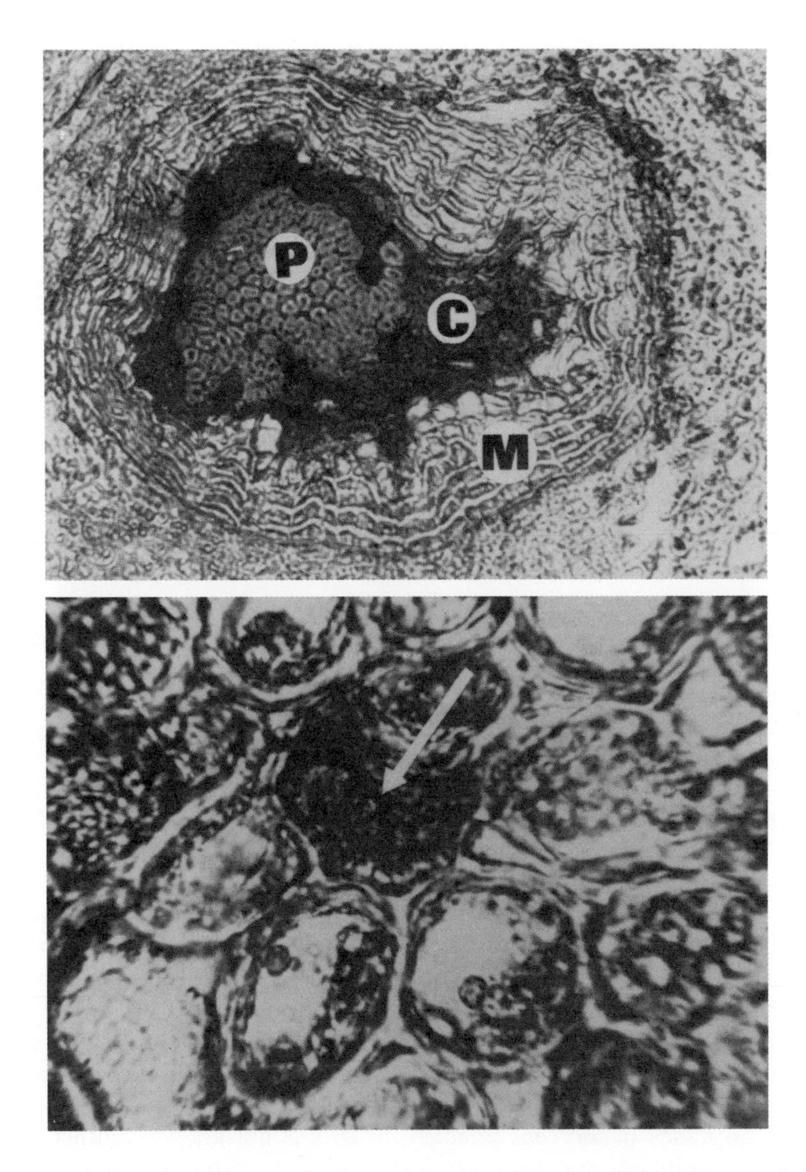

Figure 2.20 Development of Mn toxicity disorders, internal bark necrosis: P, phloem; C, affected region (cortical parenchyma) always near phloem; M, mesophyll cells of apple shoot. Arrow indicates single cell with dark amorphous mass. (Reproduced by permission from Eggert and Hayden, 1970.)

TABLE 2.14 Accumulation of Manganese in apple bark tissue (ppm)[a]

Level of K (meq liter^{-1}) in Nutrient Culture	Level of Ca (meq liter^{-1}) in Nutrient Culture		
	2	8	14
	1-year-old Bark		
2	523	301	268
5	480	328	265
8	510	335	251
	2-year-old Bark		
2	624	355	309
5	639	383	349
8	666	393	314
	Rootstock Bark		
2	651	348	284
5	721	388	430
8	886	399	321

[a] After Domoto and Thompson, 1976.

Although the bark tissue is high in Mn when IBN develops, a point accumulation of high Mn rather than an overall high Mn content of the tissue was associated with the disorder (Crocker and Kenworthy, 1973; Eggert and Hayden, 1970). Boynton and Oberly (1966) reported a 10-fold increase in bark Mn in connection to IBN development. While the bark of normal tissue contained 50 ppm Mn, the affected tissue contained 500 ppm. Very similar lesions to the high-Mn-induced IBN can also be induced by low B, and it is called low-B IBN. The difference is that the Mn-induced IBN lesions are high in Mn and the low-B-induced IBN lesions are low in B (Crocker and Kenworthy, 1973). The nature of the brown substance in the cells affected by IBN is not known; neither the fact of similarity nor that of difference between the two types of IBN was established. In solution culture experiments, low Ca was necessary for high Mn to induce IBN. If Ca was low, K aggravated the appearance of Mn-induced IBN. Low concentrations of B and P also contributed to the development of high-Mn-induced IBN (Domoto and Thompson, 1976).

The effect of Ca in reducing the accumulation of Mn is clearly shown in Table 2.14. The situation is very similar to that of Ca-related disorders. There must be disposing factors other than Mn present to develop Mn-related disorders.

The apple cultivars 'Delicious' and 'Jonathan' are most affected. Trees having IBN symptoms usually have foliar Mn levels greater than 150 ppm but could be as high as 600 ppm (Zeiger and Shelton, 1969). The degree of severity of IBN increases with increasing levels of K and age of the bark (Domoto and Thompson, 1976) and N fertilization (Rogers et al., 1965). A reduction in Mn toxicity symptoms is exerted by Ca. Calcium has influence on both Mn absorption and translocation in apple (Fucik and Titus, 1965).

Deficiency symptoms can be controlled by $MnSO_4$ as 0.3–0.5 kg per hundred liters of foliar sprays or by 0.7–5.4 kg per hundred liters of dormant sprays, which can be delayed until budbreak (Thompson and Rogers, 1945), or by a few grams of $MnSO_4$ injected into the trunk. The most effective protection against Mn toxicity is bringing the soil pH to 6–6.5. This is important before planting because when measles develop in a young orchard, growth of the trees is permanently impaired and the deficiency in growth can never be corrected.

OTHER ELEMENTS OF IMPORTANCE

Sulfur

Although S can be taken up in considerable quantity by plants from the atmosphere, the majority of S is still taken up by the roots. Sulfur is taken up as the divalent anion SO_4^{2-} by the roots at low rates and the sulfite ion is transported up in the xylem. the incorporation of S into amino acids requires reduction, which is somewhat similar to the assimilation of nitrate. Sulfur, however, could be utilized without reduction in sulfolipids of membranes. Reduced sulfur also can be oxidized in higher plants when protein degradation takes place as leaves senesce. In this reaction the reduced S of cysteine is converted to sulfate, the storage form of S in plants (Marschner, 1986).

Sulfur is a constitutent of the amino acids cysteine and methionine. Both of these amino acids are constituents of proteins and precursors of other S-containing secondary products. Sulfur is a structural constituent of coenzymes, ferredoxin, biotin, and thiamine pyrophosphate. In addition, the formation of disulfide bonds is essential in the polypeptide chain to form the characteristic tertiary structure of the proteins. Sulfur in its unreduced form is a component of sulfolipids and is thus a constituent of biological membranes (Marschner, 1986).

It is notable that although S has many essential functions in the plant cells of fruit trees, from a production point of view S is an unimportant element.

Only two areas in the world, Australia and central Washington state, have reported S deficiency. Sulfur deficiency is much like N deficiency, the leaves are pale yellow, the yellowness more noticeable in young leaves than in leaves at the bottom of the shoot (Benson, 1962). When S deficiency was induced in nutrient cultures, the roots of peaches were very fine, with a distinct brown color that distinguished them from any other treatment (Weinberger and Cullinan, 1936). The sulfur concentration of normal midterminal peach leaves on a dry-weight basis was around 0.3% from July to September regardless of sampling date (Thomas et al., 1954). Benson et al. (1963) recommended the use of terminal leaves to measure the S status in apple. According to them, the use of terminal leaves for leaf sampling is a more sensitive method for S than the use of midterminal leaves. Sulfur-deficient terminal leaves contained 100 ppm S, while the S concentration in normal leaves was 150 ppm.

Copper

In soil solutions almost all (98%) Cu is complexed with low-molecular-weight organic compounds. In the roots and in the xylem, Cu is also present in complexed form. The functions of Cu as a plant nutrient are primarily in the redox reactions of terminal oxidases. In the redox reactions Cu directly reacts with molecular oxygen, and in living cells the terminal oxidases are Cu enzymes. Copper is part of the plastocyanin, a component of the electron transport chain of photosystem I; the superoxide dismutase, which plays a role in the detoxification of superoxide radicals; the metal component in cytochrome oxidase, which is the terminal oxidase of the mitochondria; ascorbate oxidase, which catalyzes the oxidation of ascorbic acid; and phenolase and lactase, which catalyze the oxygenation of phenols. Details of these functions are described by Marschner (1986). Very little is known about the specific functions of Cu in fruit trees. There are only few locations throughout the world where Cu is insufficient for fruit trees. However, at these locations trees respond to Cu deficiency with severe dieback, which is in many ways a more severe reaction than their response to other nutrient deficiencies.

The most striking symptom of Cu deficiency is the dieback of shoots that were growing vigorously. Usually brown spots appear first on the terminal leaves followed by necrotic areas developing on the leaves and finally the upper 7–30 cm of the shoot dies and withers (Dunne, 1946). The dieback usually is not associated with chlorosis, although chlorosis associated with dieback has been reported (Boynton and Oberly, 1966). Leaf concentration of Cu on a dry-weight basis in apple is 1–4 ppm in trees with dieback and 3–12 ppm in healthy trees, in pears it is 3.2–5.1 ppm in affected trees and 10–41 in healthy trees, and in plums its 3–4 ppm in deficient trees and 7–9 ppm in healthy trees. Bould et al. (1953) concluded that the threshold effect at which deficiency develops is

<5 ppm Cu in leaf dry matter in July–August. The deficiency is easily correctable by $CuSO_4$ sprays. Copper toxicity has not been reported for temperate zone fruit trees.

Molybdenum

In higher plants Mo plays a major role only in two enzymes, nitrogenase and nitrate reductase (Marschner, 1986). Little information is available on Mo nutrition of fruit trees. Deficiency symptoms were induced in nutrient culture in apple (Fernandez and Childers, 1960) and in myrobalan rootstock seedlings (Hoagland, 1941). Symptoms start with mild uniform chlorosis of the young leaves followed by tipburn of older leaves, which progresses along the leaf margins. Leaves develop a downward cupping, and the lowest leaves abscise. In orchards only some indication of Mo deficiency was observed in plum, apricot, and apple (Benson et al., 1966; Chittenden, 1956). When leaf level of Mo is at 0.16 ppm in apple, the leaves are normal; when deficiency was visible, the leaf Mo level was 0.05 ppm (Fernandez and Childers, 1960).

Arsenic

Injury to fruit trees from soil-borne As occurs on soils where relatively large amounts of arsenate spray residue accumulated (Blodgett, 1941; Lindner, 1943). The As toxicity usually occurs when the orchard is replanted. Roots of old trees are relatively deep in the soil, and As residues do not penetrate that deep. When the orchard is replanted, the roots of young trees are in the soil zone where the As is located, whis accounts for the As toxicity.

Apricots are relatively sensitive, but plums and prunes are much more tolerant. Toxicity in apricot occurs at 6 ppm leaf concentration on a dry-weight basis, while plums do not show toxicity even at 13 ppm (Benson et al., 1966). Symptoms appear on the foliage as interveinal and marginal red or brown spots, and the leaves abscise prematurely. Young trees do not develop at the expected rate. In peach trees a high rate of N fertilization and $ZnSO_4$ reduced As toxicity (Thompson and Batjer, 1950b).

Aluminum

On acid soils throughout the world aluminum (Al) is a problem for fruit trees. Such an area is the southeastern United States, the states of Georgia, and North and South Carolina, where peaches are affected by the presence of Al in the soil. On acid soils below pH 5.5, the solubility of Al increases sharply, and below pH 4.0 all soluble Al exists as Al^{3+}. In

TABLE 2.15 Uptake of Various Elements by Apple and Peach in Presence of 4 ppm Aluminum in Nutrient Solution[a]

	Total Uptake per Seedling							
	Ca (mg)	Mg (mg)	K (mg)	P (mg)	Zn (μg)	Cu (μg)	Mn (μg)	Fe (μg)
APPLE								
Without Al	5.7	2.3	12.4	1.8	72.2	29.6	45.9	614.7
With Al	2.0	0.7	7.3	0.6	24.5	13.8	23.3	179.5
PEACH								
Without Al	10.1	2.1	20.1	2.6	66.6	26.0	61.0	104.7
With Al	4.4	1.9	12.8	0.9	49.1	13.4	38.5	454.3

[a] Nitrogen was supplied as nitrate. After Kotze et al., 1977.

this pH range more than half of the soil's cation exchange sites may be occupied by Al (Mengel and Kirkby, 1982). Under such conditions toxicity to fruit trees is likely to occur. Among fruit trees the effect of Al is studied most in peaches. The first observable effect of Al is the limitation of root growth. The epidermal cells are small with thickened cell walls. The cells of the cortex are shorter in the longitudinal axes. Obviously the growth of the tree is affected (Kirkpatrick et al., 1975).

The presence of Al in the root zone decreases the uptake of Ca, Mg, K, P, Zn, Cu, Mn, and Fe in both peaches and apples (Table 2.15) (Kirkpatrick et al., 1975; Kotze et al., 1977). It appears the level of 3–4 ppm Al in the soil solution could cause an imbalance in the nutrition of trees. The effect of Al is more severe if the trees are supplied with NO_3^- in comparison to NH_4^+ as a source of N (Kotze et al., 1977). Inhibition of Mg and Ca uptake is mainly due to cation competition for or blocking the binding sites (Marschner, 1986).

Phosphorus uptake is affected directly through the precipitation of Al phosphate at the root surface and/or in the free space (Marschner, 1986).

Apples are more resistant to Al toxicity than peaches (Kotze et al., 1977). Fruit trees (apple and peach are investigated only) can be classified as nonaccumulators of Al based on the classification of Marschner (1986). In this group of plants, regardless of their tolerance, the transport of Al into the shoots is severely restricted and the accumulation is confined to the roots (Table 2.16).

The presence of Al in the root zone decreases stomatal conductivity in peaches. The decrease in stomatal conductivity seems to be correlated with the decreased root volume, which apparently cannot supply the tree with sufficient quantity of water. Thus the tree is under water stress and its stomates are at least partially closed (Horton and Edwards, 1976).

TABLE 2.16 Accumulation of Aluminum (ppm) in Various Tissues of Peach[a]

	Aluminum in Tissue		
Aluminum in Solution	Root	Stem	Leaf
0	1572	45	121
3	1700	48	113
30	2634	56	237
100	3054	115	437

[a] After Kirkpatrick et al., 1975.

REFERENCES

Anne, P., M. Dupuis, and R. Marocke. 1952. Ann. Agron. Ser. A 3:678–679.

Asamoah, T. E. O. 1984. Fruit tree root systems: Effect of nursery and orchard management factors on nutrient and water uptake. Ph.D. Dissertation. University of London.

Askew, H. O., E. T. Chittenden, R. J. Monk, and J. Watson. 1960. N.Z. Agr. Res. 3:169–178.

Atkinson, D. 1976. Sci. Hort. 4:285–290.

Atkinson, D. 1977. Plant & Soil 49:459–471.

Atkinson, D. 1980. Hort. Rev. 2:462–490.

Atkinson, D. 1983. Compact Fruit Tree 16:1–16.

Atkinson, D. 1986. Adv. Plant Nutrition 2:93–128.

Atkinson, D., and G. C. White. 1976. Proc. 1976 British Crop. Prot. Conf.—Weeds. 3:873–884.

Atkinson, D., and G. C. White. 1980. *In* the mineral nutrition of fruit trees. D. Atkinson, J. E. Jackson, R. O. Sharples, and W. M. Waller (eds.). London: Butterworth, pp. 241–254.

Atkinson, D., and S. A. Wilson. 1979. *In* The soil root interface. J. L. Harley and R. S. Russel (eds.). London: Academic Press, pp. 259–271.

Atkinson, D., and S. A. Wilson. 1980. *In* Mineral nutrition of fruit trees. D. Atkinson, J. E. Jackson, R. O. Sharples, and W. M. Waller (eds.). London: Butterworths, pp. 137–150.

Atkinson, D., D. Naylor, and G. A. Coldrick. 1976. Hort. Res. 16:89–105.

Ballinger, W. E., H. K. Bell, and N. F. Childers. 1966. *In* Temperate to tropical fruit nutrition. N. Childers (ed.). New Brunswick, NJ: Horticultural Publ., pp. 276–390.

Bangerth, F. 1979. Ann. Rev. Phys. Path. 17:97–122.

Barney, D., R. H. Walser, S. D. Nelson, C. F. Williams, and V. D. Jolley. 1984. J. Plant Nutr. 7:313–317.

Batjer, L. P., and E. S. Degman. 1940. J. Agr. Res. 60:101–116.

Batjer, L. P., and B. L. Rogers. 1953. Proc. Amer. Soc. Hort. Sci. 62:119–122.

Batjer, L. P., B. L. Rogers, and A. H. Thompson. 1952. Proc. Amer. Soc. Hort. Sci. 60:1–6.

Batjer, L. P., and A. H. Thompson. 1949. Proc. Amer. Soc. Hort. Sci. 53:141–142.

Beevers, L., and R. H. Hageman. 1980. *In* The biochemistry of plants, Vol. 5, B. J. Miflin (ed.). New York: Academic Press, pp. 115–168.

Bell, H. K., and N. Childers. 1954. *In* Mineral nutrition of fruit crops. N. Childers (ed.). New Brunswick, NJ: Hort. Publ., Rutgers University.

Benson, N. R. 1962. Proc. Washington State Hort. Assoc. 19–22.

Benson, N. R., L. P. Batjer, and I. C. Chmelir. 1957. Soil Sci. 84:63–75.

Benson, N. R., E. S. Degman, I. C. Chmelir, and W. Chenault. 1963. Proc. Amer. Soc. Hort. Sci. 83:55–62.

Benson, N. R., R. C. Lindner, and R. M. Bullock. 1966. *In* Temperate to tropical fruit nutrition. N. Childers (ed.). New Brunswick, NJ: Hort. Publ., Rutgers University, pp. 504–517.

Beyers, E., and J. H. Terblanche. 1971. Decid. Fruit Grower 21:265–281.

Bhat, K. K. S. 1983. Plant Soil 71:371–380.

Black, J. D. F., and P. D. Mitchell. 1974. *In* Proc. 2nd Int. Drip Irrig. Conf., San Diego. Riverside, CA: University of California Press, pp. 437–438.

Blanpied, G. D., and G. H. Oberly. 1978. J. Amer. Soc. Hort. Sci. 103:638–640.

Blodgett, E. C., 1941. Plant. Disease Rep. 25:549–551.

Bohomaz, K. I. 1960. Visnyk. Kyyivs'k Univ. Ser. Biol. 3:56–66; Biol. Abstr. 41:1272.

Bould, C. 1966. *In* Temperate to tropical fruit nutrition. N. Childers (ed.). New Brunswick, NJ: Hort. Publ., Rutgers University, pp. 651–684.

Bould, C., D. J. D. Nicholas, J. A. H. Tolhurst, and J. M. S. Potter. 1953. J. Hort. Sci. 28:268–277.

Boxma, R. 1972. Plant Soil 37:233–243.

Boynton, D., and A. B. Burrell. 1944. Soil. Sci. 58:441–454.

Boynton, D., D. Margolis, and C. R. Gross. 1953. Proc. Amer. Soc. Hort. Sci. 62:135–146.

Boynton, D., and G. H. Oberly. 1966. *In* Temperate to tropical fruit nutrition. N. Childers (eds). Hort. New Brunswick, NJ: Publ., Rutgers University, pp. 1–50.

Bradford, F. C. 1924. Michigan Agr. Exp. Sta. Spec. Bull. 127.

Brewster, J. L., and P. B. Tinker. 1972. Soil & Fert. 35:355–359.

Buban, T., A. Varga, J. Tromp, E. Knegt, and J. Bruisma. 1978. Z. Pflanzenphysiol. 89:289–295.

Bunemann, G. 1962. Inst. Agr. Gembloux Hort. Ser. Bul. 1962:1162–1166.

Burrell, A. B. 1940. Cornell Ext. Bul. 428.

Butcher, D., and P. E. Pilet. 1983. Experimentia 39:493–494.

Cain, J. C. 1948. Proc. Amer. Soc. Hort. Sci. 51:1–12.

Callan, N. W., M. M. Thompson, M. H. Chaplin, R. L. Stebbins, and M. N.

Westwood. 1978. J. Amer. Soc. Hort. Sci. 103:253–257.
Catlin, P. B., and C. A. Priestley. 1976. Ann. Bot. 40:73–82.
Chandler, W. H. 1937. Bot. Gaz. 98:625–646.
Chaplin, M. H., R. L. Stebbins, and M. N. Westwood. 1977. Hort. Sci. 12:500–501.
Childers, N. F., and F. F. 1935. Cowart. Proc. Amer. Soc. Hort. Sci. 33:160–163.
Chittenden, E. 1956. Orchard. N.Z. 29:2.
Cocker, G. E. 1959. Hort. Sci. 34:111–121.
Conway, W. S., and C. E. Sams. 1987. J. Amer. Soc. Hort. Sci. 112:300–303.
Cooper, D. R., and D. G. Hill Cottingham. 1974. Physiol. Plant. 31:193–199.
Cooper, R. E., and A. H. Thompson. 1972. J. Amer. Soc. Hort. Sci. 97:138–141.
Couvillon, G. A. 1982. J. Amer. Soc. Hort. Sci. 107:555–558.
Crassweller, R. M., D. C. Ferree, and E. J. Stang. 1981. J. Amer. Soc. Hort. Sci. 106:53–56.
Crocker, T. E., and A. L. Kenworthy. 1973. J. Amer. Soc. Hort. Sci. 98:559–562.
Cummings, G. A. 1973. J. Amer. Soc. Hort. Sci. 98:474–477.
Day, L. H. 1951. Calif. Agr. Exp. Sta. Bull. 725.
Day, L. H. 1953. Calif. Agr. Exp. Sta. Bull. 736.
Domoto, P. A., and A. H. Thompson. 1976. J. Amer. Hort. Sci. 101:44–47.
Doud-Miller, D. 1983. Compact Fruit Tree 16:106–110.
Dunne, T. C. 1946. J. Western Australia Dep. Agr. (Ser. 2) 23:124–127.
Eaton, F. M., R. D. McCallum and M. S. Mayhugh. 1941. USDA Tech. Bul. 746.
Eckerson, S. H. 1931. Contrib. Boyce Thomp. Inst. 3:405–412.
Edwards, J. H., and B. D. Horton. 1981. J. Amer. Soc. Hort. Sci. 106:401–405.
Eggert, D. A., and R. A. Hayden. 1970. J. Amer. Soc. Hort. Sci. 95:715–719.
Epstein, E. 1961. Plant Phys. 36:437–444.
Faust, M. 1965. Proc. Amer. Soc. Hort. Sci. 87:10–20.
Faust, M. 1980. *In* Mineral nutrition of fruit trees. D. Atkinson, J. E. Jackson, R. O. Sharples and W. M. Waller (eds.). London: Butterworths, pp. 193–200.
Faust, M., and C. B. Shear. 1972. J. Amer. Soc. Hort. Sci. 97:434–439.
Faust, M., and C. B. Shear. 1973. Proc. Res. Inst. Pomology. Skierniewicze, Poland, Series E 3:423–438.
Faust, M., C. B. Shear, G. B. Oberly, and G. T. Carpenter. 1971. HortScience 6:452–453.
Faust, M., C. B. Shear, and C. B. Smith. 1967. Proc. Amer. Soc. Hort. Sci. 91:69–72.
Faust, M., C. B. Shear, and M. Williams. 1969. Bot. Rev. 35:169–194.
Fernandez, C. E., and N. F. Childers. 1960. Proc. Amer. Soc. Hort. Sci. 75: 32–38.
Ferree, D. C., and A. H. Thompson. 1970. Maryland Agr. Exp. Stat. Bul. A-166.

Fisher, E. G., K. G. Parker, N. S. Luepschen, and S. S. Kwong. 1959. Proc. Amer. Soc. Hort. Sci. 73:78–90.

Forshey, C. G. 1963. Proc. Amer. Soc. Hort. Sci. 83:32–45.

Frith, G. J. T., and D. G. Nichols. 1975a Physiol. Plant. 34:129–133.

Frith, G. J. T., and D. G. Nichols. 1975b Physiol. Plant. 33:247–250.

Fritzsche, R., and A. Nyfeler. 1974. Schweiz. Landw. Forsch. 13:341–351.

Fucik, J. E., and J. S. Titus. 1965. Proc. Amer. Soc. Hort. Sci. 86:12–22.

Geddeda, Y. T., J. M. Trappe, and R. L. Stebbins. 1984. J. Amer. Soc. Hort. Sci. 109:24–27.

Geisler, D., and D. C. Ferree. 1984. Hort. Rev. 6:155–188.

Givan, C. V. 1979. Phytochemistry 18:375–382.

Goode, J. E., K. H. Higgs, and K. J. Hyrycz. 1978. J. Hort. Sci. 53:307–316.

Granger, R. L., C. Plenchette, and J. A. Fortin. 1983. Canad. J. Plant Sci. 63:551–555.

Grasmanis, V. O., and G. E. Edwards. 1974. Austral. J. Plant Phys. 1:99–105.

Grasmanis, V. O., and D. J. D. Nicholas. 1966. Plant Soil 25:461–462.

Grasmanis, V. O., and D. J. D. Nicholas. 1971. Plant Soil 35:95–112.

Gyuro, F. 1974. Gyümölestermesztès alapjai. Budapest: Mezögazdasagi Kiado.

Haller, M. H., and L. P. Batjer. 1946. J. Agr. Res. 73:243–253.

Hamze, M., J. Ryan, R. Shawyri, and M. Zaabout. 1985. J. Plant. Nutr. 8:437–448.

Hansen, C. J. 1962. Western Fruit Grower 16:12–13.

Hansen, P. 1980. *In* Mineral nutrition of fruit trees. D. Atkinson, J. E. Jackson, R. O. Sharples, and W. M. Waller (eds.). London: Butterworths, pp. 201–212.

Hanson, E. J., and P. J. Breen. 1985. J. Amer. Soc. Hort. Sci. 110:389–392.

Hanson, J. B. 1984. Adv. Plant Nutr. 1:149–208.

Harley, C. P., and R. C. Lindner. 1945. Proc. Amer. Soc. Hort. Sci. 46:35–44.

Harley, C. P., J. R. Magness, M. P. Measure, L. A. Fletcher, and E. S. Degman. 1942. U.S. Dep. Agr. Techn. Bull. 792.

Harley, C. P., L. O. Regimbal, and H. H. Moon. 1958. Proc. Amer. Soc. Hort. Sci. 72:57–63.

Head, G. C. 1965. Ann. Bot. 29:219–224.

Head, G. C. 1966. J. Hort. Sci. 41:197–206.

Head, G. C. 1967. J. Hort. Sci. 42:169–180.

Head, G. C. 1968. J. Hort. Sci. 43:275–282.

Head, G. C. 1969. J. Hort. Sci. 44:175–181.

Heinicke, A. 1934. Proc. Amer. Soc. Hort. Sci. 32:77–80.

Hendrickson, A. P. 1924. Proc. Amer. Soc. Hort. Sci. 21:87–90.

Hill, R. G. 1966. Proc. 17th Intern Hort. Cong., Maryland 1:24.

Hill-Cottingham, D. G., and D. R. Cooper. 1970. J. Sci. Food Agr. 21:172–177.

Hill-Cottingham, D. G., and C. P. Lloyd-Jones. 1973. Physiol. Plant. 29:39–44.

Hill-Cottingham, D. G., and C. P. Lloyd-Jones. 1975. J. Sci. Food. Agr. 26:165–173.

Hilton, R. J., and H. Khatamian. 1973. Can. J. Plant. Sci. 53:699–700.

Hipp, B. W., and G. W. Thomas. 1968. Agron. J. 60:467–469.

Hoad, G. V., and D. L. Abbott. 1983. *In* Regulation of photosynthesis in fruit trees. A. N. Lakso and F. Lenz (eds.). Spec. Rep. New York State Agr. Exp. Sta., Geneva, NY, pp. 87–92.

Hoagland, D. R. 1941. Proc. Amer. Soc. Hort. Sci. 38:8–12.

Hoepfner, E. F., B. L. Koch, and R. P. Covey. 1983. J. Amer. Soc. Hort. Sci. 108:207–209.

Horton, B. D., and J. H. Edwards. 1976. Hort. Sci. 11:591–593.

Hooker, H. D. 1922. Proc. Amer. Soc. Hort. Sci. 18:150.

Hsiao, T. C., and A. Lauchli. 1986. Adv. Plant Nutr. 2:281–312.

Iglanov, Y. K. 1976. Izv. An. Turkm. SSR Biol. Nauk. 5:29–32.

Johnson, F., D. F. Allmendinger, V. L. Miller, and D. Polley. 1954. Phytopathology 45:110–114.

Johnson, D. S., G. R. Stinchcombe, and K. G. Scott. 1983. J. Hort. Sci. 35:317–326.

Juritz, C. F. 1913. Agric. J. Union South Africa 5:102–122.

Kallay, T. and Szücs. 1982. Proc. 9th Int. Plant Nutr. Coll. Vol. 1. Slough U.K.: Commonwealth Apr. Bur., pp. 256–261.

Kamali, A. R., and N. F. Childers. 1970. J. Amer. Hort. Sci. 95:652–656.

Kang, S. M., K. C. Ko, and J. S. Titus. 1981. Plant Physiol. 67:86.

Kang, S. M., and J. S. Titus. 1980. Physiol. Plant. 50:291–297.

Kang, S. M., and J. S. Titus. 1981. Physiol. Plant. 53:239–244.

Katzfuss, M. 1973. Arch. Gartenb. 21:637–651.

Kenworthy, A. L., and L. Martin. 1966. *In* Temperate to tropical nutrition of fruit crops. N. Childers (ed.). Rutgers University, New Brunswick, NJ: Hort. Publ., pp. 813–870.

Kessler, B. 1957. Nature 179:1015–1016.

Kessler, B. 1958. Ktavim 8:287–293.

Kidson, E. B. 1947. J. Pomology 23:178–184.

Kidson, E. B., H. O. Askew, and E. Chittenden. 1940–1941. J. Pomol. Hort. Sci. 18:119–134.

Kirkpatrick, H. C., J. M. Thompson, and J. H. Edwards. 1975. Hort. Sci. 10:132–134.

Klepper, L., and R. H. Hageman. 1969. Plant Physiol. 44:110–114.

Knee, M. 1987. Phytochemistry 17:1261–1264.

Knowles, J. W., W. A. Dosier, Jr., C. E. Ewans, C. C. Carlton, and J. M. McGuire. 1984. J. Amer. Soc. Hort. Sci. 109:440–444.

Kolesnikov, V. 1971. The root system of fruit plants. Moscow: MIR Publishers.

Korbam, S. S., and J. M. Swiader. 1984. J. Amer. Soc. Hort. Sci. 109:428–432.

Korcak, R. F. 1987. Hort. Rev. 9:133–186.

Kotze, W. A. G., C. B. Shear, and M. Faust. 1977. J. Amer. Soc. Hort. Sci. 102:279–282.

Kramer, P. J., and T. T. Kozlowsky. 1979. Physiology of Woody Plants. New York: Academic Press.

Kwong, S. S., and E. G. Fisher. 1962. Proc. Amer. Soc. Hort. Sci. 81:187–190.

Lachmo, D. R., R. S. Harrison-Murray, and L. J. Andus. 1982. J. Exp. Bot. 33:943–951.

Latimer, L. P. 1941. Proc. Amer. Soc. Hort. Sci. 38:63–69.

Lauchli, A., and R. Pfluger. 1979. *In* Potassium research—Review and trends. Berne: Intern. Potash Inst. pp. 111–163.

Leece, D. R., D. R. Dilley, and A. L. Kenworthy. 1972. Plant Phys. 49:725–728.

Letham, D. C. 1962. Exp. Cell Res. 27:352–355.

Lilleland, O., and M. E. McCollam. 1961. Better Crops Plant Food 45(4):1–7.

Lindner, R. C. 1943. Proc. Amer. Soc. Hort. Sci. 42:275–279.

Lindner, R. C., and N. R. Benson. 1954. *In* Fruit nutrition. N. Childers (ed.). Rutgers University, New Brunswick, NJ: Hort. Publ., pp. 666–683.

Lindner, R. C., and C. P. Harley. 1944. Plant Phys. 19:420–439.

Loneragan, J. F., T. S. Grove, A. D. Robson, and K. Snowball. 1979. J. Amer. Soil. Sci. Soc. 43:966–972.

Lucic, P. 1967. Contemp. Agr.–Novi. Sad. 15:31–37.

Lundegardh, H. 1945. Leaf analysis. Jena: Verlag G. Fisher.

Lyon, T. L., A. J. Heinicke, and B. D. Wilson. 1923. Cornell Univ. Agr. Exp. Sta. Memoir No. 63.

Maggs, D. H. 1964. J. Exp. Bot. 15:574–583.

Marlow, G. C., and W. H. Loescher. 1984. Hort. Rev. 6:189–251.

Marschner, H. 1986. Mineral nutrition in higher plants. New York: Academic Press.

Marschner, H., and H. Ossenberg-Neuhaus. 1977. Z. Pflanzenphysiol. 85:29–44.

Martin, D., T. L. Lewis, and K. Stockhouse. 1962. Austral. J. Expt. Agr. Anim. Husb. 2:92–96.

McGeorge, W. T. 1949. Univ. Arizona Ag. Exp. Sta. Tech. Bull. 117.

Mengel, K., and E. A. Kirkby. 1982. Principles of plant nutrition. Bern, Switzerland: Int. Potash Inst.

Miller, R. H. 1980. J. Amer. Soc. Hort. Sci. 105:355–364.

Mishra, K. A. 1984. Agr. Sci. Digest India. 4:226–227.

Mosse, B. 1957. Nature 179:922–924.

Mulder, D. 1950. Phytopath. Z. 16:510–511.

Murneek, A. E. 1930. Proc. Amer. Soc. Hort. Sci. 27:228–231.

Newman, E. L. 1969. J. Appl. Ecol. 6:1–12.

Oaks, A., W. Wallace, and O. Stevens. 1972. Plant Physiol. 50:649–654.

Oberly, G. H. 1963. Proc. New York State Hort. Soc. 149–151.

Oberly, G. H. 1973. J. Amer. Soc. Hort. Sci. 98:269–271.

O'Kennedy, B. T., M. J. Hennerty, and J. S. Titus. 1975. J. Hort. Sci. 50:321–329.

O'Kennedy, B. T., and J. S. Titus. 1979. Physiol. Plant. 45:419–474.

Oland, K. 1963. Physiol. Plant. 16:682–694.

Paliyath, G., B. W. Poovaiah, G. R. Munske, and J. Magnuson. 1984. Plant Cell Physiol. 25:1083.

Papp, J., and J. Tamasi. 1979. Gyümölcsösök talajmüvelése és tapanyagellatasa. Budapest: Mezögazdasagi Kiado, p. 372.

Peikic, B. and R. Boskovic. 1985. Yugoslov. Vocartsro. 19:437–445.

Perez, J. R., and W. M. Kliever. 1978. J. Amer. Soc. Hort. Sci. 103:246–250.

Perring, M. G., and A. P. Preston. 1974. J. Hort. Sci. 49:85–93.

Phillips, W. R., and F. B. Johnston. 1943. Sci. Agr. 23:451–460.

Poling, E. B., and G. H. Oberly. 1979. J. Amer. Soc. Hort. Sci. 104:799–801.

Poovaiah, B. W. 1986. Food Techn. 40:86.

Poovaiah, B. W., and A. S. N. Reddy. 1987. CRC Critical Rev. Plant Sci. 6:47–103.

Popenoe, J., and L. E. Scott. 1956. Proc. Amer. Soc. Hort. Sci. 68:56–62.

Priestley, C. A. 1972. Ann. Bot. 36:513–524.

Priestley, C. A., P. B. Catlin, and E. A. Olsson. 1976. Ann. Bot. 40:1163–1170.

Quinlan, J. D. 1965. Rpt. East Malling Res. Sta. 1964:117–118.

Raese, J. T., and C. L. Parish. 1984. J. Plant Nutr. 7:243–249.

Razeto, B. J. 1982. Plant Nutr. 5:917–922.

Reckruhm, I. 1974. Archiv Gartenbau 22:501–517.

Richards, D. 1980. Scientia Hort. 12:143–152.

Richards, D., and B. Cockroft. 1975. Austral. J. Agr. Res. 26:173–180.

Richards, D., and R. N. Rowe. 1977. Ann. Bot. 41:1211–1216.

Roberts, R. H. 1921. Proc. Amer. Soc. Hort. Sci. 18:143–145.

Rogers, W. S. 1939. J. Pomol. Hort. Sci. 17:67–84.

Rogers, W. S. 1968. J. Hort. Sci. 43:527–528.

Rogers, B. L. and L. P. Batjer. 1954. Proc. Amer. Soc. Hort. Sci. 63:67–73.

Rogers, B. L., L. P. Batjer, and A. H. Thompson. 1953. Proc. Amer. Soc. Hort. Sci. 61:1–5.

Rogers, B. L., A. H. Thompson, and L. E. Scott. 1965. Proc. Amer. Soc. Hort. Sci. 86:46–54.

Rogers, W. S., and G. C. Head. 1969. *In* Root growth. W. Y. Whittington, (ed.). London: Butterworths.

Rom, R. C. 1983. Fruit Var. J. 37:3–14.

Rossignol, M., D. Lamant, L. Salsac, and R. Heller. 1977. p. 483. *In* Transmembrane ionic exchange in plants. M. Thellier, A. Monnier, M. Demarty, and J. Dainty (eds.). Paris: CNRS. p. 483.

Russel, R. S., and D. T. Clarkson. 1976. *In* Perspectives in experimental biology, Vol. 2, N. Sunderland (ed.). New York: Pergamon Press, pp. 401–411.

Schultz, R. P. 1972. Proc. Soil Sci. Soc. Amer. 36:158–162.

Sharples, R. O. 1980. *In* Mineral nutrition of fruit trees. D. Atkinson, J. E. Jackson, R. O. Sharples, and W. M. Waller (eds.). London: Butterworths, pp. 17–28.

Shear, C. B. 1971. J. Amer. Soc. Hort. Sci. 96:415–417.

Shear, C. B. 1974. Plant Analysis and Fertilizer Problems. 2:427–436.

Shear, C. B. 1980. *In* Mineral nutrition of fruit trees. D. Atkinson, J. E. Jackson, R. O. Sharples, and W. M. Waller (eds.). London: Butterworths, pp. 41–50.

Shear, C. B. 1982. *In* Plant nutrition 1982. E. Scaife (ed.). Commonwealth Agr. Bur. Slough, U.K., pp. 607–612.

Shear, C. B., H. L. Crane, and A. T. Myers. 1946. Proc. Amer. Soc. Hort. Sci. 47:239–248.

Shear, C. B., and M. Faust. 1970. Plant Phys. 45:670–674.

Shear, C. B., and M. Faust. 1971. J. Amer. Soc. Hort. Sci. 96:234–240.

Shear, C. B., and M. Faust. 1980. Hort. Rev. 2:142–207.

Shen, Tsuin, and H. Tseng. 1949. Proc. Amer. Soc. Hort. Sci. 53:11–12.

Shim, K. K., W. E. Splittstoesser, and J. S. Titus. 1973. Physiol. Plant. 28: 327–331.

Shim, K. K., J. S. Titus, and W. E. Splittstoesser. 1972. J. Amer. Soc. Hort. Sci. 97:592–596.

Southwick, L. 1943. Proc. Amer. Soc. Hort. Sci. 42:85–94.

Spencer, P. W., and J. S. Titus. 1972. J. Amer. Soc. Hort. Sci. 96:131–133.

Spencer, P. W., and J. S. Titus. 1974. Plant Physiol. 54:382–385.

Stahly, E. A., and N. R. Benson. 1970. J. Amer. Soc. Hort. Sci. 95:727–729.

Sun Xiu Ping, S. Y. Wang, Yue Ao Tong, R. F. Korcak, and M. Faust. 1987. J. Plant Nutr. 10:1021–1030.

Swietlik, D., and M. Faust. 1984. Hort. Rev. 6:287–356.

Swietlik, D., and K. Slowik. 1981. Fruit Sci. Rep. 8:49–59.

Syrgiannidis, G. 1985. Acta Hort. 173:383–388.

Tamasi, J. 1986. Root location of fruit trees and its agrotechnical consequences. Budapest: Academiai Kiado.

Taylor, B. K. 1967. Austral. J. Biol. Sci. 20:379–387.

Taylor, B. K., and L. H. May. 1967. Austral. J. Biol. Sci. 20:389–411.

Taylor, B. K., B. van den Ende, and R. L. Centerford. 1975. J. Hort. Sci. 50:29–40.

Terry, N., and G. Low. 1982. J. Plant Nutr. 5:301–310.

Thomas W. 1927a Science 66:115–116.

Thomas, W. 1927b Plant Physiol. 2:109–137.

Thomas, W., W. B. Mack, and C. B. Smith. 1954. Pennsylvania Agr. Exp. Sta. Bul. 564.

Thompson, A. H., and L. P. Batjer. 1950a. Proc. Amer. Soc. Hort. Sci. 56:227–229.

Thompson, A. H., and L. P. Batjer. 1950b. Soil Sci. 69:281–290.

Thompson, S. G., and W. O. Rogers. 1945. East Malling Res. Sta. Ann. Rep. 1944:60–63.

Thorne, D. W., and A. Wallace. 1944. Soil Sci. 57:299–312.

Thorne, D. W., and F. B. Wann. 1950. Utah Agr. Exp. Sta. Bull. 338.

Titus, J. S., and S. M. Kang. 1982. Hort. Rev. 4:204–246.

Titus, J. S., and N. Ozerol. 1966. Proc. 17th Intern. Hort. Congr., pp. 161–162.

Titus, J. S., and W. E. Splittstoesser. 1969. Phytochemistry 8:2133–2138.

Titus, J. S., W. E. Splittstoesser, and P. W. Spencer. 1968. Plant Physiol. 43:619–621.

Tong, Y. A., F. Fan, R. F. Korcak, and Miklos Faust. 1986. J. Plant Nutr. 9:23–41.

Tromp, J. 1970. *In* Physiology of tree crops. L. C. Luckvill and C. V. Cutting (eds.). New York: Academic Press, pp. 143–159.

Tromp, J. 1980. *In* Mineral nutrition of fruit trees. D. Atkinson, J. E. Jackson, R. O. Sharples, and W. M. Waller (eds.). London: Butterworths, pp. 173–182.

Tromp, J., and J. C. Ovaa. 1971. Physiol. Plant. 25:16–22.

Tromp, J., and J. C. Ovaa. 1976. Physiol. Plant. 37:29–34.

Tromp, J., and J. C. Ovaa. 1979. Physiol. Plant. 45:23–38.

Tukey, H. B. 1927. New York State Agr. Exp. Sta. Bull. 541:1–26.

van Goor, B. J., and P. van Lune. 1980. Physiol. Plant. 48:21–26.

Van Steveninck, R. F. M. 1965. Physiol. Plant. 18:54–69.

Vose, 1982. P. B. J. Plant Nutr. 5:233–249.

Wallace, A., and G. A. Wallace. 1982. *In* Plant nutrition. Vol. 2, A. Scaife (ed.). Commonwealth Agr. Bur. Slough, U.K. pp. 696–701.

Wallace, T. J. 1939. Pomol. Hort. Sci. 17:150–166.

Wann, F. B., and D. W. Thorne. 1950. Scientific Monthly 70:180–184.

Weinbaum, S. A., M. L. Mervin, and T. T. Muraoka. 1978. J. Amer. Soc. Hort. Sci. 103:516–519.

Weinbaum, S. A., and P. M. Neuman. 1977. J. Amer. Soc. Hort. Sci. 102: 601–604.

Weinberger, J. H., and F. P. Cullinan. 1936. Proc. Amer. Soc. Hort. Sci. 34:249–254.

Westwood, M. N., and F. B. Wann. 1966. *In* Temperate to tropical fruit nutrition. N. Childers (ed.). New Brunswick, NJ: Hort. Publ., Rutgers University, pp. 158–173.

White, G. C., and R. I. C. Holloway. 1967. J. Hort. Sci. 42:377–389.

Wieneke, J., and F. Führ. 1973. Gartenbanwissenschaft 38:91–108.

Wiersum, I. K. 1979. Acta Bot. Neerl. 28:221–224.

Wilkinson, B. G. 1958. J. Hort. Sci. 33:49–57.

Wilkinson, B. G., and M. A. Perring. 1964. Sci. Fd. Agric. 15:146–152.

Willoughby, P., and B. Cockroft. 1974. *In* Proc. 2nd Int. Drip Irrig. Conf., San

Diego. Riverside, CA: University of California Press, pp. 439–442.

Woodbridge, C. G. 1954. Canad. J. Agr. Sci. 34:545–551.

Woodbridge, C. G. 1971. Hort Sci. 6:451–453.

Woodbridge, C. G., A. Venegas, and P. C. Crandall. 1971. J. Amer. Soc. Hort. Sci. 96:613–615.

Yoshikawa, F. T., W. O. Reil, and L. K. Stormberger. 1982. Calif. Agric. 36:13.

Young, E., and D. J. Werner. 1984. J. Amer. Soc. Hort. Sci. 109:548–551.

Young, E., and D. J. Werner. 1985. J. Amer. Soc. Hort. Sci. 110:769–774.

Young, E., and D. J. Werner. 1986. Hort. Sci. 21:280–281.

Zeiger, D. C., and J. E. Shelton. 1969. Hort. Sci. 4:213–216.

Zhou, H. J., R. F. Korcak, F. Fan, and M. Faust. 1984a. J. Plant Nutr.

Zhou, H. J., R. F. Korcak, and Miklos Faust. 1986. Scientia Hort. 27:233–240.

Zhou, H. J., R. F. Korcak, W. P. Wergin, F. Fan, and Miklos Faust. 1984b. J. Plant Nutr. 7:911–928.

3

THE USE OF WATER BY FRUIT TREES

Of all the materials used by fruit trees for growth, water is taken up in the largest amount. The bulk of the water does not remain in the tree but evaporates into the air from the leaves and other aerial parts of the tree. The evaporated water in a well-watered tree is replaced by absorption from the soil. If absorption is insufficient to replace transpirational losses, the water status of the tree is changed, its 'water potential' is lowered, and many of its life processes are affected. Processes affected by water deficit include

1. growth of the tree, by influencing cell division and expansion;
2. fruiting, by influencing flower bud differentiation; and
3. fruit quality, by decreasing carbohydrate production through altering stomatal aperture and enzyme activities of photosynthesis and respiration.

The overall water use by fruit trees has been determined only in a few cases. The transpirational coefficient, the unit of water transpired per unit of dry weight produced, for apples is 113–140 (Lenz, 1986). This is low compared to other crops, and it indicates that fruit trees, exemplified by apple, are efficient users of water. It is critical to satisfy the water requirement of fruit trees at the beginning of the growing season when fruit set occurs, during the period of flower bud formation, and when the final fruit swell occurs before harvest.

It has been known for a long time that drought lowers productivity in fruit trees. What is now beginning to be understood is that water use is directly related to productivity even under conditions when the trees are not visibly stressed from drought. Dry-weight increases in apple (Gyuro, 1974) and peach (Richards, 1976; Richards and Rowe, 1976) are proportional to the amount of water transpired. By integrating dry-weight increase and transpiration over a period of time, Richards and Rowe (1976) demonstrated that the yield of dry matter is directly proportional to actual transpiration. In their studies with peach trees, this relationship held regardless of plant size or incidental treatments. The fact that water use, directly and proportionally, affects yield makes water relations a central issue in fruit production and fruit tree physiology.

The water needs of fruit trees vary with species and even within species during different parts of the growing season. The order of water requirement of fruit trees is quince > pear > plum > peach > apple > cherry sour cherry > apricot. During the dormant period, water use by trees is low. When leaves are fully developed and fruits are on the tree, water use is much higher. Toward the end of the growing season water use decreases again. The daily water use of a 4-year-old 'Jonathan' apple tree is illustrated in Figure 3.1.

The water requirement of an orchard, to be supplied by natural rain or irrigation, includes water that evaporates from the soil surface and from free water on the plant surface in addition to that used by transpiration. Water that evaporates from the soil surface is only a small portion of the total yearly evaporation. Thus transpiration accounts for most of the water used by orchards. The actual transpirational rate of an orchard is

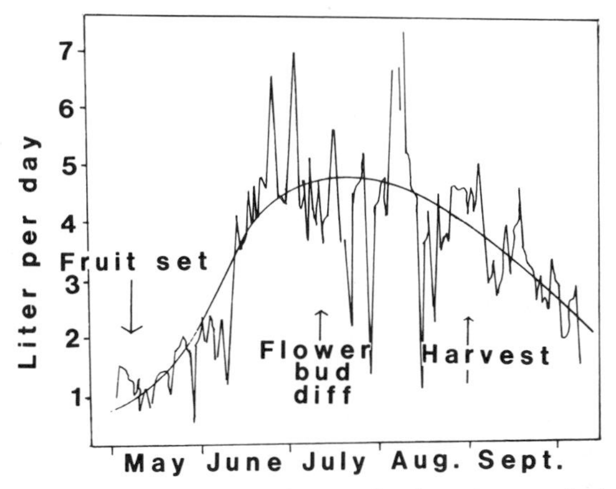

Figure 3.1 Water use of a 4-year-old 'Jonathan' apple tree during the growing season determined in a lysimeter. (Reproduced by permission from Gyuro, 1974.)

determined by a combination of environmental factors such as radiation, temperature, humidity and wind speed, and crop factors including canopy structure, leaf area, and stomatal conductance.

It is useful to distinguish between water evaporated by an orchard and that transpired from the tree. Water evaporated from the soil may be important in long-term lowering of soil water content, but it does not enter directly into determining the water status of the tree. Here we are concerned with the transpirational losses and the environmental factors affecting transpiration and how these losses affect the water status of the trees. A schematic diagram of the major effects determining the water use of a tree is presented in Figure 3.2.

Although sufficient water supply is essential for fruit production, water deficit may not be undesirable in all cases. A mild water deficit may shift vegetative growth toward reproductive growth (Chalmers et al., 1983) in peach and an increase in cold hardiness in apple (Wildung et al., 1973), characteristics that may at times be necessary.

DETERMINING WATER REQUIREMENT OF FRUIT TREES

Several methods have been used to estimate plant water requirement. The comparative merits of these methods for estimating evapotranspiration of fruit trees have been reviewed by Elfving (1982). The methods measure soil moisture status and estimate water use from the decreasing amount of ground water as opposed to determining the relationship between evaporimeter readings and actual evapotranspiration by trees and using this relationship in future determinations of water requirement.

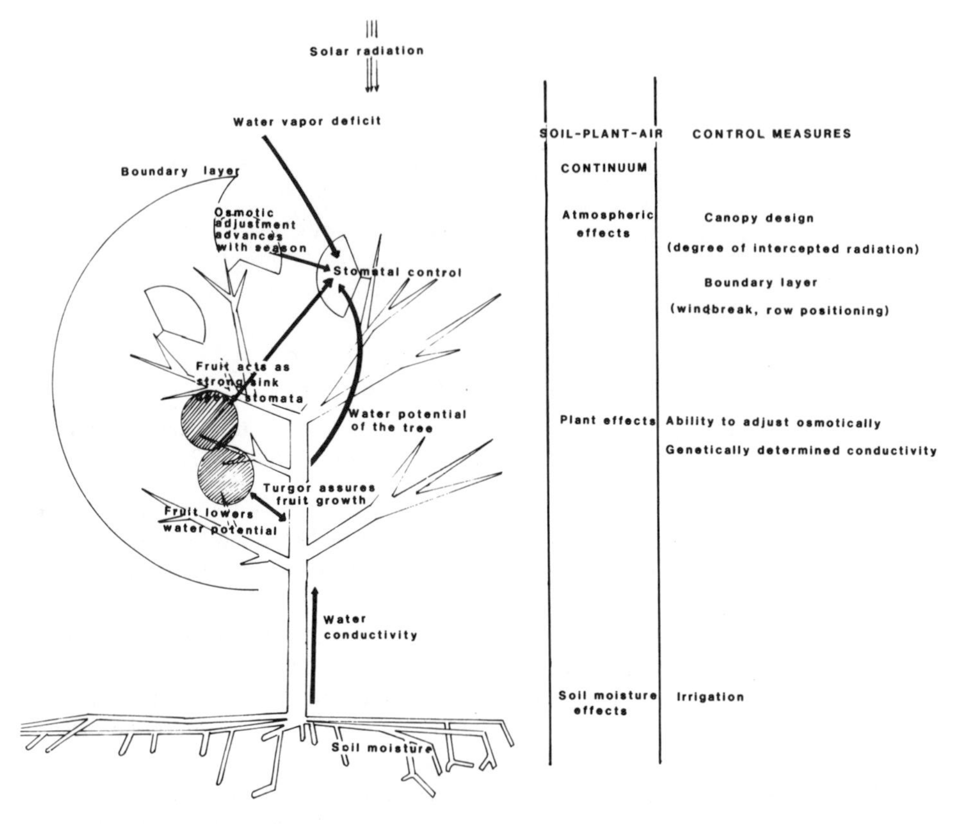

Figure 3.2 Schematic representation of major effects determining water use of fruit trees.

Measuring soil water status by its matric potential is advantageous when rainfall supplies the soil moisture or an irrigation method is used that supplements moisture to the entire orchard floor. When irrigation methods are used that wet only part of the root system, it is difficult to locate the moisture sensor in a soil zone that accurately reflects the soil moisture available to the tree. In addition, when water loss is measured by the sensor, it is difficult to interpret whether it is a downward loss or a loss to the atmosphere or whether the water is used by the tree.

The so-called Class-A pan evaporimeter (U.S. Weather Bureau Class-A evaporation pan) is popular for scheduling irrigation because of the high association between water loss and actual evapotranspiration (ET_a). Evaporation is easy to monitor, and the equipment necessary is simple. The utilization of an evaporimeter is a two-step process. The evaporation data are obtained from the Class A pan as a first step, and these data are converted into an estimated crop water use value by a conversion factor. This relative value must be further converted into actual water volume used by the plant and to be applied as irrigation.

Lysimeters are most often used for direct measurement of water use by fruit trees. Lysimeters are closed systems in which the actual water loss can be measured in a variety of ways. The actual water use by transpiration is correlated with the Class-A pan evaporation to establish crop coefficients (k_c). As a result ET_a is defined as a simple function of E_{pan},

$$ET_a = k_c E_{pan}$$

Conversion of crop water use estimates from evaporation units to volume units is dependent on the actual size of the evaporating surface (canopy of the tree). Estimation of the land area covered by transpiring canopy or total land area assigned for each plant (Black and Mitchell, 1974; Willoughby and Cockroft, 1974) or the area covered by the canopy (Kenworthy, 1972; Ross et al., 1980) all have been used for estimating evaporative surface. As the canopy rapidly expands, the k value steadily changes. This is illustrated by the changing k values during the growing season for peach and apple (Table 3.1).

Natali and Xiloyannis (1984) based water use coefficients for peach on the fruit growth stage. Coefficients were as follows: stage 1, 0.69; stage 2, 1.0; stage 3, 1.26; and after harvest, 1.24. Worthington et al. (1984) concluded that the correlation between the pan evaporation and the actual water use by peach trees was poor on a daily bases ($r = 0.61$), indicating that factors other than those associated with the physics of evaporation strongly affect water use. The correlation was better when weekly averages were calculated ($r = 0.88$).

An overall crop coefficient of 0.7 has been given by various workers for peach (Daniell, 1980; Worthington et al., 1982); the crop coefficients are 0.5 for low-density and 0.7 for high-density apple orchards when the total land area (m^2) is used in the equation. Although the water use of fruit trees determined in this way may be useful for scheduling irrigation, the crop coefficients are only very rough estimates of average water use and should be treated as such.

WATER STATUS OF FRUIT TREES

Transpiration rate and availability of water in the soil have a great influence on the water status of the tree. If water is available in sufficient quantity in the soil, the tree is usually able to satisfy the transpirational demand without being affected. However, if water is not available in the

TABLE 3.1 **Water Use Coefficients for Peach and Apple Trees as Fraction of Class-A Pan Evaporation for Tree Canopy Area**[a]

	March	April	May	June	July	August	September	October	November
APPLE									
Nonirrigated	0.55	0.75	0.90	0.95	0.95	0.90	0.86	0.83	0.70
Irrigated	0.83	0.98	1.11	1.21	1.23	1.23	1.16	0.93	0.83
PEACH									
A	—	—	—	—	0.90	0.71	0.67	0.59	—
B	—	0.66	0.74	0.98	1.07	1.26	—	—	—

[a] Apple data from Natali and Xiloyannis 1982. Peach data from (A) Worthington et al. (1984) and (B) Giulivo and Xiloyannis 1986.

quantity demanded by transpiration, then the water status of the tree changes, and the tree is under water stress. There are several physiological and morphological mechanisms developed in various plants to avoid or tolerate conditions of water stress. Few if any of these exist in fruit trees.

Traditionally, water stress in orchards has been dealt with by growing the trees in environments where the water need of the tree was satisfied by precipitation; by preserving natural precipitation by special agrotechniques, thus assuring water supply to the tree even in the absence of rain; or by applying a sufficient number of irrigations. This has been satisfactory in the past when yield was lower than that which is essential for economic production today. With the large fruit loads produced by the agrotechniques used today, transpirational rates are high and trees are likely to undergo water stress more often than in the past. Even when the trees are well watered, they react to the changing demand for water caused either by diurnal changes or longer range variations in the environment. The water status of trees was recently reviewed in detail by Jones et al. (1985).

The water status of the tree is conveniently measured by the water potential ψ. The water potential describes the state of water by its chemical potential within the system relative to that of pure water, that is, water containing no solutes and bound by no forces. This gives the capacity of the water at any point to move about in the system compared to the capacity of free water. The total water potential ψ consists of several mutually independent components: the osmotic potential ψ_s arising from dissolved solutes in the water; the pressure ψ_p or turgor potential arising from the balance of internal and external pressures and that can be described as an excess pressure inside of the cell; and the matric potential arising from capillary forces at the water–air interface. The matric potential is usually assumed negligible. For fruit trees, tree height may also introduce a component, the gravitational potential. Jones et al. (1985) partitioned the water potential ψ into pressure, osmotic, and gravitational components:

$$\psi = \psi_p + \psi_s + \psi_g$$

where ψ_p is the pressure potential, ψ_s is the osmotic or solute potential, and ψ_g is the gravitational potential. They estimated that ψ_g changes only by 0.01 MPa m^{-1} of tree height and is of little importance for fruit trees.

The total leaf water potential can be readily determined by using a pressure chamber. Many workers rely on this measure as the sole measure of the water status of trees even though it is probably only relevant to water flow and is not an indicator of physiological stress (Jones et al., 1985). Leaf water potential fluctuates with environmental demand, whereas many stress responses integrate water status over a period of time. Estimation of ψ at the surface of the roots has been proposed ot be a good general measure of the water stress of fruit trees as well as other plants (Jones, 1983). The technique is based upon the determination of stomatal conductance g_s and leaf water potential ψ_l using well-watered controls to estimate the hydraulic flow resistance in the tree. This can be used to estimate the average ψ at the root surface and as an indication of the degree of stress to which the tree is exposed.

Osmotic potential ψ_s can be estimated by the pressure–volume method modified for fruit trees by Davies and Lakso (1978) by determining the refracting index of sap expressed from frozen and thawed leaf blades (Lakso et al., 1984) and by leaf sap osmotic potential determination by psychometry (Lakso et al., 1984) or cryoosmometric methods.

Turgor potential is usually estimated as a difference between ψ_l and ψ_s.

Another measure of the water status is the *relative water content* (RWC) (at one time called 'relative turgidity'):

$$\text{RWC} = \frac{\text{fresh weight} - \text{dry weight}}{\text{saturated weight} - \text{dry weight}} \times 100$$

The RWC is a useful value. It is usually more closely related to turgor than ψ. In general, RWC and ψ_p are more relevant to the control of photosynthesis than ψ_s (Jones et al., 1985).

Yet another term often used in expressing water use is *water use efficiency* (WUE). It has been used in two ways. Physiologists calculate WUE in units such as g CO_2 g^{-1} H_2O or occasionally as mol CO_2 mol^{-1} H_2O. This is the ratio of the carbon fixed in photosynthesis to the water lost by transpiration. Agronomists usually define water use efficiency in terms of the ratio of dry matter produced as crop yield to water used in transpiration and evaporation:

$$\text{WUE} = \frac{\text{dry matter of fruit yield}}{\text{evapotranspiration}}$$

This places the emphasis on the economic unit, the amount of fruit produced.

Hydraulic Flow through the Tree

The hydraulic flow through any plant should be considered as water flow in the soil–plant–atmosphere system. The flow is determined by the transpirational rate and the ultimate source of water, the water content of the soil. There are impedances to water flow in the plant. The steady-state flow of water through the soil–plant–atmosphere system is often described by a simple catenary model (Jones, 1983). Using such a model, the steady-state leaf potential can be expressed as

$$\psi_l = \text{soil} - E_a C_{sp}$$

where soil is the bulk soil potential, E_a is the transpiration rate, and C_{sp} is the total hydraulic conductance in the soil–plant pathway. From the preceding equation, it is clear that the ψ_l is determined by the flow rate through the plant, which is in turn determined by stomatal control, hydraulic conductivity, and soil water potential.

Stomatal control has been discussed previously. The soil water potential in most situations is given. The flow of water in the soil–plant–atmosphere pathway is the determining factor for tissue ψ. The general conclusion of several authors reviewed by Jones et al. (1985) is that the conductivity of the plant only becomes significant at low soil water contents.

Landsberg and Jones (1981) estimated that in apples, the conductivity of the roots is the lowest, and root resistance accounted for 40–68% of the total plant resistance. Root conductivity to water flux is usually determined by estimating the difference between the observed conductivity of whole trees with their root in moist soil or water and the conductivity calculated when water uptake was directly through the cut base of the stem (cut under water).

Water Potential of Tissues

Leaf water potential in fruit trees generally follows the diurnal trend similar to those exhibited in other plants. Minimum values between −1.0 and −2.5 MPa usually occur in the early afternoon and coincide with the highest rate of transpiration (Jones et al., 1985). Under nonlimiting conditions when the trees are well watered, the transpiration rate is the major determining factor that could influence ψ_l, which in turn is influenced by *water pressure deficit* (VPD). Most fruit trees show no difference in minimum ψ_l in humid and arid environments. This indicates an effective control of transpiration rate under conditions of high evaporative demand. There are species differences. Judging from the diurnal trends in ψ_l, apricot is the most effective in controlling stomatal opening and transpiration followed by pear and apple (Figure 3.3). It is notable that several factors in the soil–plant–atmosphere continuum determine the water requirement of a fruit species. Practical experience indicates that pears require more moisture or more humid climate than do apples. At the same time, the daily minimum ψ_l values would indicate that pear

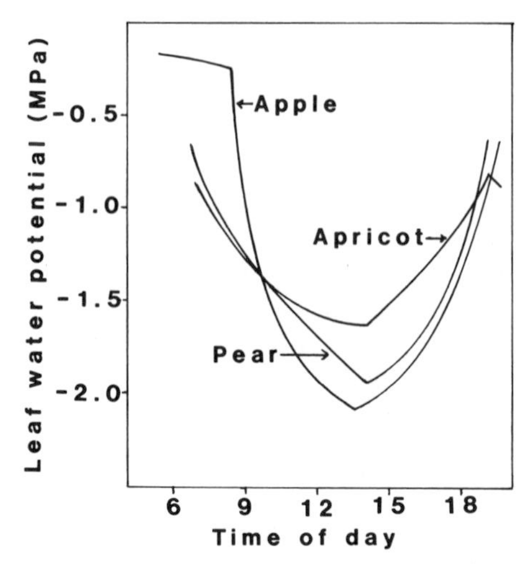

Figure 3.3 Diurnal changes in leaf water potential for apricot, apple, and pear. (Redrawn by permission from Jones et al., 1985.)

possibly exerts a better control of stomata than does apple (Figure 3.3) and thus is able to survive in more arid climates.

With severe soil water deficit, ψ_l can fall much lower than normally experienced. In apples, values for trees under water stress were recorded between −2.5 and −3.0 MPa (Jones and Higgs, 1979). The diurnal minimum decreases as the season advances. In one study (Goode and Higgs, 1973) ψ declined from −1.0 to −2.0 MPa in irrigated trees and from −1.5 to −3.0 in unirrigated trees between June and September.

Stomata close when water potential reaches a low level. Lakso (1979) measured the relationship between stomatal conductance and ψ_l. He found a straight-line relationship between g_s and ψ_l. However, the ψ_l value required for stomatal closure changed as the season advanced with about −3.0 MPa from −2.0 to −5.0 MPa. By September, apple leaves exhibited characteristics much like drought-tolerant plants. Pears also exhibit low ψ_l before stomata close. In field-grown pear leaves photosynthesis did not decrease until ψ_l approached −3.0 to −3.5 MPa (Kriedeman and Canterfold, 1971).

Leaf water potential is often lower in apple trees grafted on dwarfing rootstocks than in more vigorous ones. Olien and Lakso (1984) explained lower ψ_l in more dwarfing stock in apples by lower root conductances in the dwarfing stock. Others (Giulivo and Bergamini, 1982) attributed the difference to the higher fruit load induced by the dwarfing rootstock. In peach, differences in ψ_l were not directly related to tree size. Trees on the intermediate-sized St. Julien GF 655/2 rootstock used the most water and suffered the greatest stress (Natali and Xiloyannis, 1984; Young and

TABLE 3.2 Capacitance of Apple Tissues

Tissue	% f.w./MPa	Reference
Whole plant	9.8	Davies and Lakso, 1979
Leaves, Cox, Empire	2.5–3.0	Davies and Lakso, 1979; Landsberg et al., 1976
Golden Delicious	3.4–5.8	Jones and Higgs, 1979
Bark	7.6	Powell and Thorpe, 1977
Fruit	5.5–20.0	Landsberg and Jones, 1981

Houser, 1980). However, St. Julien rootstock is known to perform best under ample soil moisture conditions. Smart and Barrs (1973) determined minimum values of ψ_l on hot dry days for prunes and peaches. Multiple regression of water potentials against radiation, temperature, and VPD indicated that radiation was the dominant factor in controlling ψ_l, perhaps by influencing the rate of transpiration.

Various tissues of a tree store water. Release of stored water to the transpiration stream in response to decreasing water potential is a possible means to regulate water movement. Plant capacitance can provide sufficient water for 3 hr of transpiration in some herbaceous plants and up to 14 hr of transpiration in large conifers. Landsberg et al. (1976) calculated that a mature apple tree experiencing 2.0 MPa change in water potential and transpiring 20 mg $H_2O\ m^{-2}\ s^{-1}$ has sufficient storage capacity for 2 hr of transpiration.

Capacitance can be expressed as percentage of fresh weight per megapascal of change in water potential (see Table 3.2 for capacitances of apples). When water is lost from the tree and the water available through capacitance is exhausted, the branches of the tree shrink. This is illustrated by Chalmers et al. (1983), who measured the diurnal shrinkage of limbs of fruiting and nonfruiting peach trees. The degree of shrinkage was not the same on consecutive days, but in any given day the shrinkage of fruiting trees was more than that of nonfruiting trees, and shrinkage during the slow fruit growth period was less than during the rapid fruit growth period. This indicates that shrinkage reflects the water use of a tree whether it is seasonal or diurnal. Fruit and branches also shrink as the diurnal xylem potential reaches low levels. When the water potential is restored, growth takes place and stems and fruit regain or increase in volume (Figure 3.4).

Osmotic Adjustment in Tissues of Fruit Trees

A decrease in the relative water content of tissues of fruit trees at midday (Davies and Lakso, 1979; Goode and Higgs, 1973) results in passive

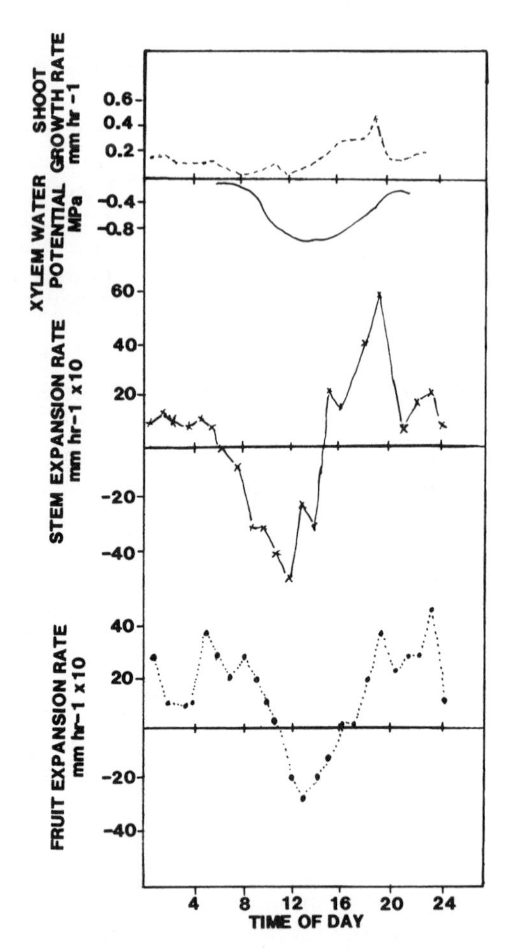

Figure 3.4 Diurnal changes in water potential, shoot growth, and shoot and fruit expansion rate. (Redrawn by permission from Landsberg and Jones, 1981.)

concentration of solutes in the symplast of the cells. The decrease in osmotic potential as the result of passive concentration can be substantial. Fanjul and Rosher (1984) reported between 0.6 and 1.0 MPa decrease in ψ_s when ψ_s was measured at full and zero turgor pressure. Such differences occurred regardless of the treatment of the trees. They measured the ψ_s of apple leaves grown in growth chambers, outdoors in pots, or in orchards and that were stressed or fully watered. In addition to the passive concentration of solutes because of water loss, there is an active accumulation of solutes resulting from decreased translocation rates compared to the production of photosynthetic carbon (Priestley, 1973).

When considering the osmotic responses of fruit trees, it is useful to

TABLE 3.3 Osmotic Adjustment in Fruit Tree Leaves

Species	Diurnal Osmotic Adjustment	Active Osmotic Adjustment
	Diurnal	
Apple	0.5–1.0	0.2–0.4[a]
Apple	1.65	0.37[b]
Apple	None	None[c]
	Long Term	
Apple	0.5–July to September	0.5[a]
Apple	2.0–2.5 over the season	2.0–2.5[d]
Apple	0.3–0.4 induced by stress	0.3–0.4[e]

[a] Goode and Higgs, 1973.
[b] Davies and Lakso, 1979.
[c] Young et al., 1982.
[d] Lakso, 1983.
[e] Fanjul and Rosher, 1984.

distinguish between passive and active adjustments. To study the active component, the passive component must be eliminated. One way to accomplish this is to convert observed osmotic potentials to corresponding values at 100% RWC (Jones et al., 1985) or to rehydrate the leaves before determining ψ_s. The second method has a disadvantage since rehydrating takes time and changes in the solute concentration during this time may produce inaccurate values.

The magnitude of the osmotic adjustment varies (Table 3.3).

Osmotic potential is dependent on molarity of the solutes and not on size of the molecules. Small molecules are more effective in adjusting the cell sap osmotically than larger ones. In fruit trees, it is likely that sorbitol is more effective in osmotic adjustment than sucrose.

Different organs show a different degree of osmotic potential and possibly osmotic adjustment. A seasonal study of young and mature apple leaves revealed that osmotic adjustment occurs during the growing season as leaves age. As the season advanced between June and September, the ψ_s of apple leaves decreased from -2.0 to -4.0 MPa when the ψ_s was measured at 100% RWC. The position of the leaf is a determining factor for ψ_s. Immature leaves have relatively high ψ_s ranging from -1.0 MPa for the newest leaves to -2.0 MPa for the still unfolding but oldest immature leaves when measured at full turgor. The mature leaves generally show ψ_s of about -2.5 MPa. Field-grown leaves also show a decrease in ψ_s in mid to late September, which may be an autumnal aging effect (Lakso et al., 1984). Young leaves of apples do not adjust osmotically up or down with changing weather conditions, whereas older leaves show a great ability to adjust osmotically (Swietlik and Miller, 1983). Young leaves do not have the enzymes to produce sorbitol in comparison to old leaves, and this may be the difference (Lakso et al., 1984).

Osmotic adjustment is very important for the trees to survive moderate drought and still carry their crop to maturity. Osmotic adjustment enables the tissue to maintain turgor, which in turn enables the tree or fruit to maintain growth. Lower ψ_s attracts water and maintains ψ_p. If ψ_l and ψ_s are lowered to the same extent, then ψ_p remains constant because $\psi_l = \psi_s + \psi_p$. This is the most essential physiological meaning of osmotic adjustment and its role in plant adaptation to water stress. Active osmotic adjustment usually occurs slowly and requires an extended period of time. Trees cannot be expected to adjust osmotically if water stress is imposed rapidly.

Turgor Potential in Tissues of Fruit Trees

The maintenance of turgor is essential for many biochemical processes in fruit trees. Processes related to turgor are expansive growth, stomatal opening, membrane hydraulic conductivity, synthesis of hormones such as abscisic acid (ABA), and distribution or concentration of gibberellins, cytokinins, and auxins. Since these compounds are involved in almost all the biochemical and physiological processes of the tree, the need for maintaining turgor potential cannot be overemphasized. The turgor of cells can be maintained by osmotic adjustment, already discussed. The magnitude of the passive compensatory osmotic adjustment depends on the elasticity of cell walls. The adjustment is small for cells with very rigid cell walls and larger for cells with thinner walls. This means that any change of cell volume as the cell loses or gains water will be reflected in the maintenance of turgor. Details of forces acting on the elasticity of cells are reviewed by Jones et al. (1985). In the overall process cell elasticity plays a relatively small role in maintaining turgor. Turgor occasionally can be excessive. When water uptake is too high and cell walls are thick, allowing little elastic adjustment, the tissue often splits. This happens in cherries and in those apples that contain insufficient concentration of Ca. Low-Ca apples usually have thicker cell walls, contain more cellulose (Faust, 1968), and are apt to split readily if the tree is watered (Shear, 1971).

Turgor pressure needs to be maintained during fruit growth in order to obtain large fruit. As was discussed, fruit imposes water stress on the tree. If fruit load is excessive and the turgor potential of the cells is low, cells of the fruit do not expand to the maximum and fruit remains small. Irrigation (i.e., maintaining turgor pressure) shifts fruit size classes in every fruit species. This is illustrated for apples (Natali and Xiloyannis,

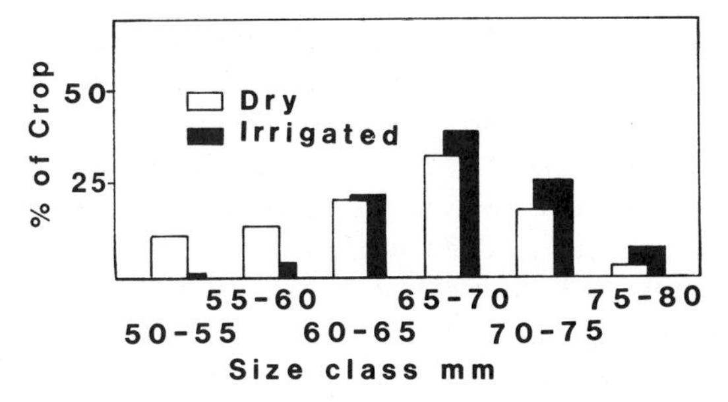

Figure 3.5 Effect of irrigation on fruit size of apple (Natali and Xiloyannis, 1982).

1982) in Figure 3.5. Some fruit species require high turgor pressure to enlarge fruit. For example, nectarines must be thinned to a higher degree (have fewer fruit per tree) than peaches to obtain similar fruit size. In the absence of data it appears that plums and apricots also require a high ψ and good water supply within the tree for the fruit to expand to a large size. Even in peach in dry conditions both number of fruit and shoot growth should be low to produce large-size fruit. This is why in such situations 'short pruning,' discussed under pruning, is practiced.

ROLE OF ROOT SYSTEM IN ABSORBING WATER

How much water the root system of a fruit tree can extract from the soil has important consequences for the physiology of the tree and the horticultural technology that assures sufficient water supply. The soil–water relations take on special importance when water is supplied by 'trickle irrigation' which only wets a relatively small portion of the root system.

Fruit trees apparently can extract significant quantities of water with only a portion of their root system exposed to moist conditions (West et al., 1970). Apple trees with quartered root systems, when each quarter was grown in separate containers, used proportionally more water: 74, 88, and 94% of the control with one-, two-, or three-quarters of the root system receiving water (Black and West, 1974). The trees apparently have the ability to distribute water laterally across the vascular tissue to all parts of the apple tree (West et al., 1970).

Root systems adapt rapidly and grow into the wet zone if only part of the root system is wetted. This was observed with peach (Black and Mitchell, 1974; Willoughby and Cockroft, 1974) and apple (Levin et al., 1979, 1980; Proebsting et al., 1977). Roots of peach proliferated four- to five-fold in the continuously wetted zone even though the other parts of the root system received water periodically from rain (Willoughby and

Cockroft, 1974). Black (1976) estimated that an increase of the root system in the wetted zone can mean over 60% of the root system is wetted with an anticipated water uptake efficiency of over 90% in comparison to trees where the entire area under the canopy is wetted.

Wetting a small part of the root system may eventually restrict the root system, and the restricted zone is filled with old roots. Atkinson and Lewis (1979) reported decreased root growth in soils containing old roots. Restriction of the root system may not be detrimental. Fruiting commenced earlier on trees where the root system was limited to about $0.69\,m^3$ wetted area in comparison to $13.5\,m^3$ when the entire root system was wetted (Proebsting et al., 1977).

As discussed before, wetting only part of the root system results in somewhat lowered total water uptake by the tree. Of course, this affects the water status of the tree. Trickle irrigation of peach trees at 50% of estimated water use as compared to irrigation at 100% depressed midday xylem pressure potentials by only 0.1–0.3 MPa. Trees of both treatments regained their water status before sunrise (0.5–0.6 MPa) compared to nonirrigated trees, which had lower before-sunrise potentials that were further lowered as the season progressed and the soil moisture was depleted (Xiloyannis et al., 1980).

Effect of Leaf Area on Water Use

Transpiration is closely related to the interception and partitioning of solar radiation. The net radiation absorbed by the leaves is a function of the available radiation and the actual interception, which depends on orchard design. Radiative properties of orchards are described in Chapter 1. The feature of orchard design that most affects transpiration is the LAI. Leaf area is the surface that absorbs radiation as well as the transpiring surface; in general, the higher the absorbing surface, the higher the transpiration rate. Such a relationship is proportional at low LAIs for almond (El Sharkawi and El Monayeri, 1976) and for apple (Atkinson, 1978), and it is likely to be proportional for other fruit crops. However, at high LAIs, evaporation from the canopy does not necessarily increase in proportion to the LAI. At high LAIs, the proportion of shaded leaves increases with increasing leaf area. The shaded leaves receive much less energy, and consequently the transpiration rate from shaded leaves is lower (Chalmers et al., 1983; Jones and Cumming, 1984; Warrit et al., 1980). For example, the transpiration of shaded peach leaves in the lowest layer of trees is less than half of those exposed to full radiation (Figure 3.6).

Reduction in radiation interception reduces total evaporation by the orchard. This can be achieved either by wide spacing or by lowering the LAI.

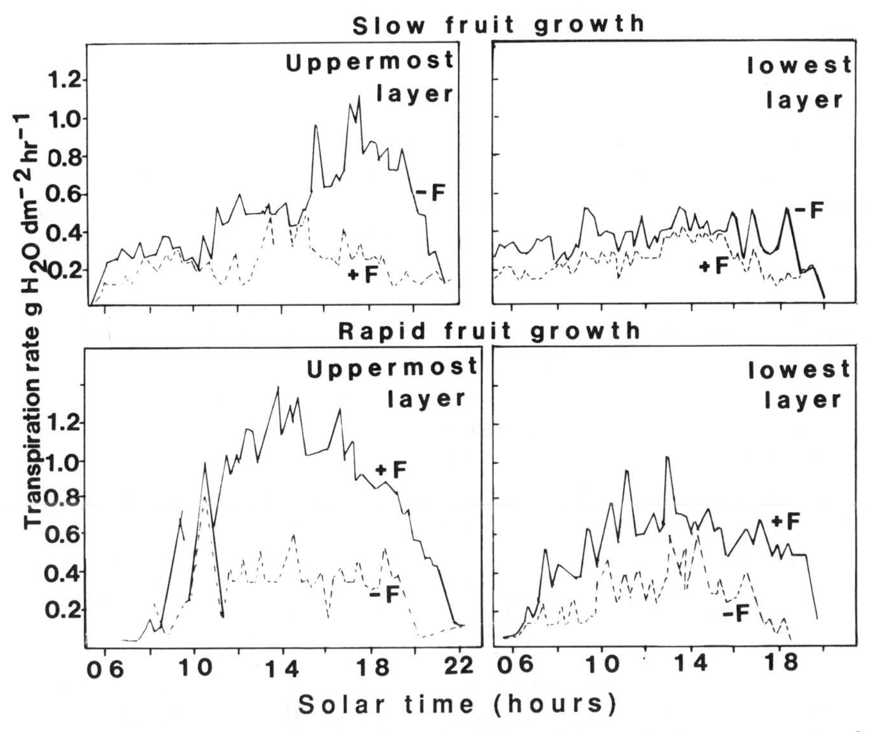

Figure 3.6 Diurnal transpirational patterns of peach leaves in the uppermost and lowest strata of the canopy during slow and rapid fruit growth. (Reproduced by permission from Chalmers et al., 1983.)

Decreased leaf area can be induced by growth regulator treatments by reducing the rate of leaf production or inducing early cessation of growth or by enhancing leaf abscission (Jones and Higgs, 1979; Lakso, 1983; Steffens and Wang, 1984). Although natural or induced leaf fall is a powerful control of transpiration, the actual leaf fall in response to water deficit may occur only after rewatering the plants exposed to severe water stress. In severe water stress, severe pruning, so-called 'dehorning,' needs to be done to decrease the water need of the trees to save them from complete dehydration. Dehorning not only reduces the leaf area but allows the root system to supply the remaining leaves with water to a higher degree. Some cherry trees died back when the trees received only 15% of pan evaporation once a week in an arid climate. Dieback effectively reduced leaf area relative to the root system (Proebsting et al., 1981). The trees recovered, but the exposed top scaffolds were severely sunburned. In peach and pear, heading back the major scaffolds to about 1.5 m also increased the root–shoot ratio and produced vigorous shoots, leading to the recovery of the tree (Proebsting and Middleton, 1980).

DIRECT CONTROL OF TRANSPIRATION

Role of Stomata

The stomata provide a dominant short-term physiological control over transpiration. The significance of this control is that stomata are able to open or close as a reaction to an increase or reduction in water content and turgor of the guard cells. The effects and interactions of external factors on guard cells are complex. Since the direct cause of stomatal closure is loss of turgor, it is not surprising that closure results from any factor that subjects the leaves to water stress through either fast transpiration or reduced soil water content.

The resistance of stomata (r_s) to the diffusion of water vapor can be measured by a transient porometer or a steady-state porometer. The transient porometer measure the time it takes the humidity in the porometer cup to increase across a preset interval and is expressed as seconds per millimeter or seconds per centimeter. The steady-state porometer, as its name indicates, measures the steady-state conditions. In addition to passing through the stomatal pathway, water vapor must pass through the laminar boundary layer of air at the leaf surface. The boundary layer has its own resistance. Thus stomatal resistance r_s derived from the mass flux of water vapor (F) from the leaf ($mg\,m^{-2}s^{-1}$) and the vapor density (concentration) difference ($\Delta\chi\ mg\,m^3$) between the substomatal cavity and the air can be represented by the equation

$$r_s = \frac{\Delta\chi}{F} - r_a$$

where r_a is the boundary layer resistance to water vapor transfer ($s\,m^{-1}$) (Warrit et al., 1980).

The resistance to vapor flux through the stomata is often expressed as its reciprocal value ($mm\,s^{-1}$) and is called *stomatal conductance* (g_s). Typical values of g_s for leaves of well-watered fruit trees vary, as shown in Table 3.4.

Naturally, when transpiration of the entire tree is concerned, one needs to consider the conductance of the entire canopy.

To determine canopy conductance, one needs to assess and sum the leaf areas and stomatal conductances for all different strata (exposed, shaded, etc.) in the canopy. Canopy conductance can be estimated from individual values of g_s (measured by porometry) and LAI values for each strata of the canopy using the relationship

$$\text{Canopy } g_s = \Sigma(g_{si}L_i)$$

TABLE 3.4 Values of Stomatal Conductance for Leaves of Well-Watered Fruit Trees

Species	g_s (mm s^{-1})	Reference
Apple	3–8 typically	Landsberg et al., 1975; Lakso, 1979; Warrit et al., 1980; Jones et al., 1983
	14–18 on occasion	Fanjul and Jones, 1982; Jackson et al., 1983
Apricot	2.7–11	Schulze et al., 1972; De Jong, 1983
Peach	1.5–7	Natali, and Xiloyannis, 1982; Tan and Buttery, 1982; De Jong, 1983
Cherry	7	De Jong, 1983
Plum	7	De Jong, 1983

where L_i is the LAI for a particular strata of the canopy and g_{si} is the corresponding mean leaf conductance for the strata (Jones, 1983).

Effect of Light on Stomatal Conductance

Reduction of CO_2 in the stomatal cavity below normal atmospheric levels causes stomatal opening. This is of major importance in causing opening in daytime and closure at night. Over a fairly wide range, the degree of opening is related to light intensity so that opening may be gradual during the first few hours of the day and closure gradual during afternoon and evening. In the orchard, apples reach maximum conductance at a photon flux density (PFD) of about 400 $\mu E\ m^{-2} s^{-1}$ (175 $W\ m^{-2}$) (Landsberg et al., 1975), whereas in leaf chambers maximum stomatal opening was achieved when the PFD was greater than 200 $\mu E\ m^{-2} s^{-1}$. There are modifying effects on stomatal conductance. The presence of fruit, discussed later, modifies stomatal opening and transpiration rate. Peach trees during periods of slow fruit growth transpire less as compared to defruited trees. During periods of rapid fruit growth transpiration rate is the opposite; trees with fruit transpire more (Chalmers et al., 1983). In apples, Hansen (1971) found similar results. Trees with fruits transpired 0.76 liters of water during the summer and 0.2 liters during fall. In comparison, transpiration rates of defruited trees were 0.38 and 0.05 liters, respectively. On a relative scale stomatal aperture of trees with fruit was 0.45 and those without fruit 0.17. For apples, Warrit et al. (1980) showed a linear relationship between internal CO_2 concentration in the stomatal cavity and ambient CO_2 concentration. They attributed much of the light response of stomata to internal CO_2 concentrations. When CO_2 is rapidly used and its concentration in the stomatal cavity is low, stomata are open. Thus it is likely that the fruit effect is manifested through its sink effect on photosynthesis and in turn on stomatal opening.

The season also seems to have an effect. In apple, stomatal conductance reaches maximum values at lower light intensities early in the season (June) than later (August). In October, leaves are insensitive to PFD levels. This indicates that the control of water loss becomes less efficient as the season advances.

Effect of Ambient Humidity on Transpiration

It is usually assumed that the vapor pressure of the air inside of the leaf is saturated. Obviously the vapor pressure of the ambient air is, in most cases, less than saturated. The vapor pressure difference between inside and outside of the leaf is expressed as vapor pressure deficit (VPD). In any given situation it is difficult to change VPD. Misting the trees can increase the water content of the air, but the effect is often very small (Kohl and Wright, 1974; Unrath, 1972). There are much larger differences in VPD if humid and arid regions are compared for the same fruit species.

Stomatal responses to the leaf-to-air vapor pressure gradient are now well established for a number of plants. Available data confirms that the VPD is an important factor in the response of fruit tree stomata to their environment. A linear relationship exists between g_s and VPD, and g_s generally decreases with increasing VPD. Landsberg and Butler (1980) examined the stomatal response in apples to humidity and temperature. They found that, especially at higher level of net radiation values, transpiration rate reached a maximum well within the limits of a linear response to VPD.

Thorpe et al. (1980) also calculated stimulated transpiration rates for apples with stomatal responses to humidity and light. They used values for VPD of 0, 1.1, and 1.7 kPa, corresponding to dew points of 20, 10, and 0°C. At high humidity (zero VPD) transpiration was lower at low light levels but transpiration was sustained at high light intensities. This indicates that the transpirational water requirement of trees can be unexpectedly high in high-energy ($\sim$600 W m^{-2}), high-humidity environments. This is an important consideration since more and more temperate zone fruit trees are planted in tropical or semitropical environments. In dry climates at larger values of VPD stomata close at relatively low levels of light intensity.

Stomatal conductance values measured with a porometer show a diurnal trend. Conductance begins to increase in the morning and reaches a maximum value before noon. Then it greatly decreases for 2–3 hr with a second peak occurring mid to late afternoon. In the past such diurnal behaviour has been explained in terms of environmental conditions, namely, by changes in the VPD.

Thorpe et al. (1980) developed an empirical model of the stomatal

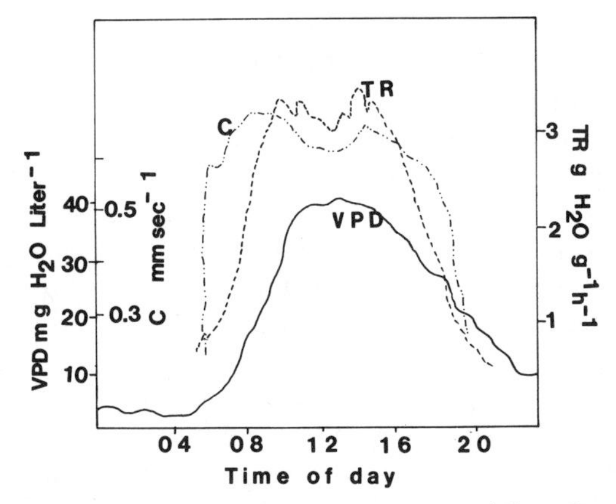

Figure 3.7 Diurnal changes in stomatal conductance (*C*) and transpiration (TR) of apricots and water pressure deficit (VPD) on a typical day (June 25) in the Negev desert. (Redrawn by permission from Lange and Meyer, 1979.)

response of apple leaves in terms of only two factors: PFD and VPD. The model was tested against mean values of g_s, total tree water used was obtained in the field; the association between measured and calculated values were good. The daily course of transpiration, stomatal opening (stomatal conductance), and VPD are illustrated in Figure 3.7. It is clear that stomatal response largely follows changes in VPD, but the two curves for transpiration and changes in VPD are not exactly alike. (The data for Figure 3.7 were obtained in extremely arid conditions and at high light intensities.) As illustrated in Figure 3.7, stomata of fruit trees are sensitive to VPD and tend to close in dry air (Hall et al., 1976; Schulze et al., 1972; West and Gaff 1976). The closing response to a decrease in humidity can be very rapid. Fanjul and Jones (1982) reported a closing response of stomata in apple within 20 s following a change in ambient humidity. However, there are other effects, reviewed by Jones et al. (1985), such as diurnal trends and water potential of the leaf, that also affect stomatal closing. Stomata also respond to temperature and irradiance independent of humidity. Thus the vapor pressure deficit is an important factor but is by no means the only one that influences stomatal movement.

EFFECT OF FRUIT ON TRANSPIRATION

The presence of fruit effectively limits the growth of the tree in apples (Avery, 1969; Hensen, 1971; Maggs, 1963) and other fruits (Lenz, 1986). This indicates that fruit may exert an influence on the water content of the tree, which is manifested in shoot growth limitation. Lenz (1986) presented evidence that apple trees that produced two crops during a 3-year period produced the least shoot growth (2.48 m^2 leaf area) but the highest rate of transpiration 180 l m^{-2} $year^{-1}$ compared with the deblossomed trees which leaf area was the largest (4.90 m^2) but had the smallest transpiration rate 81 liter m^{-2} y^{-1}. Trees producing only one crop in 3 years were intermediate. It was clear that starch accumulated to higher concentrations in the leaves of nonfruiting trees compared with fruiting trees, and the stomatal closing mechanism was regulated in such a way as to use less water. It is not clear if stomatal conductance is affected by fruiting through photosynthetic lowering of CO_2 in the substomatal spaces or if the fruit affects stomatal control through a hormonal mechanism (Lenz, 1986). The effects of various phases of peach fruit growth on transpiration (Chalmers et al., 1983) and apple fruit growth on stomatal opening (Hansen, 1971) were discussed in a previous section.

ENERGY REQUIREMENT OF TRANSPIRATION

As has been discussed, the net radiation absorbed by leaves largely determines the tissue temperature and transpiration rate and subsequently the water use by the tree. The energy balance of any individual plant can be written

$$\text{Net radiation} = \text{sensible heat} + \text{latent heat of vaporization} \times \text{flux of water vapor}$$

The amount of energy used for evaporation is called *latent heat* and is expressed either as a proportion of available energy or as the ratio between sensible heat and latent heat, called the *Bowen ratio*. Consequently, if net radiation is positive and transpiration rate is low, the temperature of the plant or a particular organ is high. Landsberg et al. (1973) calculated linear regressions of differences between bud and flower temperatures and that of air on net radiation. The regression lines were highly positive. The largest difference in bud and/or flower temperature of apple and air temperature was 4 degrees at 600 W m^{-2} net radiation level, whereas the difference at 100 W m^{-2} was negligible. Leaf temperatures could not be predicted by simple regressions, but the leaf Bowen ratio was useful in explaining variations in leaf–air temperatures. The difference between air and fruit temperature is generally higher, and differences as high as 7 degrees have been reported for sunlit fruit and 1.8 degrees for shaded fruit.

Landsberg and Butler (1980) quoted orchard measurements and calculations for energy used for transpiration. Golden Delicious apple trees grown in containers and having a leaf area of 1 m^2 on two different days with 800 $W m^{-2}$ solar radiation displayed maximum transpirational rates of 17 and 50 $mg m^{-2} s^{-1}$ at g_s values of 7 and 10 $mm s^{-1}$, which is equivalent to 42 and 124 $W m^{-2}$ energy. This gives latent heat–leaf net radiation ratios of 0.15 and 0.44, respectively (see calculation of net radiation in Chapter 1). Thus, on a day when transpiration was limited, 85% of the net radiation energy was sensible heat; on a day when transpiration was higher, 56% of the net radiation energy was sensible heat. The energy situation can be quite different if the wind speed increases and consequently the boundary layer conductance increases.

Landsberg and Butler (1980) calculated apple leaf temperatures at the net leaf radiation level of 250 $W m^{-2}$ and 20°C air temperature for various values of VPD at three boundary layer conductance levels. They illustrated that the leaf temperature will fall with increasing values of VPD with a concomitant increase in latent heat loss. However, the major effect on leaf temperature is the boundary layer conductance. A decrease in the boundary layer conductance increases leaf temperature considerably.

Expressing the energy partitioning data as a Bowen ratio gives very similar results. In general, the higher the net leaf radiation level, the lower the Bowen ratio, indicating that a smaller portion of total energy is used for transpiration as latent heat (Landsberg and Butler, 1980).

If only a small portion of total available energy is used as latent heat, the consequences to the tree are very important. such situations usually exist when the water potential of the tree is low and water supply for transpiration is limited. Upper branches, especially if exposed to radiation by pruning, often suffer 'sunburn.' Overheating destroys the bark, resulting in dieback of the affected branch. The fruit of several apple varieties, such as 'Granny Smith', is very sensitive to sunburn while still on the tree. Misting the trees occasionally so that the evaporating water cools the fruit has beneficial effects on fruit quality.

The fruit of almost every apple and plum variety is injured by the sun if left in open containers in the orchard for an extended period of time. Data are not available to indicate the temperature of fruit in such conditions.

Unrath (1972) used low-volume overtree sprinklers to control heat-induced stress in apple orchards. He compared results with undertree irrigated and nonirrigated plots. During 156 hr of operation overtree sprinkling consistently lowered all plant tissue temperatures below air temperature and lowered fruit temperature on the average by 6.7°C (Table 3.5). The lower fruit temperature resulted in higher fruit quality, firmer fruit with more soluble solids, and a larger area of the fruit surface covered with red color (Unrath, 1972).

The effect of evaporation on fruit temperature and in turn fruit temperature on fruit quality clearly indicate that in southern climates the transpiration, in addition to its obvious effects, has a very important

TABLE 3.5 Average Temperatures for Parts of Apple Tree during 156 h of Overtree Sprinkling[a]

	Mean Temperature (°C)			
Irrigation Type	Air	Bark	Leaf	Fruit
Overtree	29.6	25.6	26.1	27.6
Undertree	30.4	29.8	30.9	32.6
Nonirrigated	31.6	31.9	31.2	34.8

[a] Reproduced by permission from Unrath, 1972.

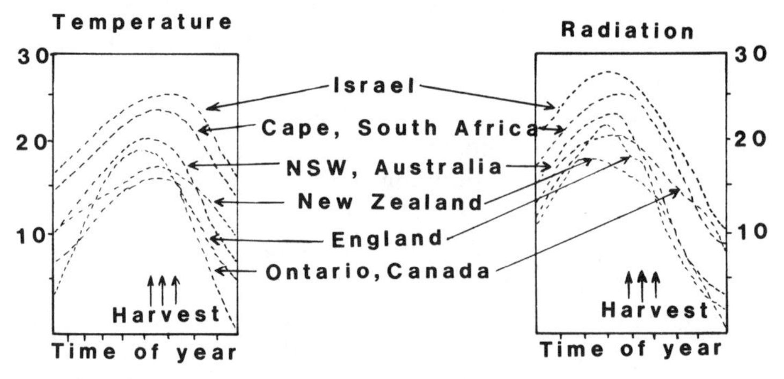

Figure 3.8 Radiation and temperature regimes during the growing season at various apple growing areas in both hemispheres. (Redrawn after Landsberg, and Jones 1981.)

role in cooling the fruit, which is essential for quality fruit production.

Much of the transpiration data obtained in connection with the water use of orchards is based on comparisons with ambient temperatures rather than radiation. Landsberg and Jones (1981) compared the total radiant energy (MJ m^{-2}) and the average monthly temperatures during the growing period for six important apple-growing regions in both hemispheres. Both the radiant energy levels and the monthly average temperatures showed bell-shaped curves during the vegetative period. During the first half of the growing season the increase in temperature lagged behind the increase in radiation, while during the second half of the season the two curves were almost identical (Figure 3.8). This indicates that although the correlation between air temperature and transpiration is not a causal relationship, it can usefully estimate the expected magnitude of water loss. The temperature-based estimation perhaps underestimates the transpirational loss of water during the first half of the season, whereas it closely estimates the loss during the second half of the season. A response for evapotranspiration based on temperature and LAI for apple, pear, and peach is illustrated in Figures 3.9 and 3.10.

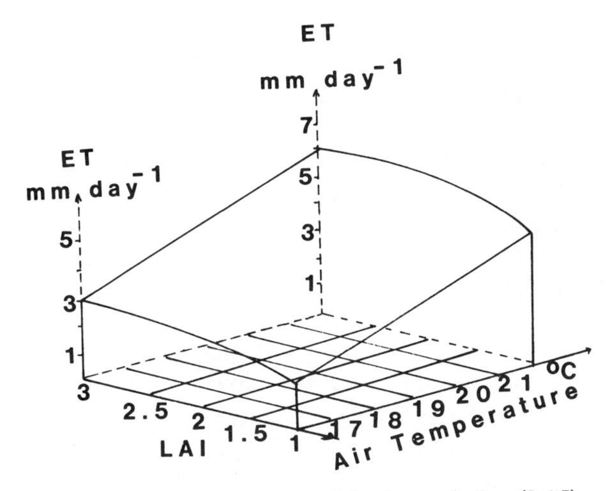

Figure 3.9 Effect of air temperature and leaf area index (LAI) on evapotranspiration of apple leaves. (Reproduced by permission from Gyuro, 1974.)

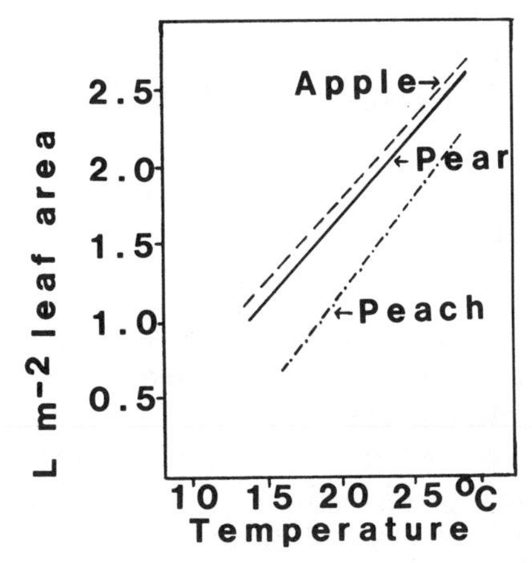

Figure 3.10 Relationships between temperature and transpiration of peach apple and pear (liters per square meter of leaf area). (Reproduced by permission from Gyuro, 1974.)

EXCHANGE PROCESSES THROUGH THE BOUNDARY LAYERS

Both individual leaves and orchards are surrounded with a layer of air through which the exchange processes take place. Landsberg and Jones (1981) modeled heat and water vapor exchange at leaf–single tree level in wind tunnels and orchards. The major components of the model are the characteristic leaf dimension, the foliage density given by the ratio of leaf area to the frontal (silhouette) area of the tree, and wind speed. The model requires that wind speed be determined at the foliage level. This has been accomplished by measuring wind speed at the canopy top and any given height and by determining a composite parameter combining the drag and momentum exchange characteristics of the canopy (Landsberg and Jones, 1981). As foliage density increases, the composite parameter for exchange also increases. The composite parameter is a relative value. In a high-density orchard for the same wind speed, this value was 1.2 when the wind direction was perpendicular to the rows and decreased to a value of 0.38 when the wind was blowing almost parallel with the rows (Landsberg and Jones, 1981). Measurements in a semidwarf orchard with planting distances of approximately 6 m in both directions and with 4 m tree height resulted in a value for the composite parameter of about 0.5 with little variation. Wind speeds at the top of the canopy are always lower than those measured in standard meteorological enclosures because of the absorption of momentum by the canopy. It is possible to estimate canopy-top wind speed from standard meteorological wind measurements. Factors for the conversion of wind speed measured by meteorological instrument to wind speed at the canopy top for hedgerow and dwarf orchard are 0.61 and 0.96, respectively (Landsberg and Jones, 1981).

Hormonal Control of Stomatal Opening

Abscisic acid (ABA) is implicated in the stomatal control of many plants. When leaves are subjected to water stress, ABA accumulates. When leaves dry at slow rates, ABA builds up before stomata close, suggesting that ABA may control stomatal opening. When water stress develops rapidly, stomata close before ABA accumulates (Salisbury and Ross, 1985). In nonirrigated peaches ABA levels were higher throughout the season than in irrigated ones, and the level in nonirrigated trees doubled during the early part of August. The nonirrigated trees obviously were under severe water stress at this time, as indicated by the ψ_l and g_s. The diurnal variation in ABA content and g_s only correlated well in nonirrigated trees. In irrigated trees, regardless of irrigation (100% or 50% of evapotrauspiration) rate, ABA concentration and stomatal opening was not correlated (Xiloyannis et al., 1980). These uncoordinated changes do not suggest a causal relationship between ABA content and stomatal closing. However, long-range buildup of ABA occurs during extended water stress. One must note that ABA exists in pools and in various forms. Thus, expecting correlations between total ABA and stomatal functions in trees may be unreasonable.

TABLE 3.6 Effect of Irrigation on Growth of Apple[a]

	Rate of Irrigation		
Characteristic	Low	Medium	High
Length of shoots, cm	17.8	28.9	52.7
Length of internodes, mm	11.7	13.3	18.6
Number of nodes per shoot	15	22	29
Ratio flower bud to vegetative bud	0.38	0.15	0.06
Production per tree, kg	44.5	61.4	55.5

[a] Reproduced by permission from Giulivo and Xiloyannis, 1988.

However, ABA may play a role in stomatal closing if plants are exposed to long-term severe water stress. Peaches planted in containers and nonirrigated for 8 days closed their stomata, and stomatal functions were not recovered for 32 hr after optimal water conditions were restored. This cannot be explained based on water movement alone (Xiloyannis et al., 1986), and the possibility exists that hormonal control is superimposed on other control measures of stomatal functions in severe water stress situations.

CONTROL OF TREE GROWTH BY LIMITING WATER SUPPLY

Maintaining high turgor potential is essential for all types of cells including those of shoots and fruits. Since shoot growth and fruit growth may occur at different times in many fruit trees, it is possible to impose a mild water stress at the time of shoot growth, thus limiting shoot expansion, in contrast to maintaining full turgor and obtaining maximum fruit growth at the time of fruit expansion.

When spring vegetative growth is too strong, a limited irrigation is recommended in apple (Gergely, 1984). Gergely (1984) recommended a delay in irrigation until the first period of shoot growth is completed. However, he warned that because flower bud initiation also occurs during this time in apple, water stress cannot be too severe. Flower bud initiation and differentiation is a photosynthate-requiring process and a high rate of water stress imposed for growth control may decrease the crop in the following year.

Giulivo and Xiloyannis (1988) summarized data on growth of apples in relation to irrigation rates. Apples grow more if the rate of available water is high (Table 3.6). Length of shoots, internode length, and number of internodes are all higher, but the proportion of buds developing to flower buds is lower. The overall production in kilogram of fruit per tree with high irrigation rate is higher, which is why growers often irrigate at a high rate. At the same time growth of the canopy shades the tree and carbohydrates become limiting for fruit growth. In addition, the formation of flower buds is low, and the tree loses productivity during the following year. Consequently, the moderate (medium) irrigation produces the highest yields (Table 3.6).

De Lotter et al. (1985) subdivided the vegetative season of the apple into four periods, each of which requires a different water supply. The first period is from full bloom to 40–50 days after bull bloom (AFB); the second period is from 40–50 days AFB to the end of the shoot growth period; the third period is from the end of shoot growth to harvest; and the fourth is after harvest. High availability of water (85% of utilizable soil reserve) in period 2 favors shoot growth while in period 3 it favors fruit growth. Deficit irrigation, when available soil water is maintained at 25%, limits fruit growth considerably during period 3, but it may favorably influence the avoidance of the development of physiological disorders in storage. Giulivo and Xiloyannis (1986) consider that maintaining 85% of available water reserves in the soil for apple and 70% for peach is optimal for cell expansion and maintaining 35% utilizable reserve in the soil for deficit irrigation, which limits cell division but still maintains stomatal opening for photosynthesis.

Tree species respond differently to deficit irrigation. Cherries developed full yield and fruit size and increase in trunk diameter with only 50% of pan evaporation supplied through high-frequency irrigation. In contrast, prunes were severely affected in all these components at the same water supply (Proebsting et al., 1981).

Chalmers and Wilson (1978) and Chalmers et al. (1975) found that in peach trees photosynthesis and translocation may occur throughout the day even though the water potential is low enough to cause the woody portion of the tree to shrink. The diffusive conductance in peach was not affected in peach leaves unless the midday water potential was lowered below −2.5 MPa (Xiloyannis et al., 1980). Thus it is possible to affect turgor pressure at least in peach without affecting photosynthesis.

Chalmers et al. (1984), using peach and pear trees, demonstrated that vegetative growth was reduced by 80 and 70% when the water allotment to the trees was reduced to 12.5 and 25%, respectively, of the rate of evaporation from the Class-A pan during early stages of fruit growth.

They based their work on the fact that only 25 and 30% of fruit growth takes place concurrently with shoot growth on 'Golden Queen' peach and pear. 'Golden Queen' peach is a long-growing-period peach with a long-stage-2 period, and essentially all fruit growth occurs during stage 3 of the fruit growth, which in this cultivar occurs well after shoot growth is completed. Shoot and fruit growth may sufficiently overlap in early or midseason ripening peaches (Figure 3.11). If shoot and fruit growth

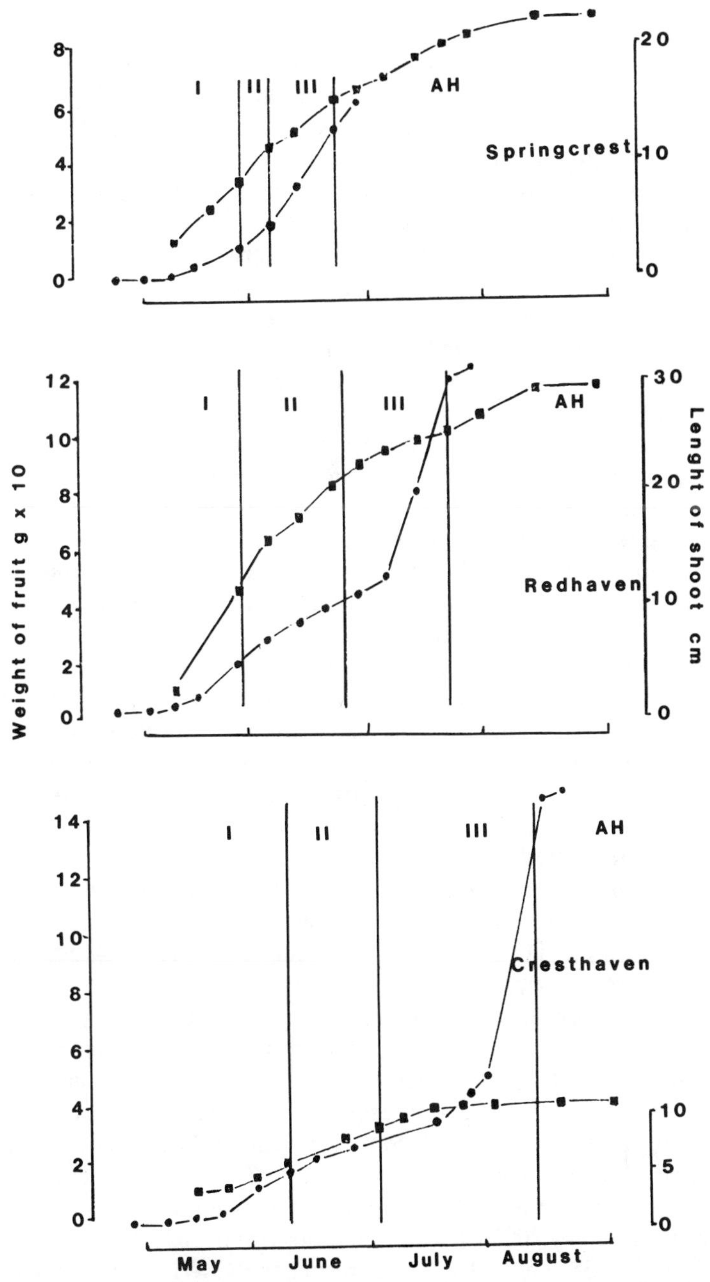

Figure 3.11 Fruit and shoot growth of peach cultivars of various ripening times: dots, fruit growth; squares, shoot growth. (Reproduced by permission from Giulivo and Xiloyannis, 1988.)

overlap, deficit irrigation is not possible. The special character of each cultivar must be taken into consideration to design the irrigation type for that particular cultivar. Nectarines or plums require a higher turgor pressure to expand. Pear fruit grows close to harvest and in this respect it is similar to a long-growing-period peach.

Almond is a considerable contrast to peach with respect to drought tolerance. Although pear and almond trees are very similar, the almond fruit does not have stage 3 in its growth pattern. Consequently, the tree does not require much water during this phase. Since both trees are relatively tolerant to water deficit during the shoot growth period (fruit growth stage 2) and the almond does not require additional water for fruit growth, the almond is more 'drought' tolerant and it can be grown on dry hills where economical peach production requires irrigation.

Spacing and water supply interact synergistically to reduce vegetative growth (Chalmers et al., 1984). Tree vigor is strongly affected by within-the-row spacing, at least in the case of peach (van den Ende and Chalmers, 1983). The likely reason for this is that the available soil water is reduced, causing a lowered water potential in the tree. If this is augmented with controlled irrigation, the result is synergistic.

As was discussed, trees with trickle irrigation when only part of the root system is wetted use somewhat less water than trees with overall irrigation. It appears that trickle irrigation also creates a mild water stress in the tree, which is beneficial. Peretz et al. (1977) reported that during 4-year trials, the leaf water potential of trickle-irrigated apple trees was lower (−1.68 MPa) than sprinkler-irrigated trees (−1.31 MPa) even though the sprinkled trees received only 90% of evaporation from the Class-A pan calculated for the entire area whereas the trickle-irrigated plots received 100% of evaporation. Corresponding to the leaf water potential, the trickle-irrigated trees were smaller [108 vs. 145 cm^2 of trunk cross-sectional area (TCA)] with a better yield efficiency (0.41 vs. 0.25 kg/cm^2 TCA), respectively.

In another experiment (Chalmers et al., 1984) furrow irrigation was compared with 100 and 50% of evaporation supplied by trickle irrigation. Trickle irrigation at 100% reduced TCA over furrow irrigation, and trickle irrigation at 50% irrigation produced even smaller trees. The yield efficiency of 100% trickle irrigation on a kg/cm^2-TCA basis was much greater than that of furrow irrigation. The 50% trickle was less than the 100% trickle although more than the furrow irrigation. Leaf water potentials for the furrow, 100% trickle irrigation, and 50% trickle irrigation were −1.31, −1.58, and −1.80 MPa, respectively. Water potentials were measured before solar noon at 1- or 2-week intervals throughout the season. It appears that a water potential of around −1.60 is best for controlling tree size and maximum productivity. At higher ψ_l the tree grows too fast; at lower ψ_l the turgor pressure is not high enough for maximum yield.

STRESS CAUSED BY EXCESS WATER

Fruit tree species possess a wide range of tolerances for water-saturated soils (waterlogging). Apricot and peach roots are considered highly sensitive, whereas pear, quince, and apple roots are relatively tolerant. 'Myrabolan' is of intermediate tolerance (Rowe and Catlin, 1971). Myrabolan required more than 15% O_2 for root growth in solution cultures, whereas apple continued root growth at 5% O_2 concentration.

Availability of O_2 to the root is not indicated simply by soil O_2 concentration but is actually limited by the rate of O_2 diffusion through the soil. Oxygen diffusion rate (ODR) is a function both of the O_2 concentration gradient between the soil and root and the path resistance to O_2 movement, largely determined by the diffusivity of O_2 in water for wet soils. Nonwaterlogged orchard soils in Maine registered on ODR rate of 0.40–0.48 $\mu g\, cm^{-2} min^{-1}$ while the ODR rate in waterlogged soils was 0.09–0.16 $\mu g\, cm^{-2} min^{-1}$ (Olien, 1987). In general, root function begins to be limited when ODR is $<0.2\, \mu g\, cm^{-2} min^{-1}$ (Olien, 1987). A physiological response, that is, a decrease in stomatal resistance, was detected in young pear and quince trees when the ODR was reduced to between 0.15 and 0.3 $\mu g\, cm^{-2} min^{-1}$ (Andersen et al., 1984).

Soil waterlogging results in a number of component stresses, which include low soil oxygen, causing anaerobic conditions for the roots; phytotoxic accumulation of reduced ions; decrease in aerobic soil microorganisms; accumulation of phytotoxic by-products of anaerobes; and attack by waterborne pathogens such as *Phytophtora* and *Phytium* (Olien, 1987) and a decreased stomatal resistance (Andersen et al., 1984).

The severity and rate of response by the tree to waterlogging not only depends on the duration of waterlogging stress but also on the genotype, transpirational demand, and time of year when trees are exposed to waterlogging stress. The time-of-year effect may be connected to soil temperature (Olien, 1987). When apricot and peach trees were flooded at 27°C, the plants wilted within 6 days; at 22°C within 10 days; and at 17°C within 16 days. Plum was more tolerant to flooding and wilted after 20, 24, and 28 days, respectively (Rowe and Catlin, 1971). Similar results were obtained by Olien (1987), summer flooding being the most severe on root loss compared with spring or fall flooding.

A greater loss of root than shoot is a common response of trees to waterlogging. In apple, summer waterlogging resulted in the greatest shoot: root ratio of 1.55 compared with spring (1.15) and fall ratios

(1.36), which were not statistically different from the nonflooded control (1.20) (Olien, 1987).

In apple, which is less sensitive to waterlogging, loss of trees from short periods of flooding is rare. Nevertheless, 60 days of flooding causes a number of effects. Trunk and shoot growth is reduced by 60 days of seasonal waterlogging, which can be ranked by severity as spring = summer > fall. Reduction in yield can be ranked as spring > summer = fall. However, severity of response within a given period can be increased if stress also occurred in the previous year (Olien, 1987).

Pears survived 1 month of flooding whereas peaches ('Halford' seedlings) died as a result of flooding. When flooding was imposed continuously for 20 months, all *Pyrus betulifolia* trees survived; *P. calleriana, P. communis*, and quince had a survival rate of 60–70%; apple rootstock MM106, 50%; and *P. pyrifolia* and *P. ussuriensis*, 30%. Whether the rootstocks were tested alone or grafted with appropriate scions did not change the survival rate (Andersen et al., 1984; Rowe and Catlin, 1971).

Since *Prunus* species are very sensitive to flooding, they have been the subject of several investigations. Rowe and Catlin (1971) determined both cyanogenic glycoside content and the portion of it that was hydrolized during waterlogging. They correlated it with the differential sensitivity observed between peach and plum. Both the cyanogenic glucoside content and the hydrolized fraction were higher in peach than in plum, indicating that they affect degree of sensitivity. The rate of cyanogenesis increased with both temperature and time. Peach and apricot roots were alike in HCN evolution whereas plum roots were lower, with release of HCN being barely detectable at 22°C. Cyanogenesis was significant in peach and apricot at as low as 17°C. When the root was exposed to O_2 levels higher than 5%, HCN was not produced (Rowe and Catlin, 1971). Even though HCN is released in considerable quantities from peach and apricot, it is not expected to cause a blockage in the respiratory mechanism. If conditions are completely anaerobic, the terminal oxidase system is inhibited regardless of the presence or absence of HCN. If oxygen is present and the root respiration is inhibited by HCN, the root of fruit trees (apple) has the capacity to channel electrons through the alternative (cyanide-resistant) pathway, and the respiration can proceed without interruption, although it produces less energy (ATP) (Wang and Faust, 1983). This is in agreement with the conclusion of Rowe and Catlin (1971), who determined from circumstantial evidence that the importance of HCN, if any, is secondary to other effects of O_2 deficiency. Nevertheless, they concluded that cyanogenesis is a highly sensitive indicator of cellular damage and relative sensitivity to waterlogging, at least in *Prunus* species.

In anaerobic roots of waterlogged plants (tomato) 1-aminocyclopropane-1-carboxylic acid (ACC) is produced 12–15 times more than in roots exposed to O_2. The ACC is readily transported to the top of the plants through the xylem, is converted to ethylene in aerobic conditions, and causes epinastic responses (Yang, 1980). Such a mechanism has not been shown in fruit trees, but it is likely to occur. Species differences in response to waterlogging could easily be caused by a differential rate of ACC production.

REFERENCES

Andersen, P. C., P. B. Lombard, and M. N. Westwood. 1984. J. Amer. Soc. Hort. Sci. 109:132–138.

Atkinson, D. 1978. Acta Hort. 65:79–89.

Atkinson, D., and J. K. Lewis. 1979. J. Photo Sci. 27:253–258.

Avery, D. J. 1969. New Phytol. 68:323–336.

Black, J. D. F. 1976. Hort. Abstr. 47:1–7.

Black, J. D. F., and P. D. Mitchell. 1974. *In* Proc. 2nd Intern. Drip Irrig. Congr., San Diego. Riverside, CA: University of California Press, pp. 437–438.

Black, J. D. F., and D. W. West. 1974. In Proc. 2nd Intern. Drip Irrig. Congr., San Diego. Riverside, CA: University of California Press, pp. 432–433.

Chalmers, D. J., R. L. Canterfold, P. H. Jerie, T. R. Jones, and T. D. Ugalde. 1975. Austral. J. Plant Phys. 2:635–645.

Chalmers, D. J., P. D. Mitchell, and P. H. Jerie. 1984. Acta Hort. 146:143–148.

Chalmers, D. J., K. A. Olsson, and T. R. Jones. 1983. *In* Water deficit and plant growth, Vol. 7, T. T. Kozlowski (ed.). New York: Academic Press, pp. 197–232.

Chalmers, D. J., and I. B. Wilson. 1978. Ann. Bot. 42:294–295.

Daniell, J. W. 1980. Fruit South 4(5):4–7.

Davies, F. S., and A. N. Lakso. 1978. J. Amer. Soc. Hort. Sci. 103:310–313.

Davies, F. S., and A. N. Lakso. 1979. Physiol. Plant. 46:109–114.

De Jong, T. M. 1983. J. Amer. Soc. Hort. Sci. 108:303–307.

Elfving, D. C. 1982. Hort. Rev. 4:1–48.

El Sharkawi, H. M., and M. El Monayeri. 1976. Plant Soil 44:113–128.

Fanjul, L., and H. G. Jones. 1982. Planta 154:135–138.

Fanjul, L., and P. H. Rosher. 1984. Phys. Plant. 62:321–328.

Faust, M. 1968. Proc. Amer. Soc. Hort. Sci. 93:746–752.

Gergely, I. 1984. *In* Alma. F. Pethö (ed.). Budapest. Mezögazdasàgi Kiadò, pp. 419–436.

Giulivo, C., and A. Bergamini. 1982. Proc. XXI Int. Hort. Cong. (Abstract) 1264.

Giulivo, C. and C. Xiloyannis. 1988. L'Jrrigazione in ortofrutti coltura. Banca Popolare. Verona. Italy.

Giulivo, C., and C. Xiloyannis. 1986. XVII Convegno Pescicolo, Cesena, Italy, March 5, 1986.

Goode, J. E., and K. H. Higgs. 1973. J. Hort. Sci. 48:203–215.

Gyuro, F. 1974. A Gyümölcstermesztés alapjai. Budapest: Mezögazdasági Kiadó.

Hall, A. E., E. D. Schulze, and O. L. Lange. 1976. *In* Ecological studies, Vol. 19, O. L. Lange, L. Kappen, and E. D. Schulze (eds.), Springer-Verlag, Berlin, pp. 169–188.

Hansen, P. 1971. Physiol. Plant 25:181–183.

Jackson, J. E., J. W. Palmer, G. C. White, and E. E. Cannadine. 1983. Ann. Rep. East Malling Res. Sta. for 1982: 34–35.

Jones, H. G. 1983. Plants and microclimate. Cambridge: Cambridge University Press.

Jones, H. G., and I. G. Cumming. 1984. J. Hort. Sci. 59:329–336.

Jones, H. G., and K. H. Higgs. 1979. J. Exp. Bot. 30:965–970.

Jones, H. G., A. N. Lakso, and J. P. Syvertsen. 1985. Hort. Rev. 7:301–344.

Jones, H. G., M. T. Luton, K. H. Higgs, and P. J. C. Hamer. 1983. J. Hort. Sci. 58:301–316.

Kenworthy, A. L. 1972. Michigan State Univ. Agr. Exp. Sta. Res. Rpt. 165.

Kohl, R. A., and J. L. Wright. 1974. Agron. J. 66:85–88.

Kriedeman, P. E., and R. L. Canterfold. 1971. Austral. J. Biol. Sci. 24:197–205.

Lakso, A. N. 1979. J. Amer. Soc. Hort. Sci. 104:58–60.

Lakso, A. N. 1983. *In* Stress effects on photosynthesis. R. Marcelle, H. Clijsters, and M. Van Poucke (eds.). The Hague: Nijhoff/Junk, pp. 85–93.

Lakso, A. N. 1984. Acta Hort. 146:151–158.

Lakso, A. N., A. S. Geyer, and S. G. Carpenter. 1984. Amer. Soc. Hort. Sci. 109:544–547.

Landsberg, J. J., D. B. B. Powell, and D. R. Butler, 1973, J. App. Ecol. 10:881–896.

Landsberg, J. J., C. L. Beadle, P. V. Biscoe, D. R. Butler, B. Davidson, L. D. Incoll, G. B. James, P. G. Jarvis, P. J. Martin, R. E. Neilson, D. B. B. Powell, M. Slack, M. R. Thorpe, N. C. Turner, B. Warrit, and W. R. Watts. 1975. J. Appl. Ecol. 12:659–684.

Landsberg, J. J., T. W. Blanchard, and B. Warrit. 1976. J. Exp. Bot. 25:579–596.

Landsberg, J. J., and D. R. Butler. 1980. Plant, Cell Environ. 3:29–33.

Landsberg, J. J., and H. G. Jones. 1981. *In* Water deficits and plant growth, Vol. 6, T. T. Kozlowski (ed.). New York: Academic Press, pp. 419–469.

Lenz, F. 1986. *In* The regulation of photosynthesis in fruit trees. A. N. Lakso and F. Lens (eds.). Symp. Proc. Publ. NY State Agr. Exp. Sta. Geneva, N.Y., pp. 101–105.

Levin, I., R. Assaf, and B. A. Bravdo. 1979. Plant Soil 52:31–40.

Levin, I., R. Assaf, and B. A. Bravdo. 1980. *In* Mineral nutrition of fruit trees, D. Atkinson, J. E. Jackson, R. O. Sharples, and W. M. Waller (eds.). Butterworths, London, pp. 255–269.

Lotter, de V. J., D. J. Beukes, and H. W. Weber. 1985. J. Hort. Sci. 2:181–192.

Maggs, D. H. 1963. J. Hort. Sci. 38:119–128.

Natali, S., and C. Xiloyannis. 1982. *In* Nuovi Orientamenti per la Coltura del Melo nel Veronese. G. Bargioni and F. Loreti (eds.). Danca Populare, Verona, Italy, pp. 145–166.

Natali, S., and C. Xiloyannis. 1984. International Conference on Peaches, Verona, July 9–14, 1984.

Olien, W. C. 1987. J. Amer. Soc. Hort. Sci. 112:209–214.

Olien, W. C., and A. N. Lakso. 1984. Acta Hort. 146:151–158.

Peretz, J., E. L. Proebsting, Jr., and S. Roberts. 1977. HortScience 12:349–350.

Powell, D. B. B., and M. R. Thorpe. 1977. *In* Environmental effects on crop physiology. J. J. Landsberg and C. V. Cutting (eds.). London: Academic Press, pp. 259–279.

Priestley, C. A. 1973. Proc. Res. Inst. Pomol. Series E 3:121–128.

Proebsting, E. L., Jr., and J. E. Middleton. 1980. J. Amer. Soc. Hort. Sci. 106:380–385.

Proebsting, E. L., Jr., J. E. Middleton, and M. O. Mahan. 1981. J. Amer. Soc. Hort. Sci. 106:243–246.

Proebsting, E. L., Jr., J. E. Middleton, and S. Roberts. 1977. Hort. Sci. 12: 349–35.

Richards, D. 1976. Ann. Bot. 41:279–281.

Richards, D., and R. N. Rowe. 1976. Ann. Bot. 41:1211–1216.

Ross, D. S., R. A. Parsons, W. R. Detar, H. H. Fries, D. D. Davis, C. W. Reynolds, H. E. Carpenter, and E. D. Markwardt. 1980. Trickle irrigation in the eastern United States. Northeast Regional Agr. Eng. Serv. NRAES-4. Ithaca, NY: Cornell University Press.

Rowe, R. N., and P. B. Catlin. 1971. J. Amer. Soc. Hort. Sci. 96:305–307.

Salisbury, F. B., and C. B. Ross. 1985. Plant physiology. Belmont, CA: Wadsworth.

Schulze, E. D., O. L. Lange, M. Evenari, L. Kappen, and U. Buschbom. 1972. Oecologia. 17:159–170.

Shear, C. B. 1971. J. Amer. Soc. Hort. Sci. 96:415–417.

Slayter, R. O. 1967. Plant water relationship. London: Academic Press.

Smart, R. E., and H. D. Barrs. 1973. Agr. Meteorol. 12:337–346.

Steffens, G. L., and S. Y. Wang. 1984. Acta Hort. 146:135–142.

Swietlik, D., and S. S. Miller. 1983. J. Amer. Soc. Hort. Sci. 108:1076–1080.

Tan, C. S., and B. R. Buttery. 1982. Hort. Sci. 17:222–223.

Thorpe, M. R., B. Warrit, and J. J. Landsberg. 1980. Plant, Cell Environ. 3:23–27.

Unrath, C. R. 1972. J. Amer. Soc. Hort. Sci. 97:55–61.

van den Ende, B., and D. J. Chalmers. 1983. HortScience 18:946–947.

Wang, S. Y., and M. Faust. 1983. J. Amer. Soc. Hort. Sci. 108:1059–1064.

Warrit, B., J. J. Landsberg, and M. R. Thorpe. 1980. Plant Cell Environ. 3:12–22.

West, D. W., and D. F. Gaff. 1976. Physiol. Plant. 38:98–104.

West, D. W., W. K. Thompson, and J. D. F. Black. 1970. Austral. J. Biol. Sci. 23:231–234.

Wildung, D. K., C. J. Weiser, and H. M. Pellett. 1973. Hort. Sci. 8:53–55.

Willoughby, P., and B. Cockroft. 1974. *In* Proc. 2nd Intern. Drip Irrig. Congr. San Diego. Riverside, CA: University of California Press, pp. 439–442.

Worthington, J. W., M. J. McFarland, and J. S. Newman. 1982. Fruit South 6(4):22–23.

Worthington, J. W., M. J. McFarland, and P. Rodrigue. 1984. Hort. Sci. 19: 90–91.

Xiloyannis, C., S. Natali, and B. Pessarossa. 1986. Riv. Ortofrutticolt. Italiana 70:107–115.

Xiloyannis, C., K. Uriu, and G. C. Martin. 1980. J. Amer. Soc. Hort. Sci. 105:412–415.

Yang, S. F. 1980. Hort. Sci. 15:238–243.

Young, E., J. M. Hand, and S. C. Wiest. 1982. Hort. Sci. 17:791–793.

Young, E., and J. Houser. 1980. J. Amer. Soc. Hort. Sci. 105:242–245.

4

FRUITING

FLOWER BUD DEVELOPMENT

Flowering in fruit trees can be divided into two major developmental processes occurring in two successive growing seasons. These developmental processes are the initiation and development of flower buds occurring during the summer and fall of one season and the flowering process itself occurring in early spring the next season.

Fruit trees must be old enough to flower. For trees developing from seeds, it takes years to reach flowering age. The nonflowering period of young seedlings is called *juvenility*. Vegetative propagation of adult trees also results in a temporary loss of ability to flower, but trees are able to regain this status soon, and it cannot be equated with juvenility. Although ontogeny plays an important role in flower development, horticulturists deal with flowering mostly in adult plants and usually as a seasonal activity.

The partially developed buds of the tree must receive a signal and subsequently undergo complex histological and morphological changes to become flower buds. The nature of the signal for initiation of flower bud development is unknown. Following initiation, the differentiation of buds occurs. The flower bud differentiation process coincides with and is influenced by the other activities occurring in the tree. Therefore, it is an extremely complicated process that requires the tree to be in the right developmental stage. Environmental conditions should be conducive to this activity, and the other processes of the tree occurring at the same time must be in harmony and not interfere with flower bud formation. In this chapter, we consider the transition from juvenile to adult phase, the differentiation process of flower buds, and all the enhancing and competing influences of tree physiology. Aspects of flowering have been reviewed previously by Nyèki (1980), Buban and Faust (1982), and Hoad (1984).

Juvenility

Seedlings of most fruit trees do not flower before they reach 3–7 years of age (Visser et al., 1976; Zimmerman, 1976). The nonflowering period after seed germination is called the juvenile period or juvenile phase. Depending on the species, plants in general may change many of their morphological characteristics signaling the end of the juvenile phase (Wareing and Frydman, 1976). Leaf shape and thickness, phyllotaxy, thorniness, ability to form adventitious roots, and other characters may change at or just before the time the plant is able to flower. In fruit trees such morphological changes are rare. Juvenile pears are usually thorny, but change in thorniness does not always coincide with commencement of flowering. The lack of signs, other than flowering, to signify the end of

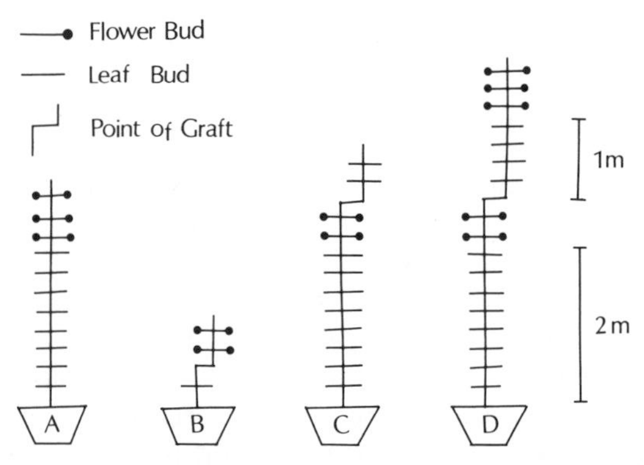

Figure 4.1 Transferring flowering in the juvenile and adult phase of *Malus hupehensis*: A, ungrafted plant; B, adult phase grafted on juvenile plant; C, juvenile plant grafted on adult plant; D, 1-m-high juvenile plant grafted on adult plant. In every case, growing point was transferred and the indicated portion above the graft grew on the nurse plant.

the juvenile period prompted Zimmerman (1972) to adopt the definition of juvenility of Stokes and Verkerk (1951) for fruit trees. According to them, the juvenile period is the time following seed germination during which the seedlings cannot be induced to flower by any means. This definition has special importance for fruit trees. It distinguishes juvenile plants from nonflowering trees propagated vegetatively. In contrast to juvenile plants, young vegetatively propagated trees can be manipulated by horticultural techniques to regain their flowering status. Consequently, they are not juvenile even though they are not flowering.

Transition from the juvenile phase to the adult phase occurs when the fruit tree reaches a certain stage. The transformed character resides in the tissue itself. Evidence for this was provided by Zimmerman (1971), who grew apples (*Malus hupehensis*) as a single stem plant in the greenhouse. The plants reached flowering age at 1.8–2 m height or at the 75th–80th node. Grafting growing tips, either from a flowering plant to a nonflowering one or vice versa, resulted in shoots that maintained the flowering status of their origin. When growing points from the midpoint (1 m) of nonflowering plants were transferred onto flowering plants, they grew the distance to reach flowering according to their location of origin. This indicated that the stage of development evolved with growth and could not be changed by the nursing plant (Zimmerman, 1971). Transferring the flowering in juvenile and adult phases of *M. hupehensis* is illustrated in Figure 4.1.

The only treatment that speeds the transition from the juvenile to the flowering phase is the rate of growth of juvenile plants. Visser et al.

(1976) compared the growth rate and flowering of apple and pear seedlings. Seedlings with a faster growth rate tended to become generative sooner than those with a slower growth rate. This relationship existed for the first 5 years after germination. However, 2 years later the relationship was no longer significant. Growth rate is important within the species to overcome juvenility, but absolute growth rate cannot overcome genetic differences between species. Trunks of pear seedlings were considerably larger than apple seedlings before flowering could occur (Visser et al., 1976). Environment, insofar as it affects growth, has considerable influence on duration of the juvenile phase of both apple and pear seedlings.

Increasing growth rate by any means decreases the length of the juvenile period. Visser et al. (1976) reported that throughout the course of 20 years, the juvenile period of genetically comparable groups of seedlings has been reduced on average by more than 3 years, from 7.4 to 4.3 years in apple and from 9.2 to 6 years in pears. The reduction corresponded to the improvements of growing conditions, especially in the nursery, resulting in seedlings that were 50–100% larger when transplanted to the orchard than before (Visser et al., 1976). A longer growing period and milder climate where N fertilization can be applied without the danger of decreasing winter hardiness are more favorable for growth and consequently for shortening the juvenile phase of fruit trees.

The nonflowering part of the juvenile tree will never flower even though the branches that have developed after the tree reached the age of maturity will flower profusely (Westwood, 1978). In a seedling orchard or in a breeding plot where mature trees are grown, the height at which flowers or fruit are located is indicative of the length of the juvenile period of that particular plant. If the flowering is located relatively close to the ground, the juvenile period of the tree has been short; if the flowering is high up in the tree, the juvenile period of that particular tree has been long.

Grafted trees in their first 2–3 years after the grafting return to the nonflowering status. This causes some confusion about the juvenility of such plants. Because flowering of grafted trees can be enhanced, by the definition described in the preceding, they are not juvenile. There appears to be a vegetative period between the juvenile and adult phases. Seedlings first go through a juvenile period, then enter into a vegetative phase, and finally go into the adult flowering status. Such a scheme described by Zimmerman (1972) and an adapted scheme are illustrated in Figure 4.2.

Seedlings may enter into the flowering status with or without passing through the vegetative phase. Flowering adult trees can alternate between flowering and vegetative phases but never return into the juvenile period. It takes 2–3 years before grafted trees regain flowering status again. Regaining flowering after tissue culture propagation is usually shorter.

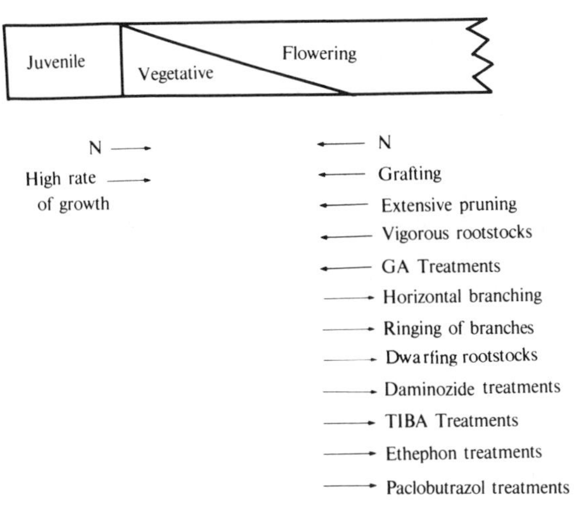

Figure 4.2 Schematic representation of juvenile, vegetative, and adult phases of fruit trees. The tree may enter from the juvenile to adult phase with or without entering the vegetative phase. Horticultural manipulation effective in influencing the transition from one phase to the other is indicated. (Modified after Zimmerman, 1972.)

Induction of Flowering

Induction of flower buds is a qualitative change that according to some is governed by hormonal balance (Luckvill, 1974) and according to others is brought about by changed distribution of nutrients inside of the apical meristem (Sachs, 1977; Williams, 1981). Regardless of the mechanism after the stimulus has been received by the bud, the meristem is programmed to form flowers. We do not know specific compounds that would act as chemical messengers to induce flower bud development.

In order for the vegetative bud to receive the inductive stimulus and undergo the preceding changes, it must be in a certain stage. Bijhouwer (1924) described the apple flower bud as a shortened axis bearing typically 21 leaf formations (Figure 4.3). At the base of the shoot, there are nine bud scales followed by three transition leaves, six true leaves, and three bracts. Only after a certain critical node number has been reached does flower induction begin first in the shoot apex itself and then in the axils of three bracts and the upper true leaves. In apple, the critical node number required before flower initiation commences is about 20 for 'Cox's Orange Pippin' and 16 for 'Golden Delicious.'

Since the growing period is limited in length in the north of the temperate zone, fruit-producing area, the rate of node production is a critical factor in determining whether a bud will produce flowers or

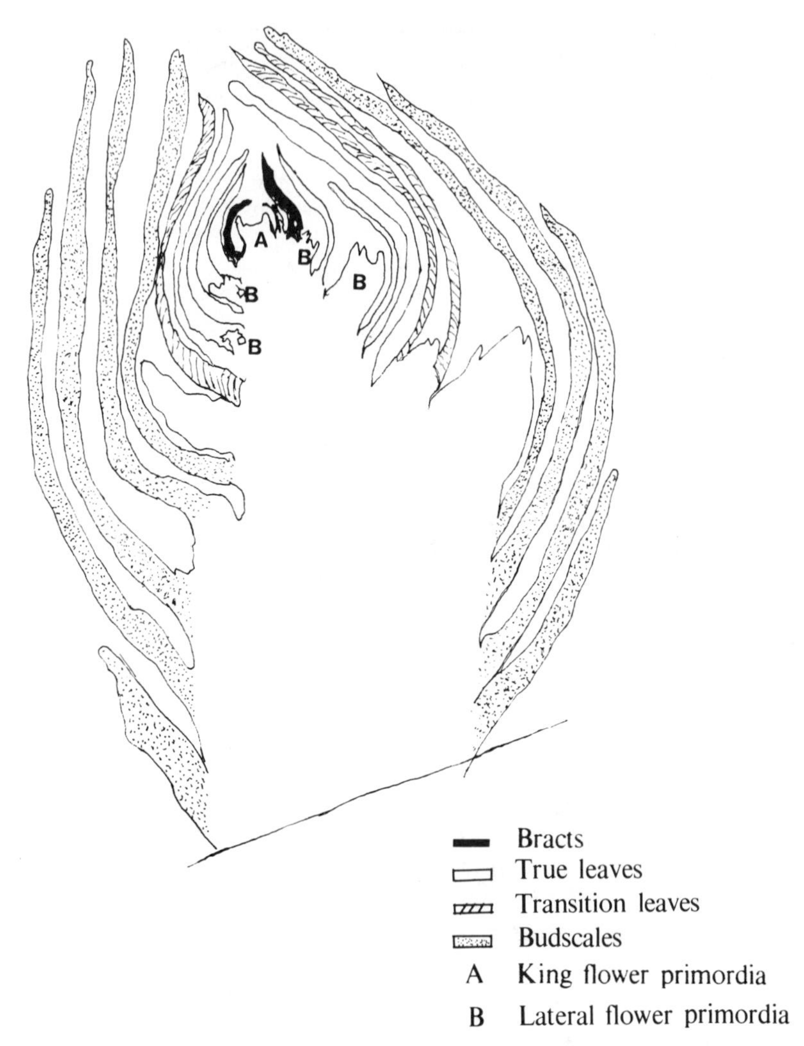

Figure 4.3 Representation of apple flower bud. It contains a bud scale, three transition leaves, six true leaves, and three bracts.

remain vegetative. The time interval between the initiation of successive leaf primordia is the plastochron. In England, the average plastochron for 'Cox's Orange Pippin' apple is about 7 days (Fulford, 1966). An apex with 6 nodes at the beginning of the season must grow for about 100 days to attain the 20-node stage for flower induction to commence (Luckvill, 1974). For this reason, at northern locations flower buds develop only on spurs, which are existing structures of the tree. At southern locations, flower buds often develop in terminal or axillary buds of the shoot, which is a new structure and has developed during the early part of the growing season.

Very short plastochrons may result in the buds' growing out as vegetative shoots in the current years, whereas with longer than 9-day plastochron the bud never reaches the stage to be receptive to flower induction. Luckvill (1974) speculated that gibberellins may extend the plastochron and inhibit flowering indirectly. This speculation is corroborated by the fact that flowers induced relatively late in the season give rise to clusters with elongated base and large spur leaves, whereas flowers initiated early in the season produce compact blossoms with small primary leaves (Abbott, 1970). Use of inhibitors of gibberellic acid (GA) synthesis, such as placobutrazol, produce more, compact flowers and small spur leaves.

From the preceding discussion, one may get the idea that initiation of apple flowers occurs sometime in August because this is the time when the bud is at the right stage for induction. Yet several treatments, discussed later in this chapter, are only effective if applied within 3 weeks of bloom. Since flower bud differentiation is clearly visible only when morphological changes begin to occur, it is not certain how much earlier induction takes place. It may be that there is an extended time interval between induction and morphogenesis.

Development of flower buds does not start uniformly throughout the tree. In apple, the terminal buds of short shoots (spurs) begin their transformation to flower bud 4–6 weeks earlier than in lateral buds (Buban, 1980). In contrast, in sweet and sour cherry trees, the transformation begins at the same time in both types of buds. In plum and apricot trees, flower bud formation starts later on secondary shoots, but buds on the secondary shoots develop faster (Bumbac, 1975).

Environmental conditions influence flower bud development. In apple in shaded areas where less than 30% of sunlight penetrates, practically no flower bud development occurs (Cain, 1973). The summer temperature may also determine the rate of flower bud initiation (Schmidt, 1971; Luckvill, 1974).

The vigor of the tree is a determining factor in the initiation and development of flower buds. Pruning, considered in another chapter, greatly influences the vigor of the tree and its flower bud formation. In general, strong pruning in apple and pear increases growth and decreases flower bud formation. However, peach must be pruned and new growth forced because flower buds are located on the new growth in this species. Nevertheless, vigor is still a determining factor in peach. Overly vigorous new growth will not produce flower buds. Feucht (1961) reported that buds of peach developed on shoots less than 200 mm^3 per internodes, whereas no flower bud developed when the volume of the internode was larger. This value is 130 mm^3 for plum.

In contrast, low-vigor trees are also unable to produce flower buds. It is discussed in other chapters that carbohydrate distribution in a low-vigor tree is such that flower bud formation may not have enough carbohydrates to proceed. Since buds are very small structures especially at the time of bud initiation and early development, this aspect of flower bud development is extremely difficult to investigate.

Gibberellins were reported to inhibit the beginning stages or perhaps the initiation of flower buds in apple (Luckvill, 1970). The seed of the developing fruit is high in GA-like substances (Dennis, 1976), and these substances were implicated in the inhibition of flower bud formation. Chan and Cain (1967) used a parthenocarpic apple variety, Spencer seedless, hand pollinated half of the tree, and adjusted the fruit load so that there was no difference between the seeded and seedless side of the tree. There was no 'return' bloom on the seeded side of the tree, whereas profuse bloom developed on the seedless side of the tree.

Gibberellins diffusing from the developing fruit are the major cause of alternate bearing (Williams and Edgerton, 1981). In a year when the apple is loaded with apples (the 'on' year), the diffusing GA prevents flower bud induction, and the following year the trees will not bloom (this is the 'off' year). During the off year, of course, there is no GA effect and flower bud development is uninhibited. This is the major reason for alternation of on and off years (biennial production). There is a varietal tendency for alternation. According to Hoad and Donaldson (1977) and Hoad (1978), the quantity of GA_3 diffusing from the seeds of alternately cropping Laxton's Superb is much greater than the steadily cropping Cox's Orange Pippin apples. There are several gibberellins produced by the apple. In addition to GA_4 and GA_7 at least 11 others have been identified (Hoad, 1984). Tromp (1982) reported that GA_7 was inhibitory to flower bud initiation and/or development but GA_4 was not. The more polar GAs apparently are more inhibitory than the less polar compounds.

In a series of experiments investigating the GA content of dwarf trees, the nonproductive trees tended to have more polar gibberellins (Grochowska et al., 1984), indicative of the differential influence of various GAs on flower bud initiation.

Compounds that counteract GA either by interfering with its biosynthesis or by its action tend to produce more flowers. Daminozide, paclobutrazol, and ethephon all produce a higher rate of bloom in the year after application (Williams, 1981).

The natural production of ethylene in flower buds also enhances the induction and/or development of flower buds. Apple wood, at locations where flower bud development is likely, contained a higher ethylene concentration compared with wood from a 1-year-old shoot that rarely produces flower buds (Klein and Faust, 1978). A variety of treatments, such as pulling the branches to horizontal position (Robitaille, 1975) and wounding, caused during the course of the so-called Lorette pruning increase both the rate of ethylene production (Klein and Faust, 1978) and flower bud formation.

In the final analysis, in apple, the presence of fruit is the most important factor in the inhibition of flower bud initiation and/or development. The inhibiting effect of fruit on flower bud formation takes place very early, within 3 weeks after bloom (Chan and Cain, 1967). Others reported an inhibiting effect only if young fruit remains on the tree for 6–8 weeks (Link, 1976; Luckvill, 1970). The time of inhibitory effect of fruit in pears was estimated as 4–6 weeks (Huet, 1972).

In peach, flower buds usually develop on 1-year-old shoots. There are three buds on each node of the peach shoot. The two side buds usually develop to be flower buds and the center bud remains a leaf bud. Since the development of buds occurs on shoots grown in the same reason, the side buds must appear on the shoot within 60 days after bloom in order to become flower buds.

The relationship of other hormones (IAA, ABA, and cytokinins) to flower bud development is not clear. Although much speculation has been advanced, the real evidence for the possible mode of action of their involvement in flower bud initiation is inconclusive or speculative. Information regarding these hormones has been reviewed by Buban and Faust (1982).

Horticulturists often believe that shoot growth must stop before flower bud formation begins. This is no experimental evidence supporting this observation in either direction. Sachs' (1977) nutrient diversion theory calls for a certain amount of nutrients, perhaps carbohydrates, available for flower bud formation activity. There is ample evidence that the shaded parts of a tree, where photosynthetic activity is low, do not develop flower buds. Perhaps cessation of shoot growth could provide the necessary carbohydrate pool for flower bud formation. Rate of photosynthesis, however, is influenced by many factors, and shoot growth must be placed into the matrix of factors, if indeed it is a factor at all.

Differentiation of the Growing Point

After flower buds receive the unknown signal for induction, detectable changes occur in the growing point. The first of these changes is the increased synthesis of DNA and RNA. Buban and co-workers (Buban and Hesemann, 1979; Buban et al., 1979) established that nucleic acid levels in 'induced' buds were higher than 'noninduced' buds. Since the presence of fruit on an apple spur prevents flower bud formation, they considered buds on spurs without fruit as 'induced' and with fruit as 'non induced' for their studies. Nucleic acid levels in apple buds determined by cytochemical methods are given in Figure 4.4. Indirect evidence obtained

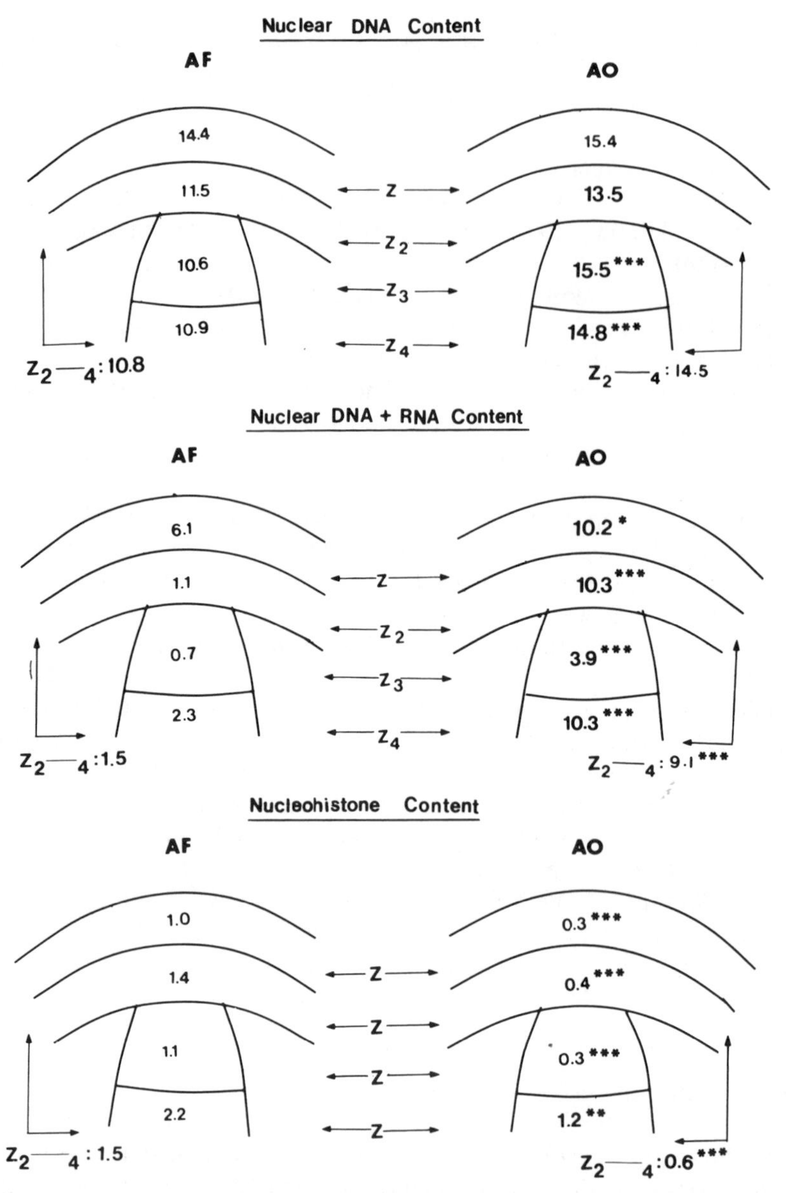

Figure 4.4 Nucleic acid and nucleohistone content in the apices of terminal buds of bearing and nonbearing spurs. significant at (*) 1%, (**) 2%, and (***) 3% level. Top: nuclear DNA content; stained by Feulgen procedure. Middle: nuclear DNA plus RNA content; staining by gallocyanine–chromealum. Bottom: nucleohistone content; staining by fast green FCF. Units are arbitrary units. Significance indicated within the same meristem zone between spurs with fruit (AF) and without fruit (AO). (Reprinted by permission from Buban and Faust, 1982.)

by inhibitor studies also indicates that nucleic acid synthesis is involved in bud development. Nucleic acid analogs that inhibit DNA or RNA synthesis inhibit flower bud formation (Buban, 1969).

Soon after cytochemical changes start, histological differentiation begins and proceeds during the following 8–10 months. The process culminates in flowering during the following season. In apple on the vegetative growing point, a well-defined zonation develops. The first visible sign of differentiation is when the flat apical meristem assumes a dome shape. Then the central meristem is partitioned, and the pith meristem develops strongly. As a result of increased mitotic activity, the reproductive growing point becomes a blocklike structure. This when a scanning electron microscope shows the beginning of flower bud development.

About 8–14 days are required from the beginning of histological differentiation to the appearance of flower meristem. The subsequent development of the flower meristem is relatively rapid. By leaf fall in a large percentage of flower buds, all parts of the flowers are present (Figure 4.5). The appearance of various tissues of the flowers proceeds from the outer structures toward the inner structures. The order of appearance of the tissues is sepals, petals, anthers, and the ovary. In apple, the apical flower develops first followed by the lateral flowers in the cluster. This order is followed by the same order in flowering. The 'king' flower of the apple opens first, which is the apical flower of the cluster.

Development of pear flowers is similar to that of the apple with the exception that the apical flower is less developed and opens last in spring (Zeller, 1960). In peach, the development of flower buds appears sufficiently different. In the triple bud, the apical dome of the two flower buds is much smaller than that of the vegetative bud. At this stage, the rapid appearance of budscales is characteristic. Budscale development is followed by a short quiescent period, and then the apical meristem is lengthened. This is followed by the thick blocklike appearance of the apical dome. In peach, this stage is considered the beginning of flower differentiation. Here, also, the development of various organs of the flower, sepals, petals, anthers, and ovary follows relatively rapidly.

There are at least three critical periods in the development of peach flower buds: (1) the appearance of side buds, (2) the beginning of flower differentiation, and (3) the development of anthers. If the plant undergoes severe stress during any of these critical periods, flower buds will not develop.

The entire process of flower bud differentiation may take 54–112 days depending on the species (Table 4.1). At any given time, there are several stages of development of flower buds in peach (Figure 4.6) and perhaps in other fruits as well.

In cherry, sour cherry, and plum, the development of flower buds

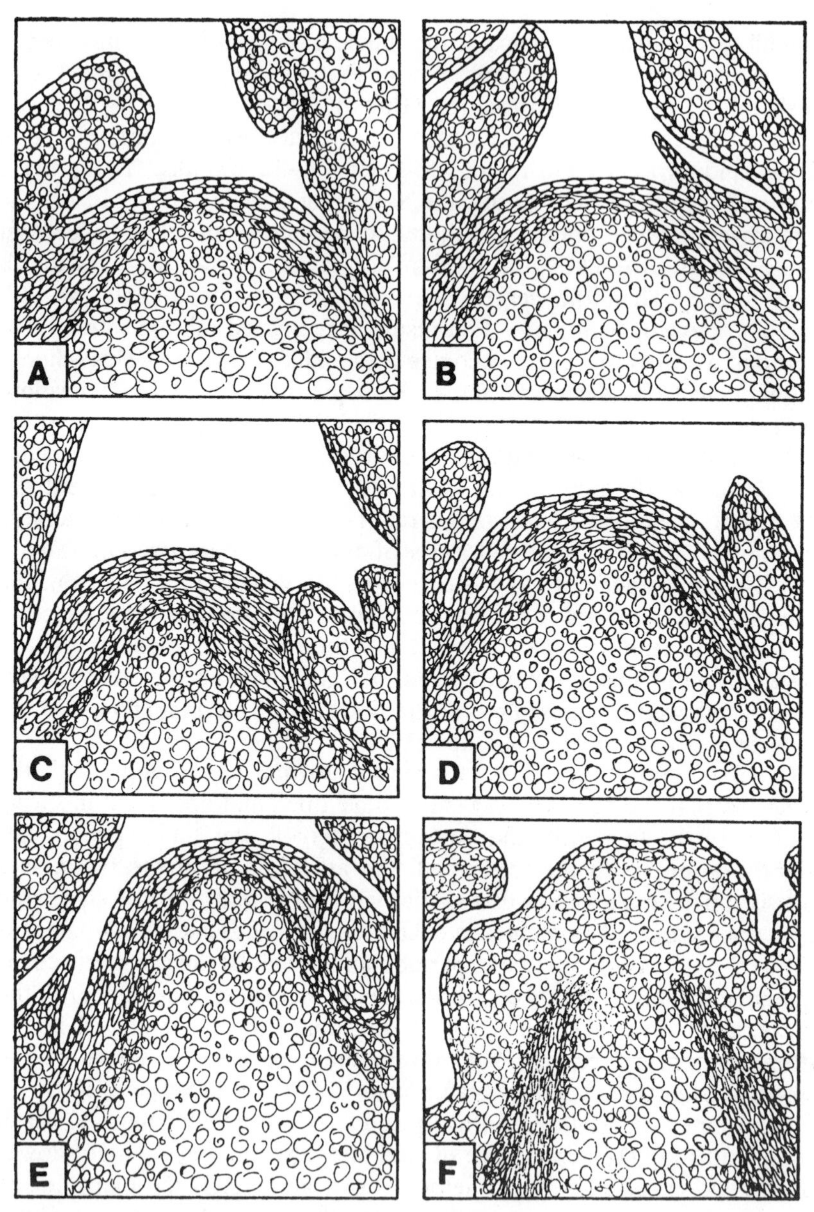

Figure 4.5 Differentiation of flower buds of apple: (A) vegetative meristem, (B) reproductive meristem, (C, D, E, F) beginning of morphological differentiation successive steps during the summer, (G, H) beginning of October, (I, J) mid-February, (K, L) mid-March. (Reprinted by permission from Pethö, 1984.)

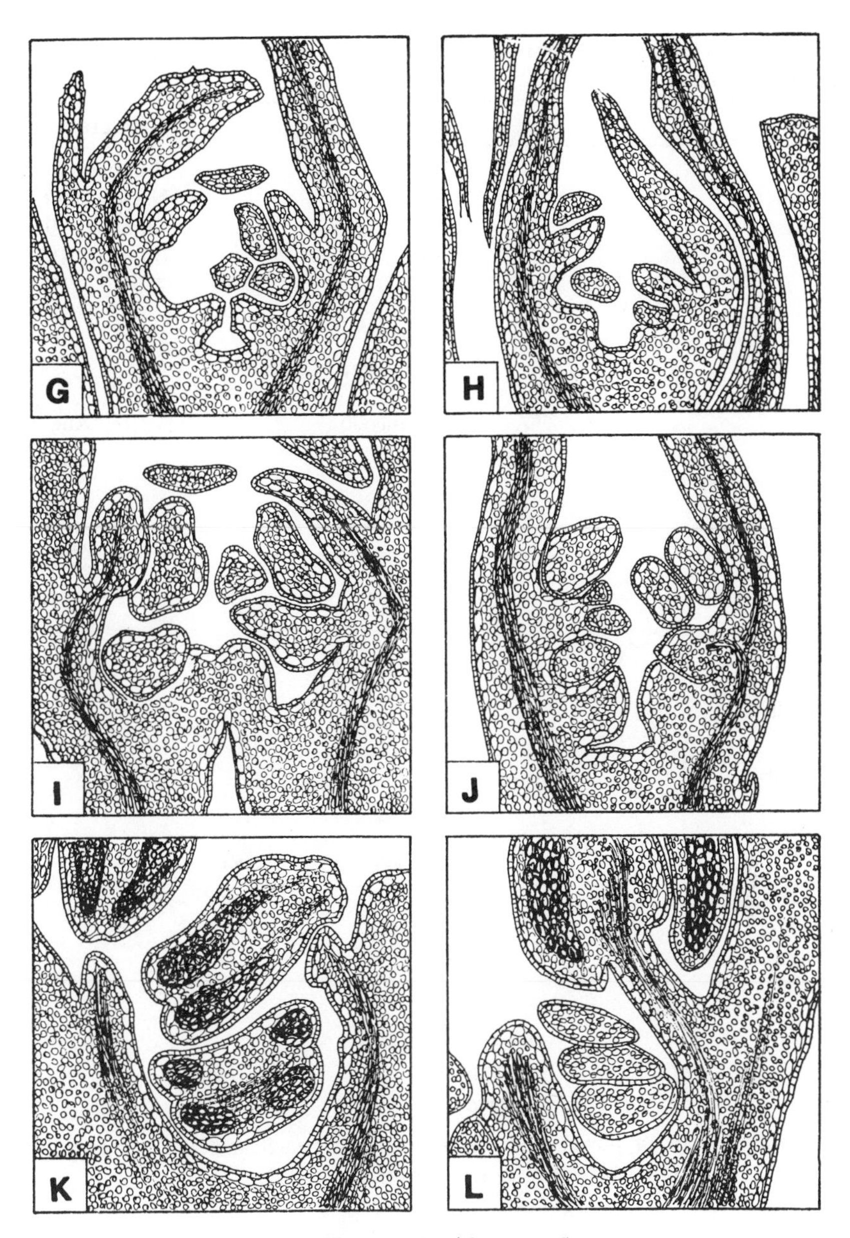

Figure 4.5 (*Continued*)

TABLE 4.1 Time Interval between Beginning and Final Stages of Flower Bud Differentiation

Species	Days	Reference
Peach	54–65	Bumbac, 1975
Sour cherry	56–100	Nyèki, 1980
Cherry	86–112	Bumbac, 1975
Plum	48–66	Bumbac, 1975

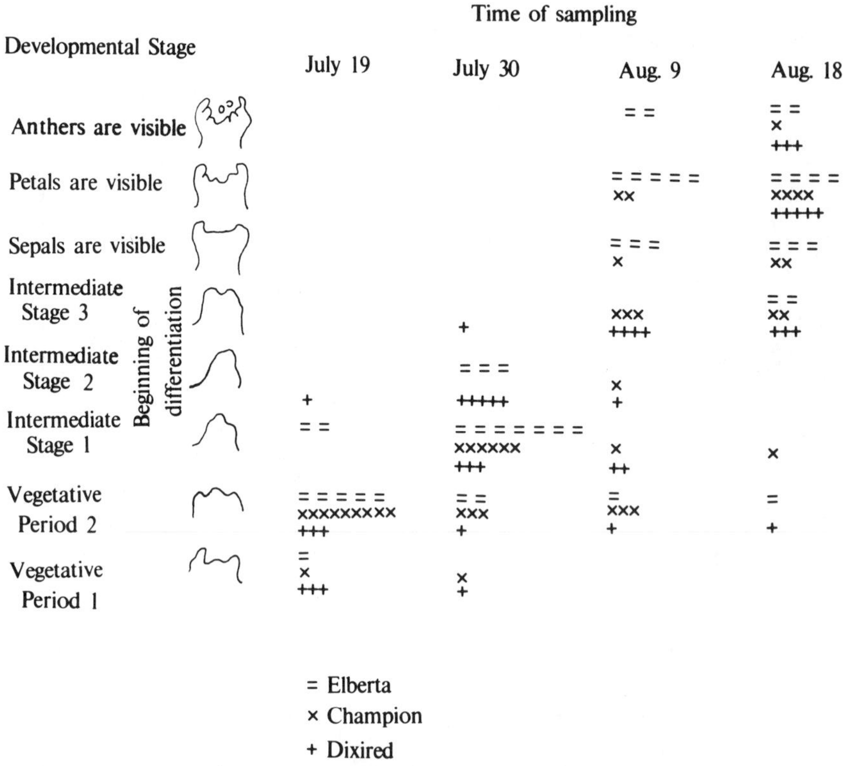

Time of sampling

Developmental Stage	July 19	July 30	Aug. 9	Aug. 18
Anthers are visible			= =	= = × +++
Petals are visible			= = = = = ××	= = = = ×××× +++++
Sepals are visible			= = = ×	= = = ××
Intermediate Stage 3 (Beginning of differentiation)		+	××× ++++	= = ×× +++
Intermediate Stage 2 (Beginning of differentiation)	+	= = = +++++	× +	
Intermediate Stage 1 (Beginning of differentiation)	= =	= = = = = = = = ×××××× +++	× ++	×
Vegetative Period 2	= = = = = ××××××××× +++	= = ××× +	= ××× +	= +
Vegetative Period 1	= × +++	× +		

= Elberta
× Champion
+ Dixired

Figure 4.6 Stages of flower bud development in peach at a norther location (Hungary). (Reprinted by permission from Nyèki, 1980.)

proceeds during the growing season only partially with the ovary only observable in October at the earliest, and it may appear perhaps as late as March. This late development of the pistillate part of the flower may be caused by the hormonal constitution of the mother plant. A large percentage of the various stone fruits may develop functional male flowers with

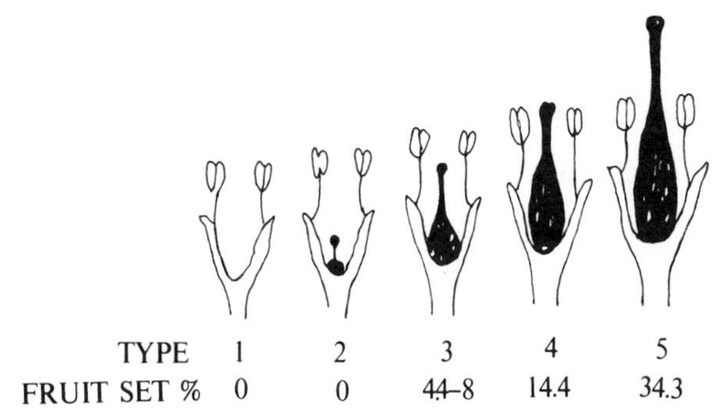

Figure 4.7 Flower types in stone fruits and their corresponding fertility. (Reprinted by permission from Nyèki, 1980.)

underdeveloped pistillate parts. Of course, this phenomenon greatly influences fruit set. The type of flowers that may develop and the corresponding fruit set are illustrated in Figure 4.7. Flowers that have underdeveloped pistils may fall off the tree even before fertilization may occur (Figure 4.8).

In a novel approach, Suranyi (1973, 1976) characterized the femininity or masculinity of the *Prunus* group. Since the size of the pistil is crucial in fruit set and consequently in the femininity of flowers, he determined the number of anthers per millimeter of pistil size. He also correlated this with self-fertility or self-sterility. A ratio of between 1.7 and 1.9 produced self-fertile individuals, whereas either lower or higher ratios produced self-sterility. Each species in the group had its own identity, *Prunus myrabolan* being the most masculine and almond the most feminine. Individual cultivars with indexes closer to the balanced self-fertile group tended to be at least partially self-fertile (Figure 4.9).

The time interval between the beginning and final stages of flower bud differentiation is relatively long and varies with species. The development of buds during fall is very important for the development of good bloom or high yield during the following year. Although one expects that, by harvest, flower buds are developed in apple, a delay of harvest greatly decreases the bloom in the following year (Figure 4.10). Premature defoliation before normal leaf fall produces similar results and greatly decreases the yield in the following year (Figure 4.11). It will be discussed in the following chapter that yield can be influenced by fruit set, which in turn can be influenced during the previous fall. Thus data on yield do not necessarily mean incompleteness of flower development. However, considering the incomplete development of female characters in *Prunus*, one must raise the question of whether fall treatments influence fruit set per se or operate through influencing functional masculinity or femininity.

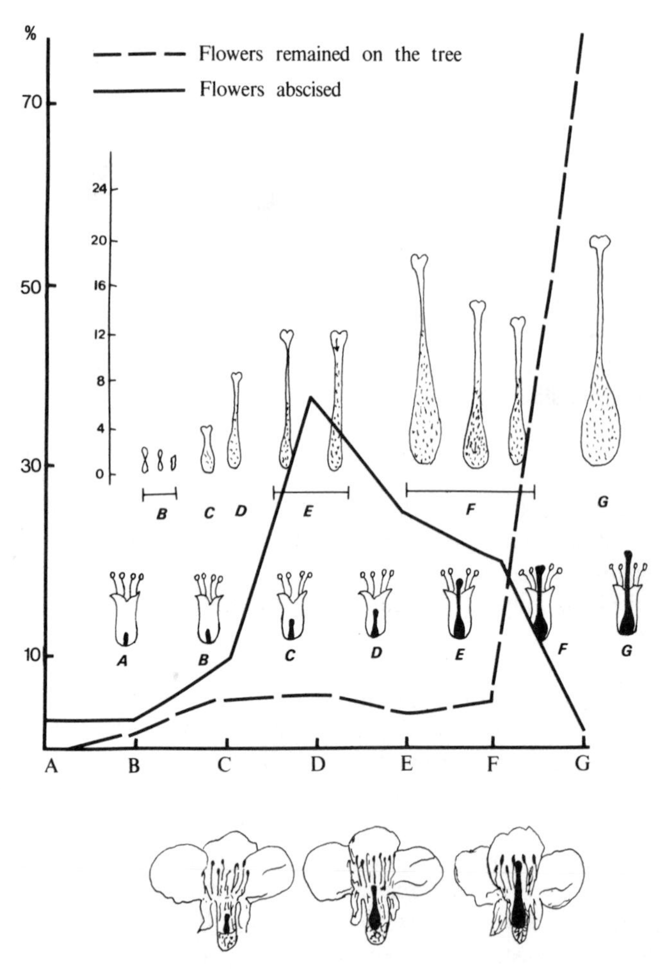

Figure 4.8 Flower types of apricot and their ability to remain on the tree. Only flowers with pistils 14 mm or larger remain on the tree and set fruit. (Reprinted by permission from Nyèki, 1980.)

The highly masculine myrabolan and mahaleb are considered evolutionally the most advanced or so-called youngest species. Their flowers are self-incompatible and small and bloom late. The next group is intermediate in sex expression and partially self-incompatible and blooms relatively early. The highly feminine group is self-compatible, flowers are large, and they bloom the earliest. These are considered evolutionally the oldest within the stone fruit group (Nyèki, 1980). The relative sex expression is obviously hormonal. Since the developmental time of sex

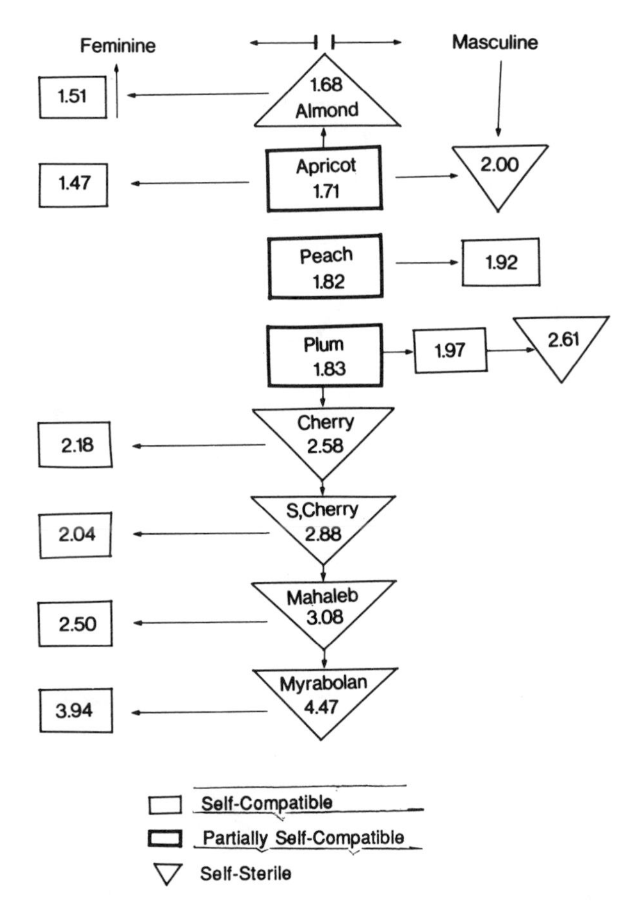

Figure 4.9 Sexual model of stone fruits. Values illustrate the number of stamens per millimeter of pistil size. (Reprinted by permission from Nyèki, 1980.)

organs in the bud can range from early October to March, the exact time when hormonal influence determines the relative sex of the flower is almost impossible to ascertain.

Buds continue to develop during the winter. Apple buds increase in size 20–25% during the months of December and January and by an additional 120–150% between mid-February and mid-March. This increase takes place before any visible bud opening (Figure 4.12). Buds that receive nitrogen fertilization during spring seem to be larger at the end of the growing season. However, fall application of nitrogen greatly accelerates the development of buds, and by March, the difference completely disappears. Apparently, fall application of nitrogen has an important role when buds are underdeveloped (Buban et al., 1979).

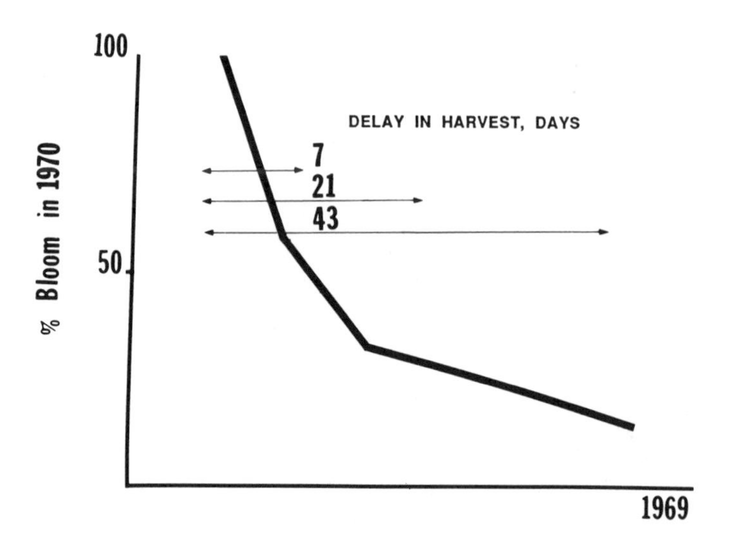

Figure 4.10 Effect of delay in harvest on the development of bloom in the following year. (Reprinted by permission from Nyèki, 1980.)

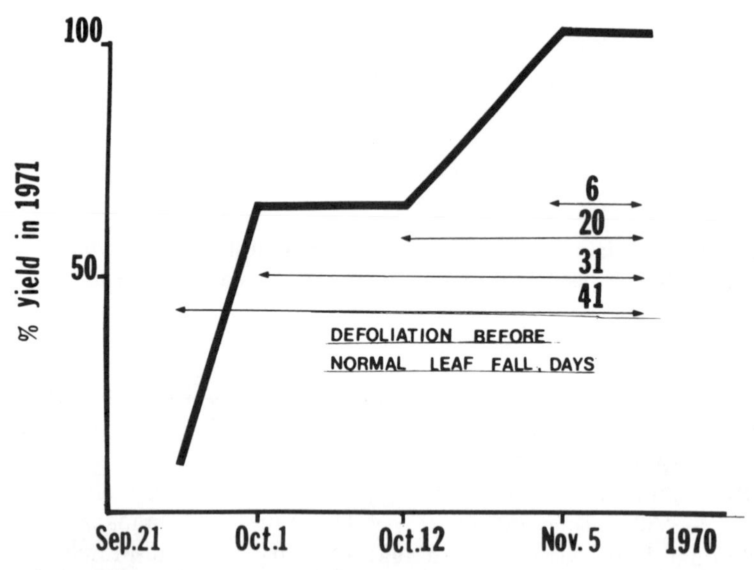

Figure 4.11 Effect of premature defoliation on yield of apples in the following year. Flowering data are not available, but fruit set, an indicative index of flower bud strength, points to insufficient flower development during fall. (Reprinted by permission from Nyèki, 1980.)

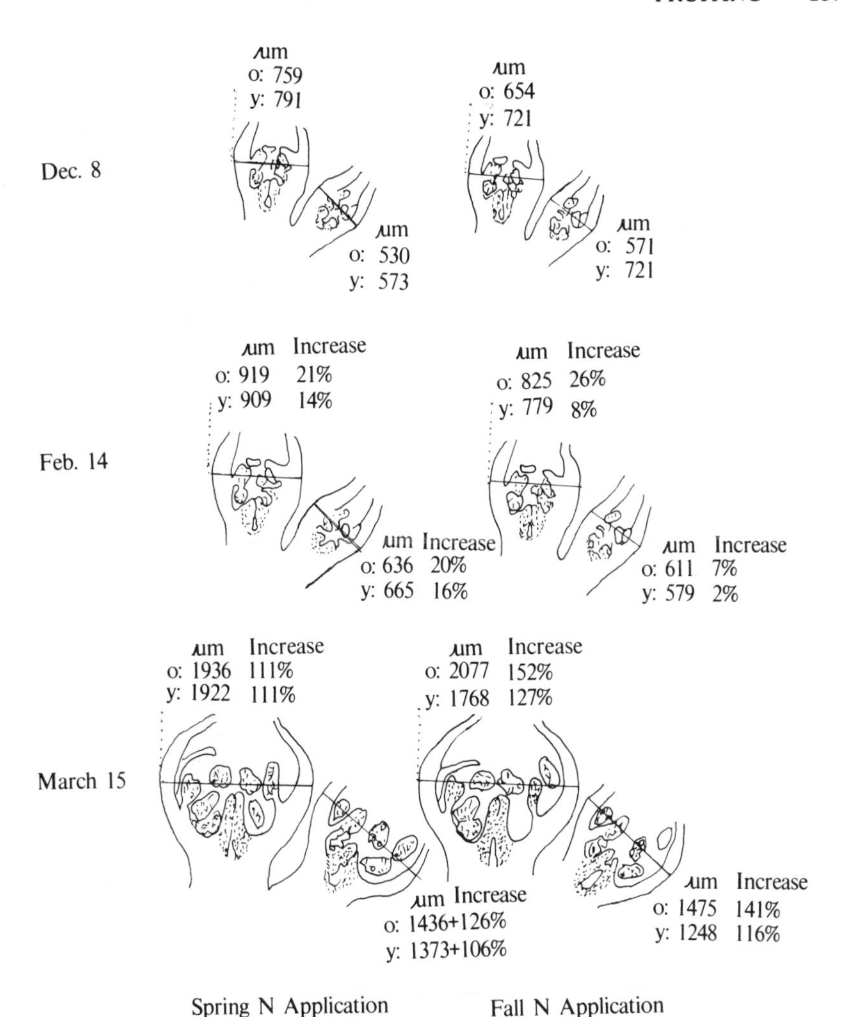

Figure 4.12 Flower bud development of apple during the winter on older and younger spurs. The center flower bud and one lateral bud are illustrated. Actual size of bud obtained at diameters marked in micrometers and percentage of increase in size since previous sampling are given for each bud: O, old spur; Y, young spur. Left: nitrogen was applied during spring. Right: nitrogen was applied during fall. (Reprinted by permission from Pethö, 1984.)

Without nitrogen or with spring application of nitrogen, the early developmental stages of flower bud differentiation are the same. The effect of summer-applied nitrogen begins to show in October when the development of stamens is initiated (Hill-Cottingham and Williams, 1967). Usually the later developmental stages, the differentiation of pistils, and formation of pollen tetrads do not take place without a sufficient quantity of nitrogen.

During the past, much emphasis has been placed on the initiation and early morphogenesis of flower buds. This is probably so because, in apples, thinning – that is, the removal of seed with the fruit – had overriding importance in flower bud development, and this procedure needed to be done early in the season. It appears that equal emphasis needed to be placed on the late events in bud development during the fall. The late phases of bud development may be more important in stone fruit than in pome fruit. It is also possible that the early phases of flower bud development are governed by hormonal events, whereas the late development of the bud depends on availability of carbohydrates and nitrogen. If, indeed, this is the case, as circumstantial evidence indicates, then growers have to adjust production practices for maximum efficiency in flower bud development. They not only have to thin the species that requires this action but must also protect the leaves until late fall for maximum photosynthetic efficiency.

TIME OF BLOOM

There are two important aspects of the time of bloom of fruit trees. Fruit trees bloom in early spring when the temperature is still relatively low and flowers may be killed by frost. Altering cultivars or developing methods that delay bloom is highly desirable to avoid frost. There can be up to 1 month in natural variation in time of bloom of a given species at a particular location (Antsey, 1966), which indicates that delaying bloom is within the realm of possibilities. Another very important aspect of the time of bloom depends on how well the trees complete dormancy. As fruit production increased in the south, it was apparent that bloom of trees was late, weak, and extended. Such blooms were undesirable not only because weak bloom produced few fruits but also because the extended bloom caused an extended harvest that required several pickings. The reasons for extended blooms were assigned to insufficient cold received during the winter. As research has progressed to understand the time of bloom, knowledge about external parameters and internal factors governing resumption of growth during the spring has accumulated.

In 1906 in California Smith observed a disastrous behavior in northern cultivars of fruit trees in mild climates (Smith, 1907). Around 1910 Chandler started a series of experiments that eventually determined that peach trees needed about 3 months of low temperature (below 9°C) to bloom. He also noticed that 'chilled' trees bloomed normally in comparison to unchilled trees whose bloom lasted 2–4 weeks (Chandler, 1925). The theory was born that trees required chilling before they could resume growth. During the 1930s, a chilling requirement for most cultivars existing at that time had been established. At about the same time, the fact that bloom can be promoted by sprays was discovered (Chandler et al., 1937). New germplasm helped the development of cultivars requiring low chilling and pushed fruit production relatively far south. In 1974, Anderson discovered that cooling buds by sprinkling irrigation in early spring delayed bloom (Anderson et al., 1975). From this work the theory emerged that trees also have a heat requirement that must be fulfilled before bloom occurs.

Studies that removed budscales (Swartz et al., 1985) and leaves (Janick, 1974; Notodimedjo et al., 1981), which forced the trees into immediate growth, pointed to the location and source of inhibitors, and studies that pointed to the importance of fall bud development (Cole et al., 1982) further complicated theories about time of bloom. Now we have to consider a series of events beginning in late fall and occurring during winter and spring, all of which influence the time of bloom.

Chilling Requirement of Trees

During late fall fruit trees enter a dormant period similar to other plants, forming resting buds. Theories that imposition of rest is caused by photoperiod in fruit trees are not well supported, although evidence is much better for other trees (Wareing and Saunders, 1971). During 'rest,' normal growth cannot be resumed whatever the external conditions may be. Comparatively little attention has been given to the various stages of dormancy in fruit trees. It is certain, however, that trees do not enter or exit from dormancy suddenly.

Lang et al. (1985) proposed a unified nomenclature for various phases of dormancy. This nomenclature can also be applied to fruit trees. Lang et al. designated the stages of dormancy as ecto-dormancy, cessation of growth is caused by physiological factors arising elsewhere in the tree; endo-dormancy, having to do with factors inside the affected structure, presumably the buds for fruit trees; and eco-dormancy, the dormancy imposed by unfavorable environmental conditions. A schematic representation of various phases of dormancy is presented in Figure 4.13.

The stages of dormancy are difficult to distinguish because there is no good physiological marker that signifies the various periods. Apple (Abbott, 1970) buds continue to grow during the entire winter. Peach buds likewise differentiate the floral parts while they are dormant. The development of buds during winter is cultivar dependent. Reports indicate that some apricot, cherry, and other cultivars of pear and peach developed very little or none when chilling presumably accumulated (Brown and Kotob, 1957; Felker and Robitaille, 1985). Pollen meiosis in peach has been used to mark the end of the rest period (Young and

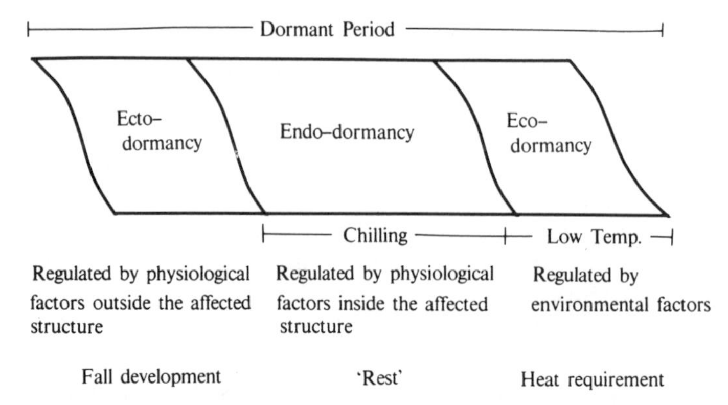

Figure 4.13 Unified nomenclature for various phases of dormancy in fruit trees. The nomenclature has been proposed by Lang et al. (1985), and portions applicable to fruit trees are presented here.

Houser, 1980), but in cherry pollen meiosis cannot be related to chilling (Felker and Robitaille, 1985). It is certain, however, that winter development does not occur at 15°C (Felker and Robitaille, 1985; Chandler and Tufts, 1933) and buds are not ready to grow at springtime. Biochemically the buds do not show rest. The respiration 'potential' of buds increases throughout the winter (Cole et al., 1982). When respiration of buds taken throughout the winter from the orchard is measured in the laboratory, the determined rate of respiration represents a potential only, because environmental conditions in the orchard limit respiration considerably. Nucleic acid and protein synthesis both indicated that buds of fruit trees do not rest, at least in this respect (Zimmerman et al., 1970).

Chandler et al. (1937) and Lamb (1948) indicated that the temperature range of 0–7.2°C was most effective for chilling. Brown (1960) and Weinberger (1954) noticed that intermittent warm periods reversed the chilling. Erez and Lavee (1971) were the first to use a weighed chill unit concept that took into account the differential effectiveness of various temperatures to which the buds are exposed. Richardson et al., (1974) refined this concept and developed the chill unit model widely used today (Table 4.2). According to their determinations, temperatures between 2.5 and 9.1°C are effective as chill units, 1.5–2.4 and 9.2–12.4°C are half as effective, and below 1.4 or between 12.5 and 15.9°C are ineffective. Reversal of chilling begins to occur above 16°C partially and above 18°C to a full extent. Using the chill unit concept, one can calculate quite accurately the time when trees are able to resume growth or the temperature regime the trees absolutely need to break buds.

In the older literature before the chill unit concept had been developed, chilling often had been calculated as chilling hours. In the more

TABLE 4.2 Conversion of Selected Temperatures to Chill Units[a]

Temperature (°C)	Chill Unit
1.4	0
1.5– 2.4	0.5
2.5– 9.1	1.0
9.2–12.4	0.5
12.5–15.9	0
16.0–18.0	−0.5
18.1–21.0	−1.0
21.1–23.0	−2.0

[a] Compiled from Richardson et al., 1974, and Shaltout and Unrath, 1983.

TABLE 4.3 Chilling Requirement Ranges for Various Fruit Tree Species

Species	Chilling Requirement (hours above 7°C)
Apple	200–2000
Apricot	250–900
Domestic plum	700–1700
Japanese plum	500–1500
Peach	200–1100
Quince	50–400
Sour cherry	600–1500
Sweet cherry	500–1300

modern literature, all values are given in the more accurate chill units.

Chilling is important only in relatively warm areas where cool temperature occurs only to a small extent. Often areas free of winter freezing temperatures still have sufficient chilling because there are sufficiently long periods in the 3–9°C range. In such areas one often finds citrus and apples in close proximity, in adjoining blocks, or divided only by a few hundred meters of altitude. In areas where citrus and apples grow together, one of the fruits often suffers. If apples are fully chilled, the fruit of citrus is usually small and very yellow and its fragrance is very strong. In contrast, where citrus is of high quality, large, and normal in color and fragrance, apples often do not receive sufficient chilling. Trees respond to temperature regimes that are hard to distinguish even with modern instrumentation.

Chilling requirements of various species differ greatly. Even within a species the range of chilling requirement is large. Chilling requirements are listed in Table 4.3. The list is partially based on Chandler et al. (1937)

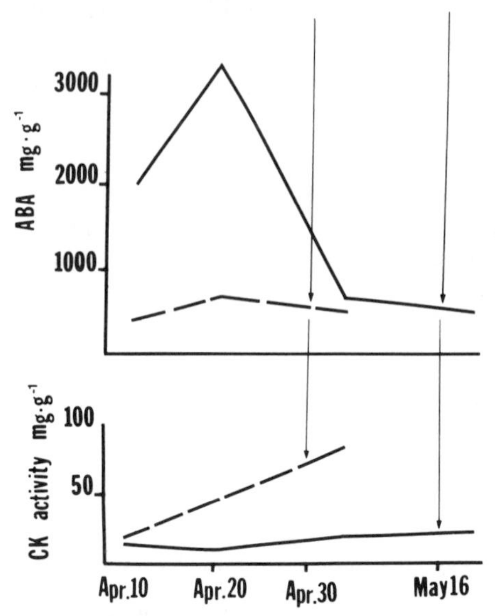

Figure 4.14 Abscisic acid (ABA) and cytokinin (CK) levels of early and late blooming apple trees during spring. Adapted from Swartz and Powell (1981). Arrows indicate bloom time of early and late blooming cultivars.

and modified by considering the recently developed low-chilling varieties.

Circumstantial evidence suggests that during rest inhibitors accumulate in the budscales and prevent the resumption of growth. Abscisic acid is the most important inhibitor causing dormancy. It is effective in delaying budbreak in cultured axillary buds of apple (Dutcher and Powell, 1972; Singha and Powell, 1978), and if injected, it can prevent budbreak in cherry (Mielke and Dennis, 1978). The highest levels of ABA negatively correlate with budbreak in apple (Seeley and Powell, 1981), sour cherry (Mielke and Dennis, 1978), and peach (Bowen and Derickson, 1978; Hendershott and Walker, 1959). There is evidence that tends to minimize the role of ABA as a major factor inducing rest in fruit trees. In apple ABA levels increased as spring growth commenced (Swartz and Powell, 1981). A decrease in ABA during winter was not due to chilling, and ABA levels could not be associated with the bloom time of cultivars (Swartz and Powell, 1981). (See Figure 4.14.)

Removal of budscales induces the growth of the bud, which strongly points to the budscales as the site of inhibitor(s) within the bud. The removal of budscales of early blooming varieties during spring has little effect on budbreak, whereas a similar action is very effective on late blooming and very late blooming varieties (Swartz et al., 1984).

TABLE 4.4 Compounds Artificially Breaking Rest in Fruit Trees[a]

Compound	Concentration Used	Remarks
Dinitroorthocresol (DNOC)	1.5% in mineral oil as dormant spray	—
KNO_3	2–5%	Promotes more flower bud opening
Thiourea	2% not closer than 2 weeks to budbreak	Promotes more leaf bud opening
Benzyladenine (BA)	500 ppm	Does not translocate, growth limited
Gibberellins	50–200 ppm	Most effective on sour cherrv and peach

[a] Some combination of the compounds used for breaking rest are synergistic. These are KNO_3 + thiourea, thiourea + DNOC, and DNOC + GA.

Experience gained in tropical areas with temperate zone fruits indicates that removal of leaves 3 weeks after harvest immediately triggers budbreak without the need for rest (Janick, 1974; Notodimedjo et al., 1981). This practice allows commercial production of apples consisting of nearly 2 million trees near Malang, East Java, and in Sri Lanka. At the latitude of 7° south with the minimum temperature of 12°C, chilling cannot occur, the cultivar used ('Rome Beauty') has a high chilling requirement, and yet no problem is experienced in budbreak if defoliation is used. This implies that the inhibitor(s) of budbreak in all likelihood is (are) produced in the leaves and transported to the budscales before leaf abscission. If leaves are removed prior to transporting the inhibitors to the buds, inhibition of budbreak does not occur.

There are several compounds used as sprays to break rest (Erez et al., 1971) (Table 4.4). The mode of action of these compounds is not known; nevertheless, they are effective in promoting budbreak and bloom in areas where chilling is insufficient.

Heat Unit Requirement during Spring

The actual time of bloom during spring greatly depends on spring temperatures. Some cultivars bloom at average temperatures of 8°C; others will not bloom until the temperature reaches at least 15°C. The heat requirement of a given cultivar is usually determined genetically. Early blooming species of pears require much less heat unit accumulation before they bloom (Antsey, 1966; Cole et al., 1982) than late blooming pear cultivars, which require a larger amount of heat accumulation (Spiegel-Roy and Alston, 1979). The actual amount of heat required often varies with the

TABLE 4.5 Pheno-climatography of Two Peach Varieties[a]

Stage Description	Redhaven (GDH)	Elberta (GDH)
Bud swell	1981	2167
Green calyx	2580	2617
Pink tip	3710	3717
First bloom	4174	4239
Full bloom	4926	5110

[a] Reproduced by permission from Richardson et al., 1975.

TABLE 4.6 Effect of Sprinkling on Stages of Apple Bloom Development

Stage Description	Unsprinkled Date	Sprinkled Date	Delay Days
Silver tip	4/4	4/4	—
Green tip	4/26	5/5	9
Half-inch green	5/2	5/21	19
First pink	5/14	6/2	19
Full bloom	5/21	6/7	18

[a] Reproduced by permission from Anderson et al., 1975.

physiological condition of the plant. Resting or partially chilled trees require much more heat accumulation than plants that have their chilling satisfied before they are able to bloom (Richardson et al., 1975; Samish, 1945; Swartz and Powell, 1981). In contrast, extra chilling given after completion of rest reduces the needed heat unit accumulation for bloom (Couvillon and Erez, 1985; Couvillon and Hendershott, 1974).

Heat requirement is most often expressed as growth degree days (GDDs) or growth degree hours (GDHs). Usually 4.5°C is the base above which the GDHs are calculated. The heat requirement of two peach varieties is given in Table 4.5.

An effective way to decrease bud temperature is to sprinkle the trees often. The heat of evaporation cools the buds as much as 10°C below air temperature. Sprinkling decreases bud development and delays bloom (Anderson et al., 1975). The magnitude of delay is illustrated in Table 4.6. During the experimental procedure from which the preceding data were obtained, an illustrative series of temperatures had been recorded as follows: Air temperature in a standard meteorological shelter was 27–28°C, wet bulb temperature was 12°C, unsprinkled bud temperature was 37–38°C, and sprinkled bud temperature was 12°C.

Sprinkling lowered the temperature in the relatively arid climate where the experiments were conducted by over 20°C, which delayed the bloom more than 2 weeks. Others obtained less spectacular differences depend-

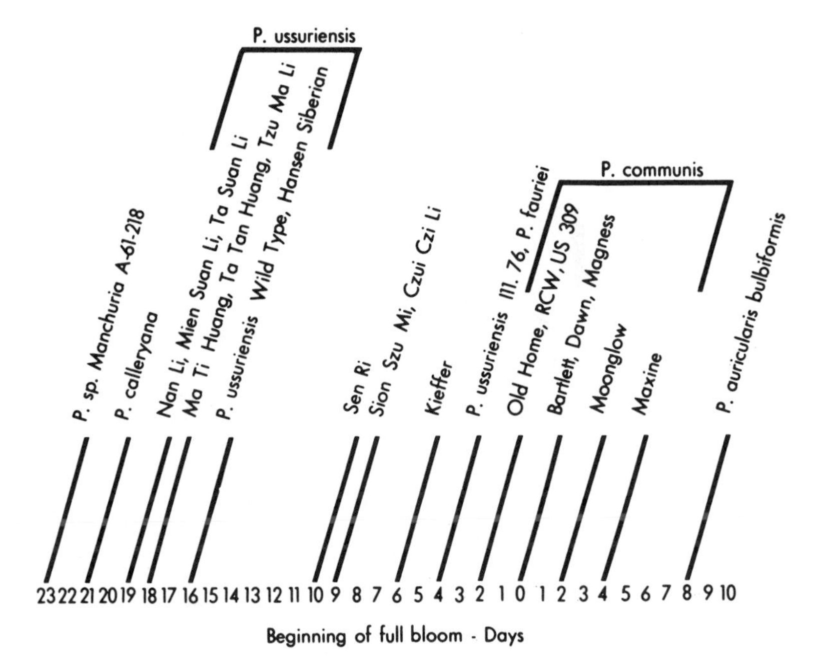

Figure 4.15 Sequence of bloom of pear cultivars at Beltsville, MD, during the spring of 1975. 'Bartlett' is designated as day 0 with regard to bloom. Those blooming before or after are designated as minus or plus day cultivars. (Reprinted by permission from Faust, et al., 1976.)

ing on bud temperature and humidity, which greatly influence evaporation. Nevertheless, the heat requirement of buds following chilling has been firmly established.

Genetic Transmittance of Time of Bloom

Time of bloom in its entire complexity can be genetically transmitted (Faust et al., 1976), but the nature of this transmittance is difficult to ascertain. Among the various pear species and cultivars the time of bloom may vary by at least 3 weeks. The flowering sequence obtained in 1975 is illustrated in Figure 4.15. Bloom times of progenies obtained by crossing cultivars with differing bloom times are illustrated in Figure 4.16. Bloom dates are expressed relative to the bloom time of Bartlett, whose bloom time is designated as day 0. When two early blooming cultivars are crossed, the entire progeny blooms early. When a -21-day cultivar is crossed with a -2-day cultivar, the progeny blooms around -10 days in the middle of the two parents' bloom times. Additional backcross to a late blooming cultivar $[(-2 \times -21) \times +2]$ results in the progeny's bloom

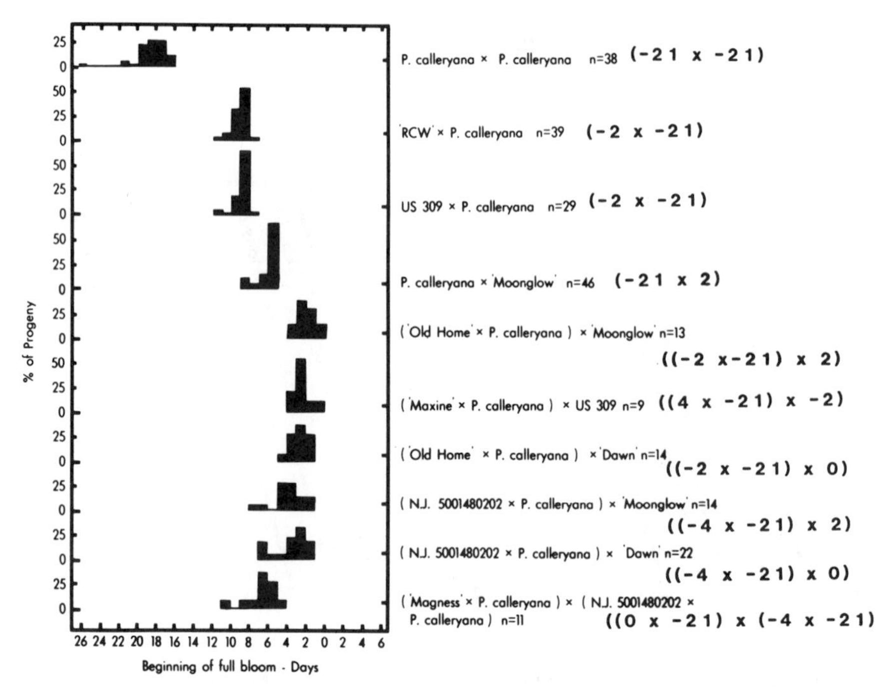

Figure 4.16 Distribution of bloom time of various progenies of pear hybrids. Days of bloom of parents in comparison to 'Bartlett' are given in parentheses. (Reprinted by permission from Faust, et al., 1976.)

close to 0 day, bringing the bloom very close to the late blooming component of the combination. Bloom time moves from early bloom to late bloom within two generations but does not extend beyond the late blooming component. It is also notable that the spread of bloom time within a progeny is very limited. This is the reason it is so difficult to develop late blooming cultivars.

Role of Cytokinins in Budbreak

There is ample information that various synthetic cytokinins, namely, benzyladenine (BA) and thiobutrazol (TBZ), are able to induce budbreak in fruit trees. Benzyladenine is successful in inducing budbreak whether it is applied to dormant trees; fully chilled buds of spur-type apple trees, which usually do not break (Williams and Stahly, 1968); or unchilled trees with or without leaves removed (Broom and Zimmerman, 1976). Shoots that result from BA-induced budbreak rarely grow to longer than 5–8 cm in apple and remain mostly of spur size. Thiobutrazol is a more effective regulator in inducing budbreak than BA (Wang and

Steffens, 1986). Shoot growth induced by TBZ is similarly shorter than that induced by BA. When TBZ-induced budbreak is combined with removal of the terminal bud, the top two to three buds of the shoot will grow and elongate normally, whereas the rest of the buds that were induced on the shoot remain short. Such results indicate that cytokinin-type compounds are able to overcome inhibitors residing in the buds and preventing budbreak, but they are not able to overcome inhibition caused by apical dominance as far as shoot growth is concerned.

The effects of BA and TBZ are not translocable. These compounds must be applied to the buds, and they only affect the bud to which they are applied. Untreated buds directly above or below the treated bud do not break.

There are some indications that late blooming cultivars of apples have lower levels of cytokinins and that this is the factor that delays bloom (Swartz and Powell, 1981). However, it is difficult to interpret the available data. The late blooming cultivar used in Swartz's studies was four times lower in cytokinin concentration, but it was also five times higher in ABA than the early blooming counterpart used in the study (Figure 4.14).

Role of Late Fall Bud Development in Budbreak

Bud development starts sometime during the summer and continues into the fall. How far bud development continues after leaf fall depends on the ability of the tissues to generate energy for further development. Bud growth is very minimal at this time, but it nevertheless continues (Abbott, 1970). There seem to be major differences between varieties in bud development. As air temperature cools, respiration decreases. Buds of some cultivars continue to respire throughout the winter largely through the so-called alternate respiratory pathway. This pathway generates only one-third of the energy that is generated by the regular respiratory chain. It appears that this pathway generates enough energy for the buds to continue their development. Cultivars that have a high rate of respiration through the alternate pathway enter into the subzero temperatures of winter with highly complex and almost fully developed buds. In contrast, buds of cultivars that have a lower rate of alternate respiration are prevented from development since the regular respiratory chain provides less energy at temperatures below 10°C, which may occur during most of November and December. Buds of these cultivars enter and exit the winter with substantially less developed buds, and bud development must occur during spring before budbreak. Such development can partially explain the heat requirement of certain cultivars.

Cole et al. (1982) compared respiratory activities of *Pyrus calleryana* and *P. communis*. *Pyrus calleryana* is an early blooming species with a

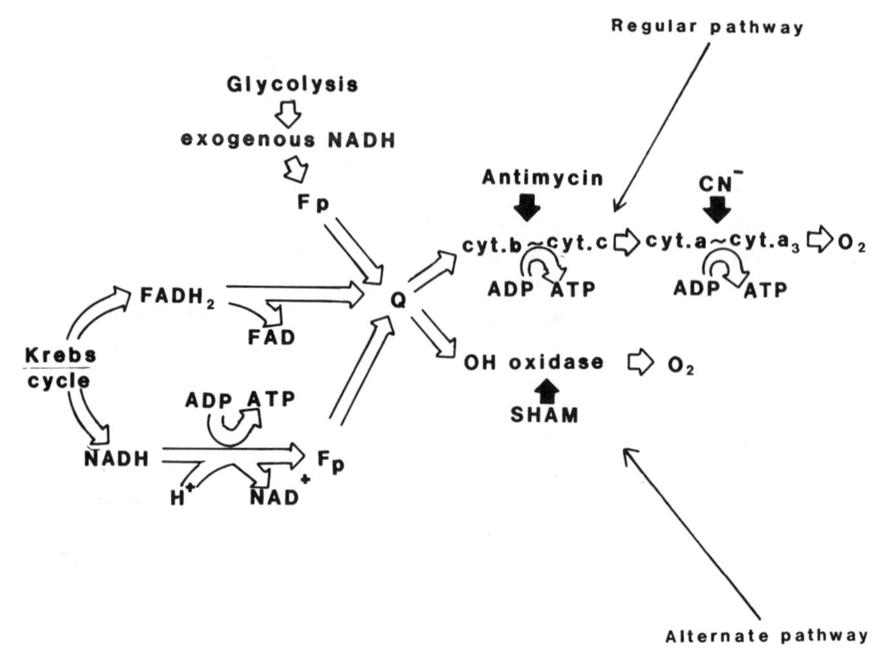

Figure 4.17 Structure of respiratory pathways of fruit tree bud mitochondria: Ep, flavoprotein; Q, ubiquinone; cyt., cytochromes; SHAM, antimycin; CN are inhibitors of pathways.

high rate of bud development during the fall. In contrast, *P. communis* enters into the winter with relatively small undeveloped buds and blooms much later than *P. calleryana*. The rate of respiration in the alternate pathway is much higher in *P. calleryana* than in *P. communis*. The hypothesis that respiration of early blooming species is carried on largely through the alternate respiratory pathway during periods of cool temperature was confirmed with a number of pairs of fruit species. Each pair contained early and late blooming cultivars of the same species. The alternate path (cyanide resistant) of the respiratory chain is illustrated in Figure 4.17.

Events Occurring during the Period When Trees Have No Leaves

The anomalous designation that when trees have no leaves they are 'dormant' or in a state of 'rest' implies that little activity is going on in their tissues. Naturally during the winter temperatures are low and this slows enzyme activity. Nevertheless, a purposeful sequence of biological events occurs in trees that greatly determines their spring activity. Late fall buds develop greatly in those cultivars that possess alternate pathways

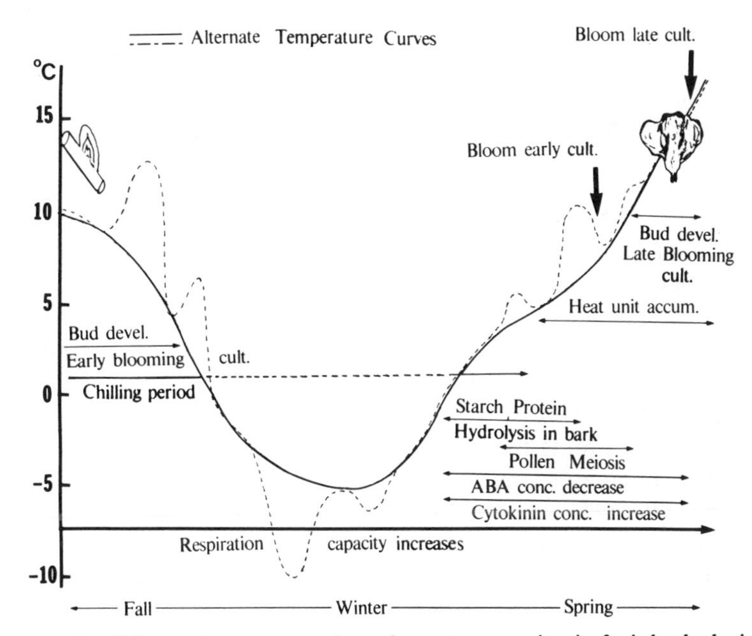

Figure 4.18 Schematic representation of events occurring in fruit buds during the period when the tree has no leaves. Dotted and solid lines indicate alternate possibilities for temperature. Dotted and solid line combination for chilling indicates times when temperature is favorable (solid line) or not (dotted line) for chilling.

of respiration. The tree then enters into rest. Although during this time growth cannot be resumed, the bud tissues are not inhibited from synthesizing nucleic acids and proteins (Zimmerman et al., 1970). Hydrolysis of starches (Grochowska, 1973) and storage proteins (Titus and Kang, 1982) also occurs. The chilling requirement must be satisfied before ABA concentrations in the bud tissues decrease, cytokinin concentrations increase, and meiosis of pollen cells can be observed. Then the tree needs a shorter or longer period of warm temperature before it formally resumes growth. When the expansion of flowers begins, respiration returns to the regular respiratory pathway (Cole et al., 1982). In cultivars that do not possess the ability to generate energy through the alternate pathway of respiration, some or none of the winter activities described here can occur. These cultivars have a longer heat requirement, and this heat requirement cannot be satisfied with temperatures much below 10°C. This is why the cooling of buds of early blooming species such as apricot does not delay bloom and late blooming cultivars are very responsive to temperature; they bloom only when temperatures increase to around 15°C and bloom can be delayed by cooling sprinkling. The schematic diagram in Figure 4.18 illustrates bud activity during dormancy.

FRUIT SET

Fruit set plays an important role in modern fruit production. A large yield of fruit can only be expected if conditions for pollination and fruit set are favorable. This involves a complex series of events, none of which can be limiting to the overall process. Given a source of viable pollen, the first step is the transfer of pollen to a receptive stigma. Transfer may be natural or artificial. Next, germination and pollen tube growth must occur. These processes are affected by environmental factors, mineral nutrition, and the genetic makeup of the tree. Concomitantly, the mature embryo sac must develop at the base of the style. Then successful fertilization must occur followed by growth of the embryo. If all these steps occur, the fruit remains on the tree. The overall process is called 'fruit set.' In general, events involving fruit set can usually be grouped into three major categories: the flowering process itself, circumstances that influence pollination and fertilization, and conditions conductive to fruit set without fertilization. All processes connected with fertilization are affected by regulators, nutrition, and rootstock type, which can all be controlled by the grower. For these reasons understanding the details of fruit set can make the difference between success and failure in orchard production.

The Biology of Bloom

During spring, depending on the species, trees are ready to bloom. Factors influencing the time of the bloom are discussed in this chapter concomitantly with dormancy. Bloom is the process in which the sepals and petals of the flower slowly enlarge and move apart and the stigma(s) and stamens are exposed. When 12–15% of the flowers of a tree are open, this time is the beginning of bloom; when 95–100% of the flowers are open, this time is the end of bloom. Cultivars of a species usually do not bloom at the same time. In any given species we distinguish three of five groups considering bloom: very early, midseason, and late blooming cultivars. If one cultivar is expected to pollinate another, the two cultivars must bloom together, or at least have the bloom period overlap sufficiently to allow for pollen transfer.

As a rule, the more northern the location is, the later the bloom is. When the bloom is late, the bloom period is compressed, and in a compressed bloom differences between cultivars in time of bloom tend to be minimized. Thus, quite often, a flowering sequence determined at a northern location does not apply to southern locations because the spread between the blooming cultivars is much greater. Complicating the decision as to which cultivars should be planted together is the characteristic of certain cultivars whose bloom is unstable. Some years they bloom

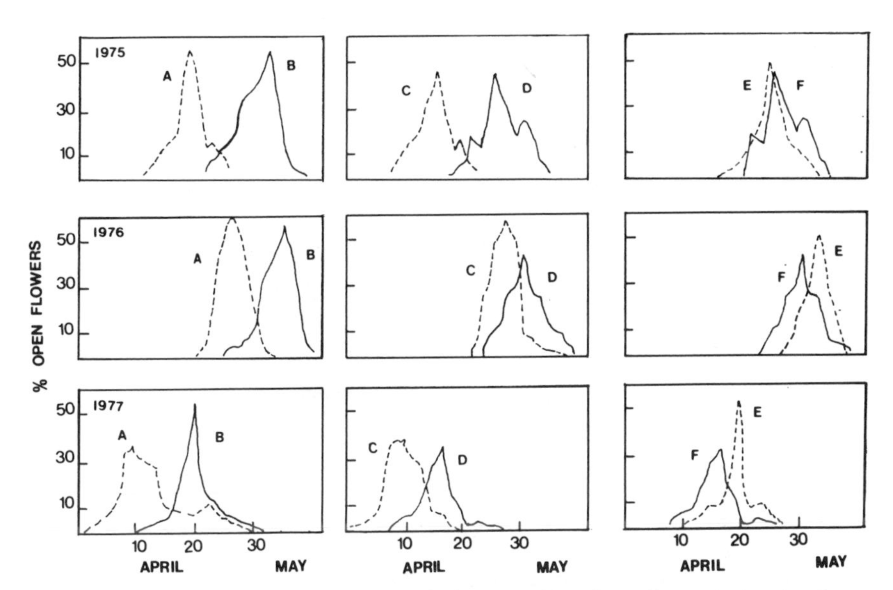

Figure 4.19 Flowering fenograms of three pairs of apple varieties in three different years at a northern location. First pair ('Eva', 'Golden Delicious') consistently flowers apart, the third pair ('Starking,' 'Golden Delicious') always flowers together, and the second pair ('Red Astrachan', 'Jonathan') is variable. A = 'Eva'; B = 'Golden Delicious'; C = 'Red Astrachan'; D = 'Jonathan'; E = 'Starking'. (Adapted from Nyèki, 1980.)

earlier and some years later regardless of the absolute date or relative sequence (Figure 4.19). The reason for the uneven shift in flowering is explained by the differential sensitivity of the various stages of bloom to temperature stimuli. A 6-year record of apricot bloom is illustrated in Figure 4.20. In some years the stages of bloom development were compressed; in other years they were spread apart. The stage between the red tip and the balloon stage is the one most likely to be extended, with the stage between beginning and end of petal fall as the second most variable stage.

The need for certain cultivars to bloom together in order to be effective pollenizers is obvious. At present we have no control over the time of bloom nor can we influence two cultivars to bloom exactly together. Only experience and testing can be suitable guides for determining which cultivars bloom consistently together or how the bloom of a certain cultivar may change. To assure that good pollenizers bloom together with the main variety, some advocate the planting of three cultivars, one that usually blooms slightly before and another that blooms slightly after the main cultivar. Nyèki (1980) illustrated such possibilities for the 'Meteor Early' sour cherry cultivar. In the year illustrated in Figure 4.21, the

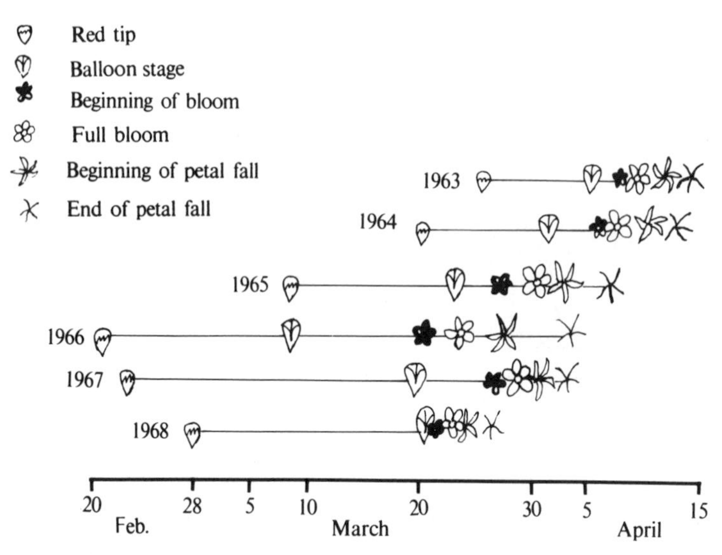

Figure 4.20 Change in rate of bloom development of apricot flowers during six consecutive years. (Reprinted by permission from Nyèki, 1980.)

bloom lasted for 2 weeks with the majority of flowers opening in 3 days. The best pollen donor cultivar to pollinate 'Meteor Early' is designated as No. 4 among the choices as pollenizer cultivars. This cultivar opens its flowers at the same time as 'Meteor Early'. If the bloom of the pollenizer variety started a day earlier or a day later, then cultivar Nos. 3 or 5 would be considered pollenizers. Considering the extremes, cultivars designated with 0 and ∅ are not good pollenizers because they barely overlap with the beginning or the end of the full bloom of 'Meteor Early'. There are lists of cultivars in every species developed for certain locations that can be used as guides for developing planting plans as far as time of bloom is concerned, but they need not be discussed here.

The Process of Pollination

The majority of fruit trees are self-incompatible. Self-incompatibility is defined as the inability of fertile hermaphrodite seed plants to produce zygotes after self-pollination. Usually self-pollen is genetically inhibited from germination or pollen tube growth so that fertilization is prevented. Only a few fruit cultivars are self-fertile and can produce viable seed with pollen from the same cultivar. Cresti et al. (1978) surveyed a number of cultivars for the type of fruit set. Results are summarized in Table 4.7.

In apples and sour cherries, some degree of self-incompatibility has been observed in cultivars generally classified as self-fruitful. In the self-

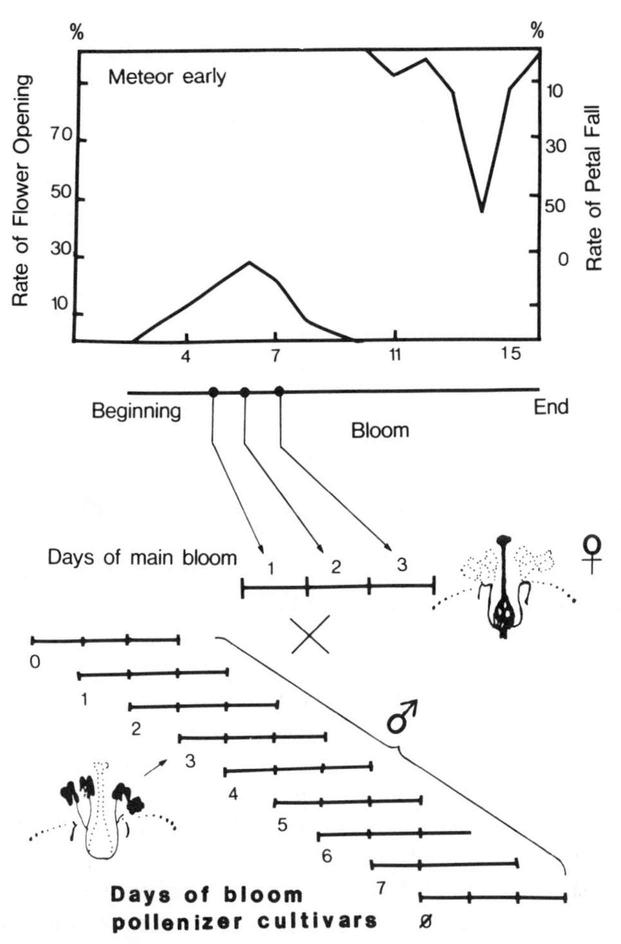

Figure 4.21 Scheme for determining pollenizer varieties. (Reprinted by permission from Nyèki, 1980.)

fruitful cultivars, cross-pollination usually sets heavier crops than self-pollination, although self-pollination also produces large enough crops that these cultivars are classified as self-fruitful. In sour cherries, Nyèki (1980) considers those varieties that set less than 1% fruit with self-pollen as self-incompatible, those that set 1–10% as partially self-fertile, and those that set fruit more than 10% as self-compatible. Some varieties such as 'Meteor Early' set fruit as a rate of 26–40% with self-pollen, 'Erdi Prolific' 15–20%, and 'Favorit' 12–24%. Some varieties are poor pollenizers even in cross-compatible situations. The utility of a pollenizer variety depends on the rate of pollen production. In apple an average anther contains 3500 pollen grains with a variation of over 6000 for

TABLE 4.7 Type of Fruit Set in Various Temperate Zone Fruit Trees

	Number of Cultivars		
Species	Self-compatible	Self-incompatible	Parthenocarpic
Apple	5	121	22
Pear	3	178	46
Peach	Most	2	—
Prune	26	26	—
Plum	3	Most	—
Cherry	1	26	—
Sour cherry	17	29	—
Apricot	15	4	—

[a] Reproduced by permission from Cresti et al., 1978.

'Delicious' and 'Stayman' and only 400 for 'Winesap'. Triploid apples contain 2.6–8.5 times more pollen than the average for the species. Plum anthers may contain 400–1800 grains and peach anthers 700–1300 depending on the variety. Although triploid cultivars contain more pollen, the pollen is less viable and it has less value as a pollenizer. Temperature has a great influence on the amount of pollen produced. After long cold winters few pollen grains are produced. Cold temperature during early spring not only can lower the number of pollen grains but also can decrease the vitality of those grains formed. High temperature during spring often results in sterile pollen. There are varietal differences in sensitivity to temperature. 'Red Astrachan' develops sterile pollen if spring temperatures reach 20°C, whereas 'Cox's Orange Pippin' is unaffected at 32°C (Nyèki, 1980). The quality of pollen depends on whether the pollen is spreading or sticks in clumps. The latter type spreads only with difficulty and insects cannot spread it widely. Bees, the most important pollinator insect, have a definite preference. They do not visit the pear variety 'Magness', which is male sterile and has no pollen. Although, if pollinated, it sets fruit, since very few bees visit it, it is a shy producer. Compatibility, quantity and quality of pollen, and bee preference are all known characteristics for most cultivars. Thus favorable conditions involving these characteristics should be secured before planting when the orchard is planned.

Effective Pollination Period

The concept of *effective pollination period* (EPP) was introduced by Williams in 1966 (Williams, 1970). In diploid cultivars of apple and pear, differentiation of ovule usually coincides with flower opening, and thereafter fertilization is possible. After pollination it takes a certain time for

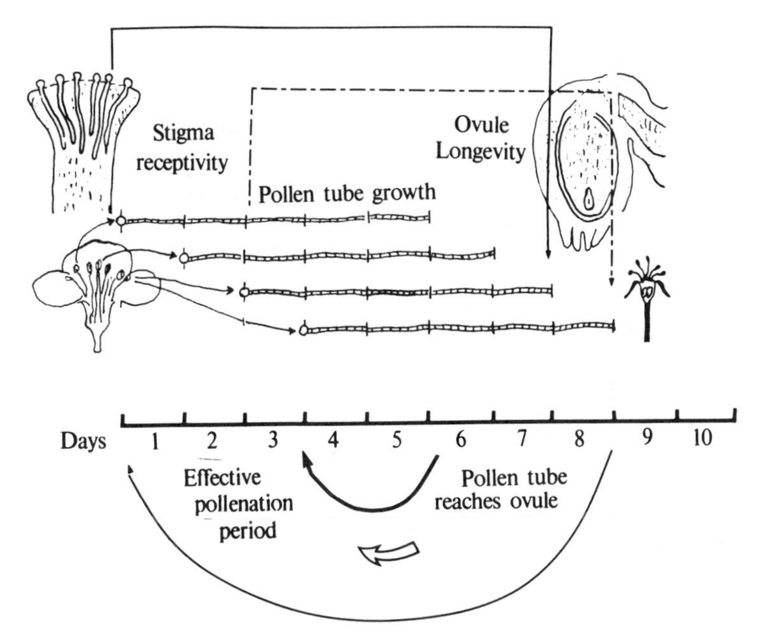

Figure 4.22 Effective pollenation period.

the pollen tube to reach the embryo sac. The transfer of male gametes, the fertilization, can take place only at this time. The mature ovule has a limited life. Effective pollination is limited to a period immediately following the opening of the flower. The duration of this period equals the longevity of the ovule minus the time required for pollen tubes to reach the embryo sac (Figure 4.22). Pollination that occurs after the EPP cannot result in fertilization because by the time the pollen tube reaches the ovule, degeneration has taken place in the placental tissues. Several modifying factors extend or shorten the time of pollen tube growth or ovule longevity. The relationship is modified in triploid varieties where the egg apparatus matures 2–3 days after the flower opens and ovule longevity is extended. Late summer nitrogen applications also effectively extend ovule longevity and EPP (Williams, 1970).

The ovules may degenerate soon after the flower opens or be viable for a longer period of time. Ovules of the 'Delicious' apple cultivar develop late and are known to degenerate soon (Hartman and Howlett, 1954). Ovule degeneration occurs later in the terminal flowers than in the lateral flowers. The ovule longevity of 'Delicious' was estimated as 5 days compared with that of 'Jonathan' (7–8 days), 'Calvil' and 'Melba' (10–12) days, and 'Dzsojsz' (14 days). In 'Windsor' cherries the rate of ovule degeneration is high. More than half of the ovules degenerated within 2

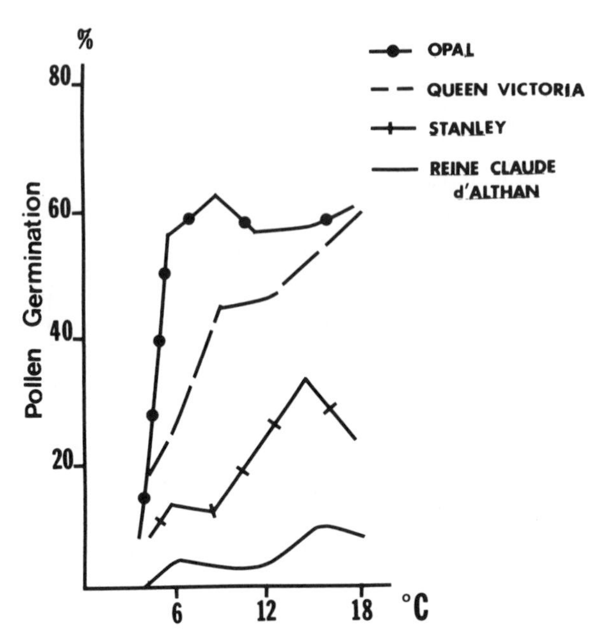

Figure 4.23 Effective of temperature on pollen germination of various plum varieties. (Adapted by permission from Keulemans, 1984.)

days and about 80% after 4 days in this variety. Even at the time of flower opening the 'Schmit' cherry variety contained only 38% of viable egg cells. Similar degeneration was found in the apricot variety 'Constant' (Nyèki, 1980).

Effect of Temperature on Fruit Set

Both the germination and growth rate of compatible or partially compatible pollen are dictated by temperature. The germination of pollen is usually not linear with temperature and greatly depends on the variety. Germination response to temperature of four plum varieties clearly indicates this point (Figure 4.23). In contrast, pollen tube growth responds to temperature in a much more linear fashion (Figure 4.24).

Williams and Wilson (1970) developed a temperature response index to allow the estimation of time required for pollen tube growth for apples. According to their concept, the daily growth index of the pollen tube should be compounded daily, considering the daily mean temperature, from 50% and from 90% bloom until 100% is reached. Daily indexes are determined by the length of pollen tube growth in percentage of stylar base. Indexes are listed in Table 4.8. Thus in a period with daily

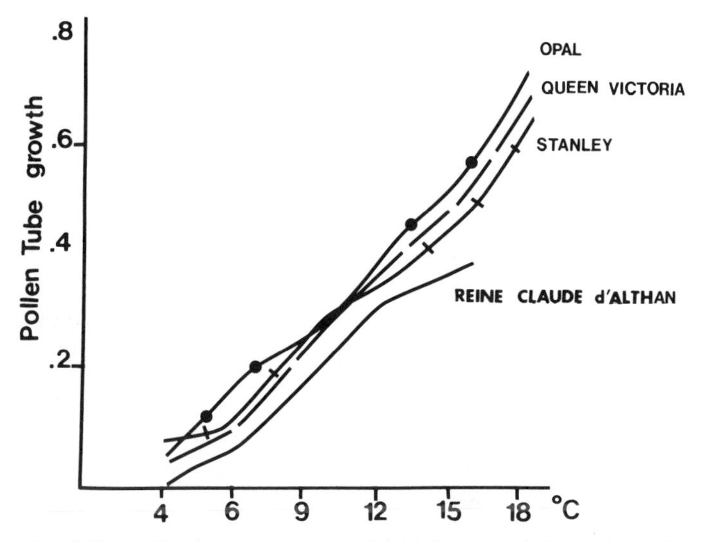

Figure 4.24 Effect of temperature on pollen tube growth in various plum cultivars. (Adapted by permission from Keulemans, 1984.)

TABLE 4.8 Apple Pollen Tube Growth Index

Mean daily temperature, °C	5	6	7	8	9	10	11	12	13	14	15
Pollen tube growth index (%)	8	9	10	11	12	14	17	20	25	35	50

temperatures of 10, 12, 10, 10, and 14°C pollen tube growth is expected to be 14 + 20 + 14 + 14 + 35 = 97%. Consequently, the pollen tube growth in this particular situation requires 5 days.

Events Important in Fruit Set

After and perhaps before pollination and fertilization, an important sequence of physiological events occurs if the fruit is to set and develop. The first prerequisite for a good set is the development of the so-called strong flower buds that takes place during the previous fall and requires a certain level of photosynthate and nitrogen supply. This is followed by the second important requirement, a certain temperature range during and soon after bloom to assure good pollination, pollen tube growth, and fertilization. The third requirement is for after fertilization, when the young developing fruit requires a relatively high level of photosynthate supply. If any of these factors are not satisfied, a poor fruit set will result, which usually means that soon after bloom most of the young fruit will fall. In general, from fruit set studies alone it is difficult to ascertain which

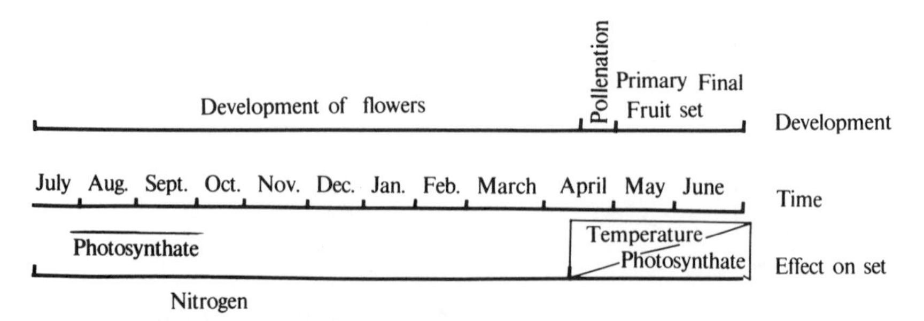

Figure 4.25 Factors influencing fruit set. (Reprinted by permission from Nyèki, 1980.)

factor is the major contributor to a good fruit set or which is responsible for a poor set. Good fruit growers usually try to assure satisfactory levels of factors under their control, namely, adequate N supply and protection of leaves for good photosynthate supply, especially during the spring. A schematic of the important controllable factors in fruit set over time is illustrated in Figure 4.25.

Early Fruit Drop

Early shedding of fruit is a regular feature of fruit set. The magnitude of fruit drop is a concern to the orchardist, and activities are usually directed to minimize the drop. In apples Murneek (1933), after an extensive study, described four waves of fruit drop at about 12–14-day intervals. The first drop occurs shortly after petal fall. It is by far the largest and tends to overlap with the second drop. The third and fourth drops are more conspicuous because the size of the dropping fruit is much larger. These two drops are collectively called the 'June drop' by the orchardist because they occur at the end of May or in early June. The first two drops occur in all varieties under all conditions. With some varieties, such as 'Delicious', these drops are very pronounced. The first drop is usually heavy after self-pollination and it is less after cross-pollination. June drop involves a complete abscision process including formation of ethylene. Prevention of ethylene production with ethylene synthesis inhibitors usually decreases fruit drop (Greene, 1980) but usually does not increase the setting of useful fruit. The fruit that has been prevented from dropping by ethylene synthesis inhibitors does not grow to full size, and in this sense fruit set cannot be improved by preventing abscision of inferior fruit. To be effective in inhibiting June drop, inhibitors must be applied soon after bloom, and the addition of hormones to the inhibitor mixture enhances their effectiveness (Figure 4.26).

The growth rates of many fruits fall behind the normal fruit soon after

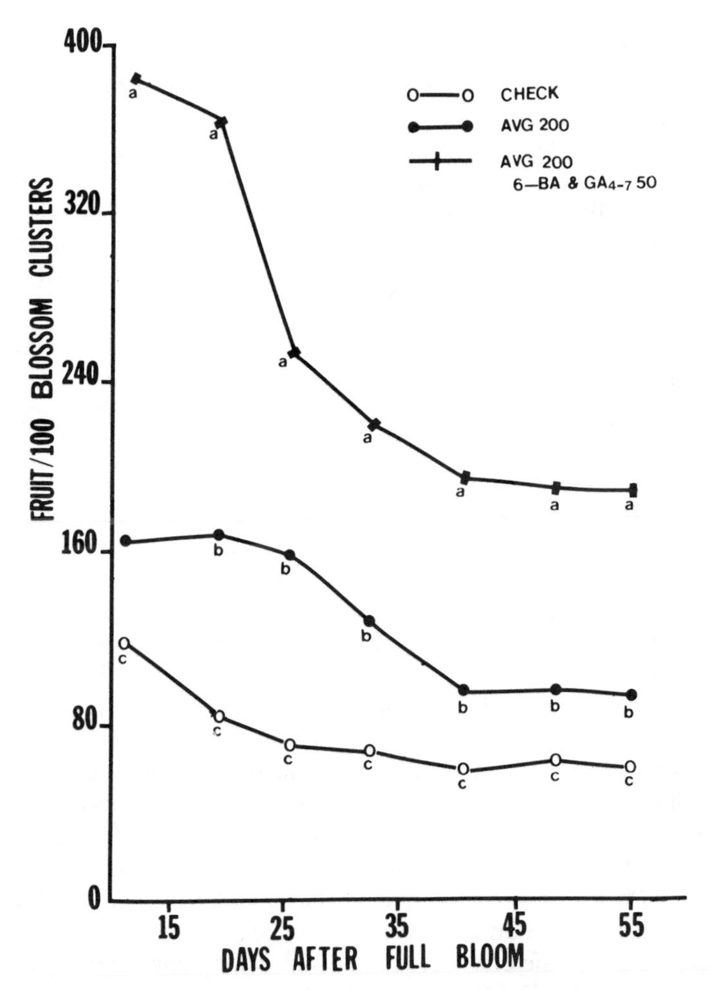

Figure 4.26 Effect of aminoethylvinylglycine (AVG) and growth regulators on fruit retention of apples. (Reprinted by permission from Greene, 1980.)

or perhaps beginning at the time of fertilization. However, there are fruits, especially peach, that are indistinguishable from other fruit and that still fall in June. Miller (1986), investigating the ethylene production of peach fruit, found that before June drop ethylene production was generally high and decreased to almost trace levels after June drop. Before the decrease the standard deviation of the mean for ethylene production was very large, indicating that some fruit produced ethylene to a much larger extent than others. The high ethylene production in some fruit easily explains the eventual drop of these fruits. What is not clear is the factor that triggers the high rate of ethylene production.

Parthenocarpic Fruit Set

Fruit can set and grow to full size without fertilization. Production of fruit without pollination is termed *parthenocarpy*. Although a small degree of parthenocarpy is observed in many fruit species, it is most widespread in pears among the temperate zone fruits. The rate of parthenocarpic fruit set is genetically determined among the pear varieties, but it is greatly influenced by environmental conditions. Pear varieties can be classified into four groups based on their ability to set parthenocarpic fruit:

1. Not parthenocarpic: 'Pap Pear'.
2. Parthenocarpic fruit is shed at the time of June drop: 'Hardy'.
3. Variably parthenocarpic: 'Bosc', 'Clapp Favorit', 'Diel', 'Madame du Pois', 'Oliver Serres', 'Bartlett'.
4. Consistently parthenocarpic: 'Arabitka', 'Hardenpont', 'Passe Crassane', 'Pringall'.

'Bartlett' is highly parthenocarpic in California, but this cultivar requires cross-pollination in the eastern United States to set fruit. Griggs and Iwakiri (1954) indicated that temperature was the determining factor in the parthenocarpic fruit set of 'Bartlett'. In most fruit the seed is the source of GA, perhaps in certain varieties the pericarp is also able to produce GA, and the GA produced by the pericarp is sufficient to induce and maintain fruit growth. The GA production is sensitive to temperature, which may explain the dependence of parthenocarpy on high temperature around bloom. The possible production of GA in the pericarp is supported by the fact that parthenocarpic fruit of a given variety usually ripens 1–1.5 weeks later than the seeded fruit. High-GA fruit or fruit treated with GA ripens later than untreated comparable fruit.

Opinions about the effect of frost in stimulating parthenocarpy are not uniform. Lewis (1942) stated that frost stimulated parthenocarpy, but Karnatz (1962) could not find such an effect. Mittempergher and Roselli (1966) found that varieties able to set fruit parthenocarpically are also likely to set parthenocarpic fruit after frost. In addition, several other varieties are able to set some fruit parthenocarpically after frost. They listed 'Conference', 'Sakesbirne', and 'Goodale' as such varieties.

Use of Hormone and Growth Regulator Sprays To Increase Fruit Set

That the set of unpollinated, self-pollinated, and cross-pollinated flowers of various fruit species can be promoted by the application of hormones or growth regulator mixtures is well established. However, when orchards have been sprayed commercially with hormone mixtures or single compounds, increases in yield have not always resulted. The setting response may depend on the time of application, the position of the flower within the cluster, and the age of the wood on which the cluster is situated (Goldwin, 1981). In apple the rootstock on which trees are grafted also may make a difference in the setting response to growth regulator sprays (Blasco et al., 1982). Goldwin (1983) reported that the set of hormone-induced parthenocarpic fruitlets was always significantly greater in the absence of pollinated fruitlets in the same cluster. He suggested that the presence of fertilized embryos created competing sinks that reduced the ability of the unpollinated flowers to respond to the hormone stimulus. In experiments where both fertilized fruit and parthenocarpically set fruit induced by hormones were present, the parthenocarpic fruit eventually abscised during June drop. The concept of flower quality within a cluster is not new. Such differences have been observed by Howlett (1927, 1931) and Detjen (1929) in open pollinated sets of apples and may be one of the major reasons for the June drop discussed previously. It is conceivable that in commercial orchards hormone-setting sprays are not successful unless cross-pollination is minimal, and it is unlikely that the induction of parthenocarpy contributes to improved yields in orchards with reasonable quantities of pollinated fruit.

The importance of the time of treatment in achieving success in fruit setting with growth regulator treatments has been stressed by Schwabe and Mills (1981), who reviewed most of the available literature. In 'Agua de Aranjuez' (Spadona) pear, the highest fruit set was achieved when hormonal GA_3 sprays were applied at the balloon stage of the flower compared with time of anthesis or petal fall. In this variety embryo sacs are not mature at anthesis, and they mature in an unsynchronous manner during the following 5 days. Petal fall occurred 12 days later. Parthenocarpic fruit set was produced with immature embryo sacs, and from this it appears that mature embryo sacs are not necessary for parthenocarpic fruit set (Herrero, 1983). Application of growth substances to the flower may stimulate the development of ovules, which in turn may produce growth substances to sustain fruit growth (Nitsch, 1970). Accepting this hypothesis may explain several effects arising from the importance of the time of application of growth regulators; the effect of rootstocks, which may create different hormonal environments; or effects of temperature during bloom, which may stimulate GA synthesis.

A variety of substances were applied to fruit trees usually as sprays to influence parthenocarpic fruit set. Treatments, concentrations used, and results obtained are summarized by Schwabe and Mills (1981). The compounds included gibberellins; cytokinins such as BA, kinetin, and Zeatin; auxins such as NAA and TIBA; and mixtures of compounds including GA_3 + NAA and GA_3 + DPU + NOXA; (DPU, diphenylurea; NOXA, z-naphthoxyacetic acid). Although apples synthesize several gibberellins, the gibberellin applied for the purpose of setting fruit may make a difference. Bukovac and Nakagawa (1967) compared the effectiveness of various gibberellins on the fruit set of apple. The order of effectiveness was GA_4, GA_2, GA_7 > GA_{13}, GA_{14} > GA_1, GA_3, GA_5, GA_{10}. All fruit treated with GA_6, GA_8, or GA_9 abscised within 5–6 weeks after treatment.

Schwabe and Mills (1981) expressed the opinion that in most situations more than one hormone is needed for inducing parthenocarpic fruit set. The role of auxin-type growth regulators in mixtures is to prevent abscission of young fruitlets, while cytokinins and possibly gibberellins may be involved in stimulation of cell division in fruit tissues. The role of gibberellins may also extend to promote the expansion of cells leading to the production of fruit of normal size.

FRUIT GROWTH

Rate of Growth

In general, fruit growth is divided into two main periods: cell division and cell enlargement. Both processes are involved in determining the rate of fruit growth in combination with the air space formation that occurs as intracellular spaces enlarge within the fruit. At the beginning, cell division dominates growth; from about 50 days after bloom, cell enlargement is the most important aspect of the overall enlargement of fruits. Air space formation occurs to a different degree and at different times in the various fruits. The size of fruit growth greatly depends on the cultivar. Early cultivars with a short period of fruit growth are generally smaller than fruit of later ripening cultivars of the same species that has a much longer period for growth.

The growth curve of pome fruit is sigmoidal. Stone fruits have a double-sigmoidal curve. Growth curves are illustrated in Figure 4.27. Apples and most stone fruits are harvested when they are mature. Maturity is interpreted for the present discussion as fully grown. In contrast, pears are harvested during the exponential portion of the fruit growth. If growers harvest pears a few days later, the harvested quantity is much greater because fruit grows rapidly at the time of harvest. Of course, later harvest decreases the storage quality of the fruit; thus determination of harvest time is a compromise between harvested size and storage quality. If left on the tree, pear also would complete the sigmoidal curve.

Stone fruits have a different growth pattern. The three distinctly different portions of the curve are designated Stages I, II, and III. Stage I is the cell division stage; Stage II, the slow growth period of the double-sigmoidal curve, coincides with the period of pit hardening, during which

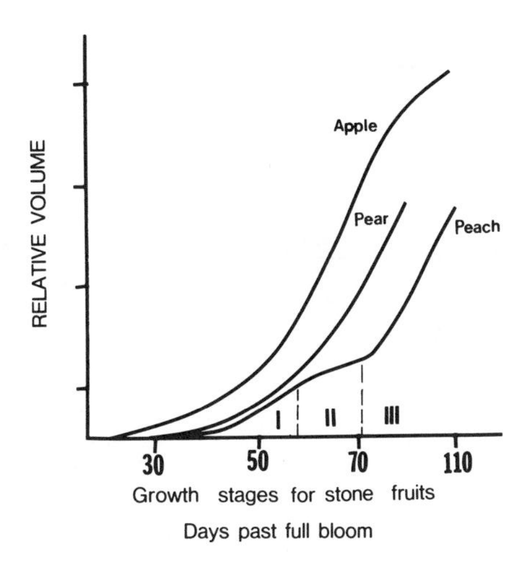

Figure 4.27 Seasonal growth curves to apple, pear, and peach fruit.

lignification of endocarp (stone) proceeds rapidly while mesocarp (flesh) growth is suppressed. Embryo development in peach starts about 50 days after bloom and is completed within 30 days. Mesocarp growth in early cultivars proceeds either simultaneously or immediately after embryo development is completed. Later maturing cultivars have a shorter or longer lag period before growth of the mesocarp is resumed in Stage III. The length of the lag period determines the time of maturity of the fruit (Tukey, 1933b). The growth pattern of cherry is identical to that of peach (Tukey, 1933a). The length of the lag period significantly modifies the appearance of the growth curve. The growth curve of early ripening cultivars is almost sigmoidal with only a little hint of the pit-hardening period. In contrast, the late ripening cultivars show a well-defined double-sigmoid curve (Figure 4.28).

Nectarines, mutants of peach, also have less defined sigmoidal growth curves (Fogle and Faust, 1975). The reason for such curves is not known. However, nectarines have no, or much smaller, air spaces in the fruit as opposed to peach. Obviously, the development of air spaces significantly contributes to the size of fruit. It is possible that the slow final swell of nectarine fruits is determined by the lack of air spaces in the fruit.

Role of Cell Division in Fruit Growth

The period from anthesis to the end of cell division is about 4 weeks for cherry, plum, and peach, 4–5 weeks for apple, and 7–9 weeks for pear (Westwood, 1978). Others calculate the period of cell division as percentage of total time required for fruit growth and maturation. Calculating

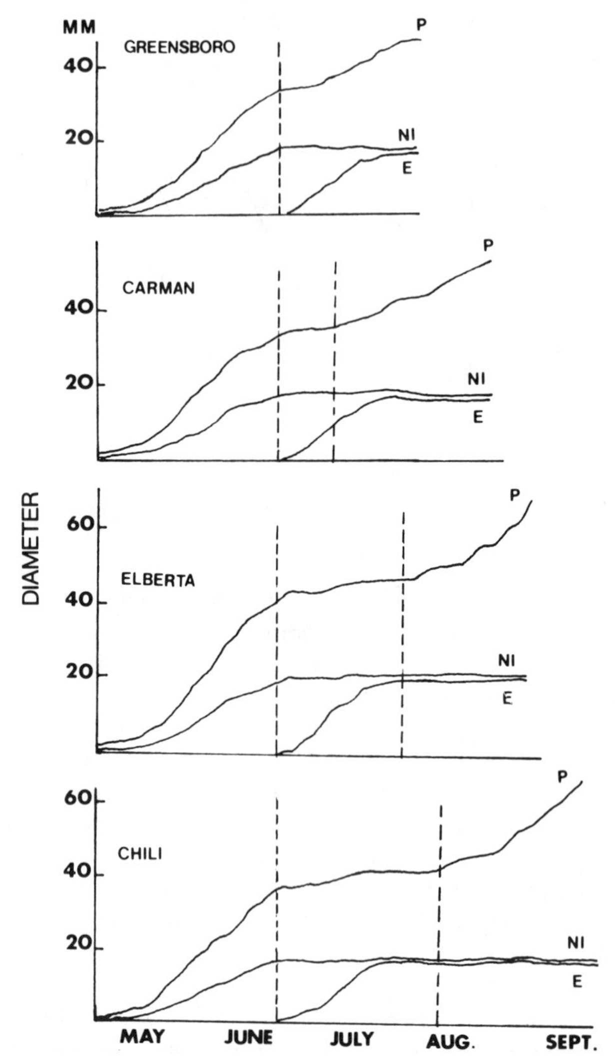

Figure 4.28 Growth curves from full bloom to fruit ripening of pericarp (P), nucellus and integuments (NI), and embryo (E) of four varieties of peaches ripening at different seasons. (Reprinted by permission from Tukey, 1933b.)

this way, the first 20% of the time of fruit growth is cell division in cherry, plum, peach, and apple but about 40% in pear. While the cell division period usually refers to cell division in the pericarp, one must recognize the variety of tissues in a fruit. Usually cell division in the skin continues almost throughout the entire period of fruit growth.

There are many millions of cells in a fruit. Martin et al. (1964) deter-

TABLE 4.9 Number of Cells per Apple Fruit ($\times 10^{-6}$)

	Australia	United States
'DELICIOUS'		
Small	—	46.6–50.1
Large	—	62.2–73.1
Very large	80.0	115.6
'JONATHAN'		
On year	46.0	64.0
Off year	38.0	—

mined the cell number of apples in Australia and Westwood et al. (1967) in western United States. Their results indicate that large fruit usually have more cells than small fruit (Table 4.9), and thinning the fruit early stimulates cell division. Apples grown in the United States contained more cells and the eventual cell size was smaller than in the Australian fruit of the same cultivar, where the final size was almost twice the size attained in the United States.

Natural cytokinins of the apple fruit play an important role in determining the eventual cell number of the fruit (Williams and Letham, 1969). The characteristic lobes of 'Delicious' apples do not develop under conditions where cell division is limited by warm temperatures soon after bloom. Treatment of 'Delicious' with cytokinin-type compounds (BA) induces lobes. Consequently, the inference has been made that the lobes of 'Delicious' are associated with the cell division of the fruit. The cell number of 'Golden Delicious' fruit that remained after thinning with NAA in comparison to unsprayed fruit indicates that the thinning spray removed the fruit that had fewer cells, which in turn also indicates a range in cell number within the same tree.

Induced by unknown factors some cells in the flesh of apple restart cell division after cell division has been completed. At this time the cell walls of the original cells are fully formed and normal cell division is not possible. The daughter cells remain within the confines of the original mother cell wall until it ruptures. Once initiated, proliferation occurs throughout the season (Miller, 1980). Such abnormal cell proliferation causes a disorder called cork spot and is prevalent in several cultivars including 'York Imperial' (Miller, 1980).

Fruit and Cell Enlargement

As indicated previously, cell enlargement follows the cell division period. Westwood (1978) lists factors that tend to increase cell size: few cells per fruit; light fruit set; adequate soil moisture; king bloom fruit in apples; excess nitrogen; late thinning; strong fruiting spurs, which in many ways is synonymous with high nitrogen; and excessive thinning, which essentially gives the same results as light fruit set and both involve a large leaf–fruit ratio.

As cells enlarge in apple, the pectin between the cells apparently degrade and the cells roll around with smaller and smaller portions of their walls connected. This allows the intercellular spaces to enlarge, and as a result the density of the fruit decreases. Air spaces are an important feature of fruit growth. Fruit with large air spaces develops to a large size with less carbohydrate input from the tree than a denser fruit. Moreover, fruit that floats in water can be handled in a water-floating handling system, which assures convenient, bruise-free handling. In contrast, fruit with a high density would sink, and water floating cannot be used for this fruit. The specific gravity of the fruit is a good measure of air spaces within the fruit. The specific gravity of apple decreases rapidly between 30 and 70 days after bloom and remains constant thereafter. The specific gravity of a fully developed apple ranges from 0.79 to 0.89 depending on size. The larger the fruit is, the lower the specific gravity is (Westwood, 1978). In a 100 g apple, the total air space occupies about 19% of the volume of the fruit. In a 400 g apple, the air space volume is 27% of the total (Westwood et al., 1967). The time of decrease in specific gravity indicates the time when intercellular spaces develop in the fruit. In peach and pear the extent of air space development is much less than in apple, and it occurs gradually during the entire period of growth (Westwood, 1962). The density of pear and peach is higher than 1 at the beginning of the season and decreases to slightly below 1 (0.99).

Fruit growth follows a diurnal pattern very similar to the diurnal pattern of the water potential of the tree. The magnitude of midday depression in fruit size (diameter) in apple is about 0.4 mm, and recovery occurs by about 1700 hr. Growth occurs during the night, and the magnitude of enlargement is about three times the magnitude of midday depression, 1.2 mm (Eggert and Miterall, 1967; Tukey, 1964). The fruit growth of apple is illustrated in Figure 4.29. The requirement of high turgor pressure for cell enlargement (Hsiao, 1973) and the previous discussion on the water status of the trees fully explain why growth occurs during the night. This is the time when turgor pressure conducive for cell enlargement exists in the fruit.

Fruit enlargement apparently greatly depends on the availability of gibberellins in the fruit. Apples treated with GA have elongated fruit (Williams and Stahly, 1969). Treatment with inhibitors of gibberellin synthesis results in small fruit (Stinchombe et al., 1983) and/or flattened fruit (Webster and Crowe, 1969). Warm conditions soon after bloom result in flat, oblate, fruit, whereas cooler conditions during bloom result

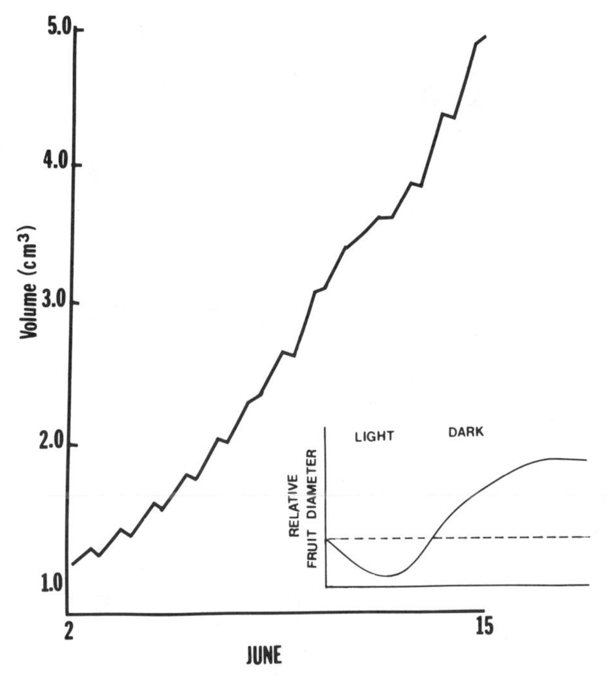

Figure 4.29 Growth of apple fruit during a 2-week period in June. Insert: diurnal change.

in more elongated fruit. Several mutations of the Delicious apple cultivar produce naturally more elongated fruit and lobed fruit than others. The requirement of cool temperature around bloom to produce long, large fruit seems not to correspond to the characteristic of 'Delicious' that in cool summer conditions when temperature does not exceed 20°C fruit size remains very small and for all practical purposes the fruit is unmarketable. This characteristic is even transmitted to the 'Empire' cultivar, which is a cross between 'Delicious' and 'McIntosh'. 'McIntosh' is able to produce large fruit in cool climates. It is likely that the length of fruit of 'Delicious' is cytokinin-related, which is produced in cool springs, whereas the overall size, which depends on cell enlargement and in turn on GA, is dependent on warm temperature.

The need for adequate levels of gibberellins is also expressed by the growth rate of epidermal cells of apples. If apples do not elongate early in their development, the surfaces often crack, which leads to the development of a disorder called *russeting*. The wax structure of apples greatly differs (Faust and Shear, 1972b). The wax on 'Golden Delicious' is unlike the wax of most cultivars. Normally wax is arranged in small platelets on the cuticular surface. The wax of the 'Golden Delicious' cultivar is amor-

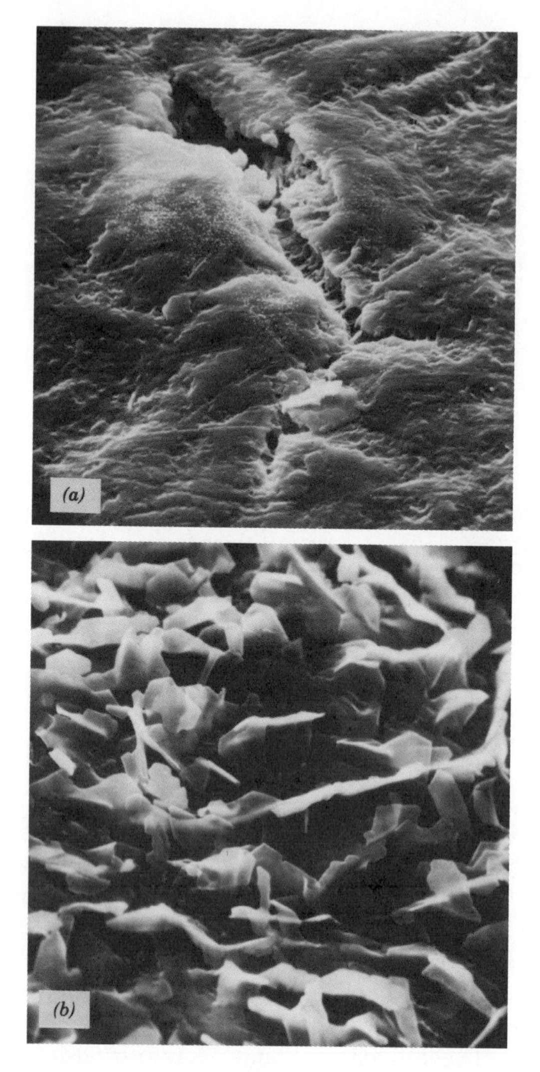

Figure 4.30 Wax patterns of (a) 'Golden Delicious' (×650) and (b) 'Delicious' (×2500) apples.

phous (Figure 4.30). The role of the wax is to protect the epidermal cells from excess uptake of water from their surface. When the fruit enlarges, the cuticular layer of the amorphous wax often cracks, water if present in the atmosphere enters into the epidermal cell, and the excess turgor pressure causes one or more cells below the crack to burst. Dead cells in the epidermis have been reported by Skene (1965). The neighboring cells are activated by the injury and form a layer of pheloderm on the surface of the fruit. Because of the appearance of the newly formed cork cells, the fruit becomes 'russeted'. In contrast, a fruit surface with a platelet type of wax does not crack; the wax platelets simply move apart and the spaces between the platelets are filled with new platelets (Faust and Shear, 1972a). The two types of waxes are illustrated in Figure 4.30. Spraying the fruit with GA_4 or GA_7 prevents russeting (Wertheim, 1982) while it elongates the fruit. The opposite effect is achieved by treating the trees with paclobutrazol, an inhibitor of gibberellin biosynthesis. Naturally water is necessary to cause excess turgor, which in turn causes the cells to burst. In dry climates russeting does not occur because nights are cooler, fruit elongates more, and water does not accumulate on the surface of the fruit unless overhead sprinkler irrigation is used.

The growth of stone fruits differs substantially from the growth of apple. A stone fruit resembles a sphere. Yet its growth is not constant in all directions. It enlarges more in one direction than another before it finally becomes a sphere. At full bloom the greatest diameter of the fruit is its length, while the cheek and suture diameters are approximately the same. During Stage I the suture diameter enlarges the most, during Stage II the fruit length increases the most, and during Stage III, depending on the cultivar, the cheek diameter increases the most. If enlargement of the cheek diameter is limited during Stage III, the appearance of the fruit of a given cultivar is flat. Several apricot cultivars have relatively flat fruit. Cherry is almost spherical. In plums there are elongated and round fruited cultivars. During Stage III, fruit size is largely caused by cell enlargement. Chalmers et al. (1983) estimated that Stage III is the time when 80–82% of the fruit growth takes place. Hsiao (1973) emphasized that cell expansion depends directly and physically upon turgor pressure, making this stage of growth of stone fruits in general and peaches in particular very sensitive to water supply.

The entire double-sigmoid growth curve of stone fruits can be shifted by increasing night temperatures during Stage I. Lilleland (1933) discovered that warming the atmosphere during the night in controlled conditions enhanced the growth of apricots, and this enhanced growth shifted the time of transition from Stage I to Stage II and Stage III without changing the time required for each stage (Figure 4.31).

Enhancing the early growth of stone fruits, especially peaches and cherries and to a certain extent plums, creates a special problem called 'split pit', associated with rapid rate of growth. The endocarp in the affected fruit separates into two halves or shatters to pieces, making the consumption of the fruit very inconvenient. Pit splitting has been associated with high N, high rate of irrigation, large fruit size, and early ripening cultivars (Woodbridge, 1978). Embryos of early ripening peaches and cherries either abort or fail to complete their development (Davidson, 1933; Tukey, 1933b). These cultivars are most prone for

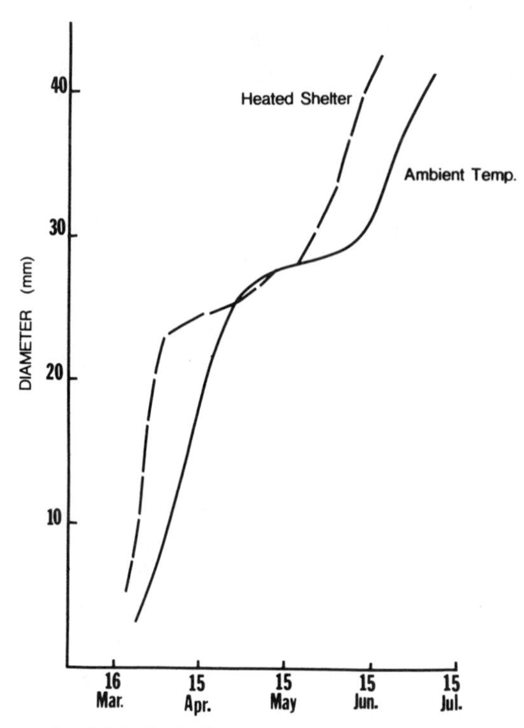

Figure 4.31 Growth of Blenheim apricot fruit in heated shelter and in ambient air. (Reprinted by permission from Lilleland, 1933.)

splitting the pit. The causes and detailed mechanism for the occurrence of this growth imbalance in the growth of fruit are not known.

Fruits are not uniform structures. Not only are there many types of tissues (vascular bundles, seed, stone, mesocarp cells, epidermal cells, etc.) but there are also various concentration gradients of nutrients, carbohydrates, proteins, and supposedly other compounds. Some of these concentration gradients are of utmost importance because either certain metabolic disorders develop where certain ingredients are the lowest or other disorders depelop where the ingredients are the highest. Concentration gradients of mineral elements in apple are illustrated in Figure 2.7. The peel and core usually have high concentrations of mineral elements and protein, whereas the flesh has much lower concentrations. This usually corresponds to the cell size found in these tissues. Occasionally the skin of nectarines cannot grow at the same degree as the flesh. This causes cracks on the skin (Fogle and Faust, 1976). Skin cracks, however, cannot be associated with the overall growth. Fogle and Faust (1976) examined pairs of early to late ripening cultivars of nectarines and found that the fruit growth rate measured by the diameter of the fruit is not

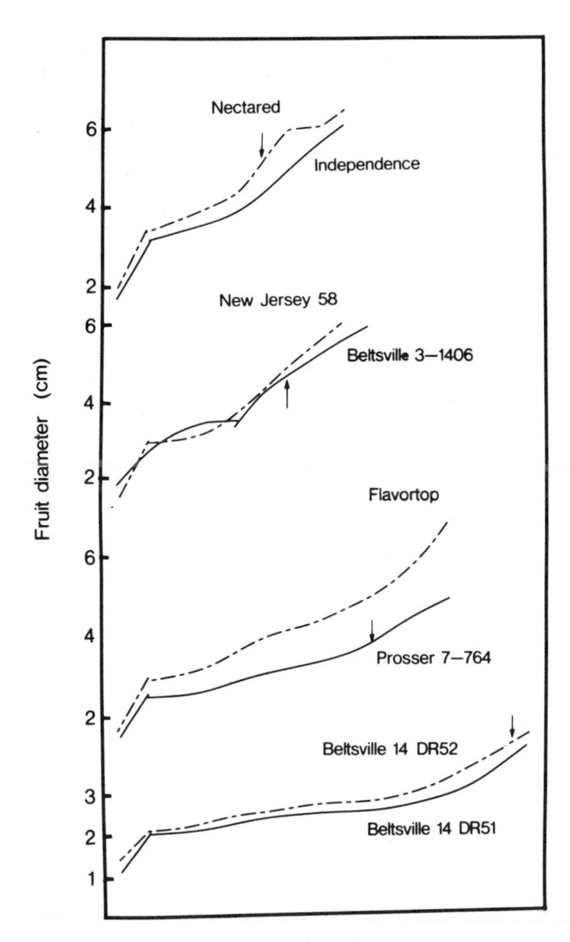

Figure 4.32 Growth rates, dates of field cracking from scanning electron micrograph of fruits of four contrasting pairs of nectarine clones. Arrow indicates the time and the fruit cracked. (Reprinted by permission from Fogle and Faust, 1975.)

correlated with cracking. The resistance or susceptibility to skin cracking is a cultivar dependent characteristics (Figure 4.32).

THINNING OF FRUITS

Under optimum conditions a tree will set a large number of fruits. It has been recognized by the earliest writers that a tree cannot support all its fruit to grow to commercially desired size and quality, produce sufficient number of flower buds for the following year, be able to support root growth, and accumulate enough reserves to be hardy to withstand the temperature stress during winter. The only adjustable activity among these processes is fruit growth. Fruit number can be reduced; the fruit can be thinned. The thinning process is considered by some to be the adjustment of leaf–fruit ratio to a desirable level. In some fruit species this is done by removing the excess fruit by chemical sprays; and term used for this process is *chemical fruit thinning*. In other fruit species the removal of fruit is more difficult and must be done mechanically or by hand. Chemical thinning can be accomplished at bloom time or during the early postbloom period. Bloom time thinning is easier to accomplish, but the fruit is often exposed to frost after bloom, and in many locations it is undesirable to thin until the frost danger is passed. Thinning serves slightly different purpose in different species. In stone fruit, without proper thinning large fruit size is almost impossible, especially with early ripening cultivars. In apples, in addition to a moderate increase in fruit size, the main purpose of thinning is to partially remove the source of GA, the seeds, which prevent flower bud formation. Thus in apples the role of thinning to maintain annual production is more important than the concomitant increase in fruit size. There are many physiological considerations to achieve successful thinning. These considerations are discussed in this chapter.

In 1729, Langley recognized that apricots and peaches were too numerous on the tree and he needed to thin them, taking away the least promising ones and preserving the best. During the 1920s, blossoms were thinned with small hand scissors or other small tools developed to ease the hand labor requirement then connected with thinning (Edgerton, 1973). Russel and Pickering (1919) were the first to recognize that the biennial bearing habit of apples could be controlled if fruit was removed at bloom time. Bagenal et al. (1925) recognized that 'lime sulfur' sprays used for controlling pests also induced fruit drop, but it was left to Auchter and Roberts (1934) to conduct the first experiments for selecting chemicals to thin fruit. The tar distillate they used (Auchter and Roberts, 1935) was a caustic chemical that burned the flowers but caused minimum injury to the vegetative tissues. In 1940 sodium, 4,6-dinitro-*o*-cresylate (DNOC), the first commercial preparation as a bloom thinner was introduced. A year later, in 1941, Burkholder and McCown reported success by using 1-naphthalene acetic acid (NAA) and 1-naphthaleneacetamide (NAAm) as postbloom thinners of apples. In 1958 carbaryl was tested (Batjer and Westwood, 1960), and the last compound, ethephon, was introduced into the thinning picture in 1969. While the chemical thinning of apples made considerable progress, the chemical thinning of peaches lagged behind.

Chemicals Used for Thinning

The main blossom-thinning chemical, sodium 4,6,dinitro-*o*-cresylate (DNOC), is a contact chemical. It is caustic and burns the flower parts, especially the pollen deposited on the stigma (MacDaniels and Hildebrand, 1940), and obviously fertilization and fruit set cannot take place. The chemical should be applied when a portion of flowers are still not open or they already have been fertilized and are not affected by the spray. For this reason DNOC needs to be applied 1–3 days after the trees reach full bloom. When bloom is extended, the timing of DNOC application is difficult. It is most often used at concentrations of 160–480 g/100 L of water. If rain occurs during bloom, the chemical can be reactivated and may cause excessive burning. This is why this chemical is not used in apple-growing areas where rain during bloom is frequent, including eastern United States (Williams and Edgerton, 1981).

Postbloom thinners can be used in all fruit-growing regions, and the compounds involved are NAA, NAAm, carbaryl, and ethephon. The Chemical formulas of the thinning compounds are given in Figure 4.33. There are cultivar differences in response to postbloom thinners. NAA is used on 'Delicious' at 2 to 5 ppm concentrations. Because it causes many fruits to remain on the tree but retards their growth, resulting in so-called buttons hanging on the tree, NAAm is not used on 'Delicious'. However, 'Golden Delicious' and some other cultivars are successfully thinned with NAAm at concentrations of 17–35 ppm. Some late summer and early fall cultivars are seldom thinned with NAA satisfactorily. If applied close to bloom, NAA causes clumping of fruit and distortion of the foliage, and if applied about 2 weeks after bloom, the growth rate of fruit is decreased. Delayed postbloom applications may also cause premature ripening and splitting of the small fruits (Hoffman, et al. 1955). For these reasons NAAm is preferred with several cultivars. Carbaryl is used on 'Delicious' at concentrations of 30–60 g/100 L of water, on spur-type 'Delicious' at 60–120 g/100 L of water, and on 'Golden Delicious' at 120–180 g/100 L of water. The increasing concentration range on the three cultivars illustrates the degree of resistance certain cultivars exert toward shedding their fruit. Postbloom thinners are applied 10–25 days after bloom. Pears are successfully thinned with NAA or NAAm at the rate of 15–20 and 10–15 ppm, respectively. The most effective time for thinning pears is 15–20 days after bloom (Williams and Edgerton, 1981).

A combination of sprays is often used to achieve the desired degree of thinning. On difficult-to-thin cultivars a two- or three-spray program is often necessary. The program usually starts with a DNOC spray at full bloom; if little thinning occurs, another spray is applied with 17–5 ppm NAAm 7–10 days after bloom. If thinning is still not satisfactory, a carbaryl spray can be applied 20–22 days after bloom. Such a program

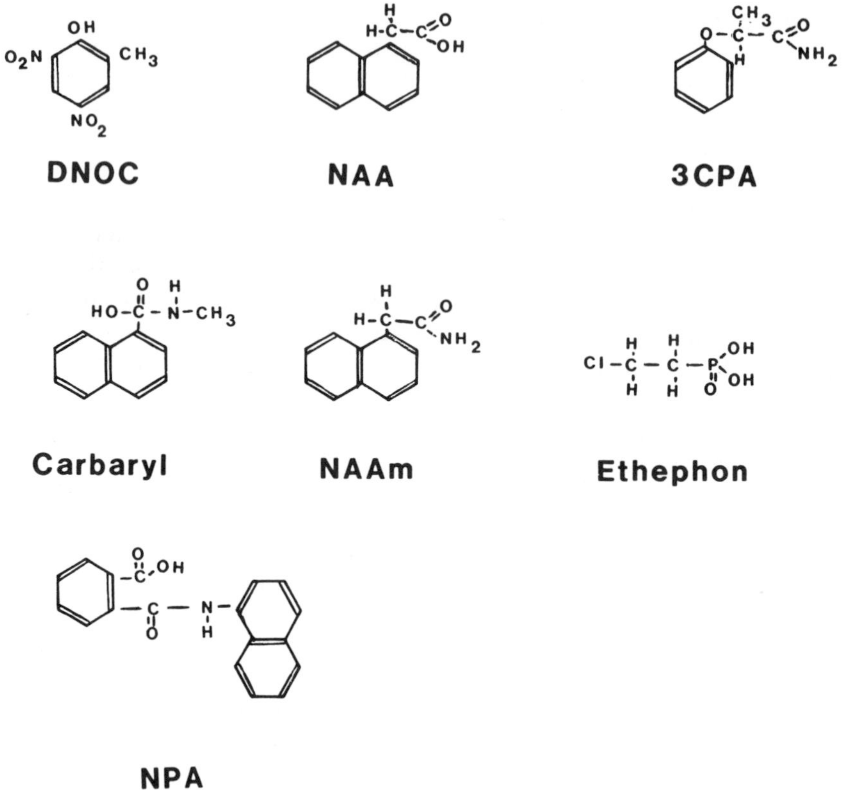

Figure 4.33 Chemicals used for thinning: DNOC, sodium 4,6-dinitro-*o*-cresylate; NAA, 1-naphthalene acetic acid; 3CPA, 2-(chlorophenoxy) propionamide; carbaryl, 1-naphthyl-*N*-methylcarbamate; NAAm, 1-naphthaleneacetamide; ethephon, 2-(chloroethyl)phosphonic acid; NPA, *N*-1-maphthyl phthalamic acid.

assures the desired level of thinning and prevents overthinning. On easy-to-thin cultivars such as Delicious one postbloom thinning spray is usually satisfactory.

Factors Influencing the Effectiveness of Thinning

Several factors affect the success of thinning. Most factors can be grouped into three categories: those affecting absorption of chemicals, those connected with the vigor of the tree, and those affecting the photosynthetic activity of the leaves during the period of fruit set. These factors are not surprising if one considers the conditions required for fruit set. If conditions are conductive to good fruit set, the same conditions create difficult-to-thin situations. In contrast, conditions that make fruit set poor also make thinning easy. Based on empirical data, Williams and Edgerton (1981) compiled a list of conditions that provide easy-to-thin and difficult-to-thin conditions. These conditions are reproduced in Table 4.10.

Conditions affecting the absorption of NAA and NAAm have been determined to considerable detail. The longer the time required to dry the spray and the higher the drying temperature, the more absorption occur. In practice, however, these two conditions rarely occur together. High temperature often creates short drying periods. Consequently, in the field most uptake occurs in wet, cool conditions (Westwood and Batjer, 1958). The absorption of NAA also increases when trees are exposed to frost a few days before the spray (Westwood and Batjer, 1960). Absorption is influenced also by the thickness of the cuticle, the wetting agent applied, the accompanying ions in the spray solution, and other factors all reviewed by Swietlik and Faust (1984).

Mechanism of Action of Thinning Sprays

Sprays of DNOC prevent germination of pollen deposited on the stigma (MacDaniels and Hildebrand, 1940). They also inactivate pollen tubes that have already grown halfway down the style of the pistil (Hildebrand, 1944). The set is reduced considerably when DNOC sprays are applied 1–3 days after full bloom at the time when most flowers are fertilized. Williams and Edgerton (1981) believe that the thinning action, at least partly, indirectly results from the temporarily altered physiology of the tree when both the flower and leaf tissues absorb the chemical, which induces physiological stress.

Using NAA may induce abscission of the young fruit immediately or delay the process and the fruit falls somewhat later (Batjer and Billingsley, 1964; Struckmeyer and Roberts, 1950; Teubner and Murneek, 1955). It is unquestionable that NAA is a promoter of ethylene synthesis (Yang, 1980), and this action is operating in apple fruit as well (Walsch et al., 1979). Application of the ethylene synthesis inhibitor aminoethoxyvinylglycine (AVG) completely eliminated fruit drop in several cultivars (Williams, 1979). Luckvill (1953) proposed that the NAA-induced abscission of fruit is mediated through the abortion of the seed in some of the fruit, which eventually abscise. Evidence has accumulated that seed abortion is not essential for the abscission of young fruit (Marsh et al., 1960). Especially work with carbaryl underlined this point. Carbaryl reduces the number of viable seeds in some but not all cultivars but thins all cultivars equally (Batjer and Thompson, 1961).

Williams and Batjer (1964) found that all of the carbaryl applied to the leaf or the fruit is deposited in the vascular tissue, and no activity was detected in the seed. These two facts, the lack of the growth regulator in the seed and the lack of correlation between seed abortion and fruit

TABLE 4.10 Conditions Affecting Ease of Fruit Thinning with Chemicals

Trees are easy to thin when:	Trees are difficult to thin when:
1. Fruit spurs on the lower, shaded inside branches are low in vigor.	1. Fruit is set on spurs in well-lighted areas of tree (tops and outer periphery).
2. Moisture or nitrogen supply is inadequate.	2. Trees are in good vigor with 12–18 in. of terminal growth with no mineral deficiencies.
3. Root systems are weakened by disease or physical damage.	3. Older trees in good vigor have a mature bearing habit.
4. Bloom is heavy, especially after previous heavy crops.	4. Light bloom or light fruit set occurs with the exception of young trees.
5. Young trees have many vigorous upright branches.	5. Trees have horizontal fruiting branches.
6. Thinners are applied to self-pollinated or poorly pollinated fruit.	6. Insects are active on cross-pollinated cultivars.
7. Fruit set is heavy on easily thinned cultivars such as Delicious.	7. Limbs and spurs have been slightly girdled following moderate winter injury.
8. Cultivars tend to have a naturally heavy June drop.	8. Biennial bearing trees are in the off year.
9. Fruit sets in clusters rather than as singles.	9. Fruit sets in singles rather than in clusters.
10. Bloom period is short and blossom thinning sprays are used.	10. Cultivars such as Golden Delicious and heavy setting spur types are to be thinned.
11. High temperature is accompanied by high humidity before or after spraying.	11. When ideal fruit growth occurs before and after time of thinning.
12. Blossoms and young leaves are injured by frost before or soon after spray application.	12. Low humidity causes rapid drying of the spray and decreased absorption occurs before and after spraying.
13. Foliage is conditioned for increased chemical absorption by prolonged cool periods.	13. Cool periods follow bloom without any any tree stress.
14. Rain occurs before or after spray application.	14. Endogenous ethylene production is low.
15. Prolonged cloudy periods reduce photosynthesis before or after application of chemicals.	15. Bloom is light and a high leaf–fruit ratio exists.

[a] Reproduced by permission from Williams and Edgerton, 1981.

abscision, suggested that seed abortion is not the primary cause of fruit abscision. That the growth regulator was found in the vascular bundles and the fertilized ovule needs a high rate of carbohydrate support to develop and the fruit to set prompted Williams and Batjer to propose that carbaryl interfered with transport and this is the major reason for its thinning action. Independently Teubner and Murneek (1955) concluded that embryo abortion and fruit abscission were two unrelated phenomena that may or may not occur at the same time.

Influence of Chemical Thinning on Alternate Bearing

The most outstanding feature of thinning chemicals on apples is their effect on alternate bearing. Previously we discussed that flower buds do not develop if gibberellin level is too high in the bud, and one of the major sources of GA is the seed of the developing fruits. Thinning sprays reduce fruit set early in the growing season, which also means the elimination of many seeds, reduction of GA levels, and reduction of the interference caused by GA in bud formation. For this reason thinning must be applied earlier than 60 days after bloom to have a positive effect on flower bud formation. This period sometimes needs to be much shorter. Spur-type 'Golden Delicious' trees must be thinned within 20 days after bloom to influence flower initiation (Williams and Edgerton, 1981).

Chemical thinning agents have been successfully combined with growth regulators such as daminozide and ethephon to help overcome biennial bearing (Williams and Edgerton, 1981). These growth regulators also can be considered as inhibitors of GA action. Another growth regulator, paclobutrazol, a GA biosynthesis inhibitor, is very active in promoting the flower bud formation of apples without thinning. This compound does not enter into the fruit, and seed GA synthesis is not affected by paclobutrazol (Steffens, personal communication, 1985). Consequently, it must be concluded that the overall GA level of the tree is important in preventing flower bud formation. Seeds are important only as a major contributing source of the overall GA level. Lowering the GA level of the tree has a positive effect on flower bud formation regardless of whether lower concentrations result from eliminating some GA produced by the seed or inhibiting GA produced by the tree. The combination of thinning sprays and daminozide have the combined effect of reducing the seeds present on the tree and modestly inhibiting the tree-produced GA. Thus the combined effect is more effective in overcoming biennial bearing and yet keeps the maximum amount of fruit on the tree.

The view described is underlined by the fact that pears in California do not need to be thinned in order to initiate flower buds. Bartlett pears are parthenocarpic at this location. In contrast, thinning is essential at other locations where the pears have seeds and presumably the overall level of GA in the tree is much higher. Thus adjustments need to be made, and that most commonly can be done by thinning. A detailed study of the GA levels of the tree after thinning and after combination sprays is not available.

Thinning Stone Fruits

The success of thinning with apples led to the early expectation that a similar success with peaches would soon follow. Peach thinning is much more difficult. Proebsting (1981) identified two characteristics that may be responsible for the lack of success in thinning peaches: Peach flowers tend to open at the same time, and the bloom development is not gradual as it is in apples. Peach fruit have only 1 seed from 2 ovules in comparison with apples, which have 5–10 seeds in 10 ovules. Proebsting believed that this greatly influences the chance of overthinning peaches, which is indeed often the case. Peach trees are not alternate bearers, and thinning is needed for size improvement only. However, thinning always reduces yield, and the economic incentive to adopt difficult-to-accomplish programs may not be great.

In a good year an 'Elberta' peach tree matures about 180 kg fruit, but this fruit only has an average diameter of 5.8 cm. To increase the average fruit size to 6.7 cm, the fruit must be thinned. If fruit is thinned at bloom, yield needs to be reduced to 90 kg to obtain the desired size; if thinning is delayed to 60 days after bloom, yield needs to be reduced to 45 kg to attain 6.7 cm size by harvest (Proebsting, 1981). The difference could be as much as 12.5 tons/ha, and for this reason early thinning of peaches is crucial to the success of the grower.

The spray DNOC can thin peaches successfully. However, peach bloom is much earlier than apple bloom, and thinning must be done when the effects of frost and fruit set are not known. For this reason growers do not adapt the program widely.

Naphthyl phthalamic acid (NPA) thins successfully if applied within 1 week after bloom, but it has not been adopted by the industry. Two other chemicals, 3-chlorophenoxy-propionamide (3-CPA) and ethephon, can also be used at the time of cytokinesis, when the endosperm becomes cellular. The effect of these chemicals has not been consistent with respect to rate, timing, and cultivar response. There is evidence that both yield beyond the level needed to increase fruit size by hand thinning. This would indicate that both chemicals are growth retardants and reduce the rate of growth of the fruit (Proebsting, 1981). Reduction of growth rate of prunes with ethephon has been demonstrated (Weinbaum et al., 1977). Ethephon also causes leaf yellowing, early leaf drop, and gummosis, but these undesirable effects can be overcome by application of GA_3 at 50–100 ppm with the ethephon (Young and Edgerton, 1979).

The importance of photosynthesis for fruit set and its partial inhibition as a means of fruit thinning in peach was shown by Byers et al. (1984). Shading of limbs (80–90%) from 31 to 41 days after bloom effectively thinned peaches. Applying terbacil, a photosynthetic inhibitor, 35 days after bloom effectively reduced photosynthates to very low levels after application. The rate of photosynthates within 2 weeks returned to slightly over half of the control rate. Inhibiting photosynthesis to such a rate overthinned peaches showing the importance of photosynthesis in fruit set. Fruit that remained per cm^2 of limb cross-sectional area was 13 for control; 5.6 for hand-thinned; 2.4 for 500 ppm terbacil-treated; and 0.1 for those exposed to 92% shade. (The ideal fruit humber would have been between 5 and 6.)

REFERENCES

Abbott, D. L. 1970. In Physiology of tree crops. L. C. Luckvill and C. V. Cutting (eds.). London: Academic Press.

Anderson, J. L., G. L. Ashcroft, E. A. Richardson, J. F. Alfaro, R. E. Griffin, G. R. Hanson, and J. Keller. 1975. J. Amer. Soc. Hort. Sci. 100:229–231.

Antsey, T. H. 1966. Proc. Amer. Soc. Hort. Sci. 88:57–66.

Auchter, E. C., and J. W. Roberts. 1935. Proc. Amer. Soc. Hort. Sci. 30:22–25.

Bagenal, N. B., W. Goodwin, E. S. Salmon, and W. M. Ware. 1925. J. Ministry Agriculture (Great Britain) 32:137–150.

Batjer, L. P. 1967. Agr. Chem. Dig. 9:6–8.

Batjer, L. P., and H. D. Billingsley. 1964. Washington State Agr. Exp. Sta. Bull. No. 651.

Batjer, L. P., H. D. Billingsley, M. N. Westwood, and B. L. Rogers. 1957. Proc. Amer. Soc. Hort. Sci. 70:46–57.

Batjer, L. P., and B. J. Thompson. 1961. Proc. Amer. Soc. Hort. Sci. 77:1–8.

Batjer, L. P., and M. N. Westwood. 1960. Proc. Amer. Soc. Hort. Sci. 75:1–4.

Bijhouwer, J. 1924. Meded. Landbhoogeseh. Wageningen, 27:1.

Blasco, A. B., S. M. El Soufaz, and J. E. Jackson. 1982. J. Hort. Sci. 57: 267–275.

Bowen, H. H., and G. W. Derickson. 1978. Hort. Sci. 13:694–696.

Broom, O. C., and R. H. Zimmerman. 1976. J. Amer. Soc. Hort. Sci. 101: 28–30.

Brown, D. S. 1960. Proc. Amer. Soc. Hort. Sci., 75:138–147.

Brown, D. S., and F. A. Kotob. 1957. Proc. Amer. Soc. Hort. Sci. 69:158–164.

Buban, T. 1969. Bot. Közlemenyek 56:251.

Buban, T. 1980. In Gyümölesfak viràgzàsbiologiàja ès termèkenyülese. T. Nyèki (ed.). Budapest: Mezogazd. Kiado.

Buban, T., and M. Faust. 1982. Hort. Rev. 4:174.

Buban, T., and C. V. Hesemann. 1979. Acta Bot. Acad. Sci. Hung. 25:53.

Buban, T., I. Zatyko, and I. Gonda. 1979. Kertgazdasag 11:17.

Bukovac, M. J., and S. Nakagawa. 1967. Experimentia 23:865.

Bumbac, E. 1975. Inst. Cerect, peutra Pomicult. Pilesti. 4:111–128.

Burkholder, C. L., and M. McCown. 1941. Proc. Amer. Soc. Agr. Sci. 38: 117–120.

Byers, R. E., C. G. Lyons, Jr., T. B. Del Valle, J. A. Barden, and R. W. Young. 1984. HortScience 19:649–651.

Cain, J. 1973. J. Amer. Soc. Hort. 98:357.

Chalmers, D. J., K. A. Olsson, and T. R. Jones. 1983. In Water deficits and plant growth, Vol. III, New York: Academic Press.

Chalmers, D. J., and B. van den Ende. 1975. Australian J. Plant Phys. 2: 623–634.

Chan, B., and J. Cain. 1967. Proc. Amer. Soc. Hort. Sci. 91:63.

Chandler, W. H., 1925. Fruit growing. Boston: Houghton Mifflin.

Chandler, W. H., M. H. Kimball, G. L. Philip, and W. P. Tufts. 1937. Calif. Agr. Exp. Sta. Bull. 611.

Chandler, W. H., and W. B. Tufts. 1933. Proc. Amer. Soc. Hort. Sci. 30: 180–186.

Cole, M. E., T. Solomos, and M. Faust. 1982. J. Amer. Soc. Hort. Sci. 107: 226–231.

Couvillon, G. A., and A. Erez. 1985. J. Amer. Sci. Hort. Sci. 110:47–50.

Couvillon, G. A., and C. H. Hendershott. 1974. J. Amer. Soc. Hort. Sci. 99:23–26.

Cresti, M., B. Donni, and M. Devreoux. 1978. 'La fertilita nelle piante da frutto.' In S. Sansavini (ed.). Bologna: Societa Orticola Italiana Sectione Frutticoltura.

Dennis, F. G. 1976. J. Amer. Soc. Hort. Sci. 101:629–633.

Detjen, L. R. 1929. Proc. Amer. Soc. Hort. Sci. 25:153–157.

Dutcher, R. D., and L. Powell. 1972. J. Amer. Soc. Hort. Sci. 97:511–514.

Edgerton, L. J. 1973. In Shedding of plant parts. T. Kozlowski (ed.). New York: Academic Press.

Eggert, D. A., and A. E. Mitchell. 1967. Proc. Amer. Soc. Hort. Sci. 90:1–8.

Erez, A., and S. Lavee. 1971. J. Amer. Soc. Hort. Sci. 96:711–714.

Erez, A., S. Lavee, and R. M. Samish. 1971. J. Amer. Soc. Hort. Sci. 96: 519–522.

Faust, M., and C. B. Shear. 1972a. Hort. Sci. 7:233–235.

Faust, M., and C. B. Shear. 1972b. J. Amer. Sci. 97:351–355.

Faust, M., C. B. Shear, and C. B. Smith. 1967. Proc. Amer. Soc. Hort. Sci. 91:69–72.

Faust, M., M. Zimmerman, and T. van der Zwet. 1976. Hort. Sci. 11:59–60.

Felker, C. F., and H. Robitaille. 1985. J. Amer. Soc. Hort. Sci. 110:227–232.

Feucht, W., 1961. Gartenbauwissenschaft 26:206.

Fogle, H. W., and M. Faust. 1975. J. Amer. Soc. Hort. Sci. 100:74–77.

Fogle, H. W., and M. Faust. 1976. J. Amer. Soc. Hort. Sci. 101:434–439.

Fulford, R. M. 1966. Ann. Bot. 30:209.

Goldwin, G. K. 1981. J. Hort. Sci. 56:352–354.

Goldwin, G. K. 1983. Acta Hort. 149:161–171.

Greene, D. 1980. J. Amer. Soc. Hort. Sci. 105:717–720.

Griggs, W. H., and B. T. Iwakiri. 1954. Hilgardia 22:643–678.

Grochowska, M. J., 1973. J. Hort. Sci. 48:347–356.

Grochowska, M., G. J. Buta, G. L. Steffens, and M. Faust. 1984. Acta Hort. 146:125.

Hartman, J. D., and F. S. Howlett. 1954. Bull. Ohio Agr. Exp. Sta. 745:1–64.

Hendershott, C. H. and D. R. Walker. 1959. Proc. Amer. Soc. Hort. Sci. 74:121–129.

Herrero, M. 1983. Acta Hort. 149:211–216.

Hildebrand, E. M. 1944. Proc. Amer. Soc. Hort. Sci. 45:53–58.

Hill-Cottingham, D., and R. Williams. 1967. J. Hort. Sci. 38:242.

Hoad, G. V. 1978. Acta Hort. 80:93.

Hoad, G. V. 1984. Acta Hort. 149:13.

Hoffman, M. B., L. J. Edgerton, and E. G. Fisher. 1955. Proc. Amer. Soc. Hort. Sci. 65:63–70.

Hoad, G. V., and S. Donaldson, 1977. Annu. Rpt. Long Ashton Res. Sta., 1977:39.

Howlett, F. S. 1927. Proc. Amer. Soc. Hort. Sci. 23:307–315.

Howlett, F. S. 1931. Ohio Agr. Sta. Bull. 483:89–93.

Hsiao, T. C. 1973. Ann. Rev. Plant Phys. 24:519–570.

Huet, J. 1972. Phys. Veg. 10:529.

Janick, J. 1974. Hort. Sci. 9:13–15.

Karnatz, A. 1962. Erw. Obstb. 4:31–33.

Kevlemans, J. 1984. Acta Hort. 149:95–101.

Klein, J. D., and M. Faust. 1978. HortScience 13:164.

Lamb, R. C. 1948. Proc. Amer. Soc. Hort. Sci. 51:313–315.

Lang, G. A., J. D. Early, N. J. Arroyave, R. L. Darnell, G. C. Martin, and G. W. Stutte. 1985. HortScience 20:809–811.

Lewis, D. 1942. Proc. Royal Soc. B. 131:13–26.

Lilleland, O. 1933. Proc. Amer. Soc. Hort. Sci. 30:269–279.

Link, H. 1976. Mitt. Obstbau 20:57.

Luckvill, L. C. 1953. J. Hort. Sci. 28:25–40.

Luckvill, L. C. 1970. *Physiology of tree crops*. In L. C. Luckvill and C. V. Cutting (eds.). London: Academic Press.

Luckvill, L. C. 1974. Proc. 19. Int. Hort. Cong. 3:237.

MacDaniels, L. H., and E. M. Hildebrand. 1940. Proc. Amer. Soc. Hort. Sci. 37:137–140.

Marsh, H. V., F. W. Southwick, and W. D. Weeks. 1960. Proc. Amer. Soc. Hort. Sci. 75:5–21.

Martin, D., T. L. Lewis, and J. Corny. 1964. Australian J. Agr. Res. 15:905–919.

Mielke, E. A., and F. G. Dennis. 1978. J. Amer. Soc. Hort. Sci. 103:446–449.

Miller, A., 1986. Ph.D. Dissertation. University of Maryland, College Park, MD.

Miller, R. H. 1980. J. Amer. Soc. Hort. Sci. 105:355–364.

Mittempergher, L., and G. Roselli. 1966. Riv. Ortoflorofruttic. Ital. 50:555–567.

Nitsch, J. P. 1970. In The biochemistry of fruits and their products. A. C. Hulme (ed.). New York: Academic Press. pp. 427–471.

Notodimedjo, S., H. Danoesastro, S. Sastrosumarto, and G. R. Edwards. 1981. Acta Hort. 120:179–186.

Nyèki, J., Gyümölcsfak viragzasbiologiaja es termekenyülése. Budapest: Mezögazdasagi Kiado.

Pethö, F. 1984. Alma, Mezögazdaségi Kiado, Budapest.

Proebsting, E. L. 1981. In Tree fruit growth regulators and chemical thinning. R. B. Tukey and M. Williams (eds.). WA: Washington State University Press.

Richardson, E. A., S. D. Seeley, and D. R. Walker. 1974. HortScience 9: 331–332.

Richardson, E. A., S. D. Seeley, D. R. Walker, J. L. Anderson, and G. L. Ashcroft. 1975. Hort. Sci. 10:236–237.

Robitaille, H. A. 1975. J. Amer. Soc. Hort. Sci. 100:524.

Russel, H. A., and S. Pickering. 1919. Science and fruit growing. York: Macmillan.

Sachs, R. M. 1977. Hort. Sci. 12:220.

Samish, R. M. 1945. J. Hort. Sci. 21:164–179.

Schmidt, S. 1973. Arch. Gartenbau 21:587.

Schwabe, W. W., and J. J. Mills. 1981. Hort. Abst. 51:661–698.

Seeley, S. D., and L. E. Powell. 1981. J. Amer. Soc. Hort. Sci. 106:405–409.

Shaltout, A. D. and C. R. Unrath. 1983. J. Amer. Soc. Hort. Sci. 108:951–961.

Singha, S., and L. Powell. 1978. J. Amer. Soc. Hort. Sci. 103:620–622.

Skene, D. S. 1965. Rpt. East Malling Res. Sta. 1964:99–101.

Smith, R. E. 1907. Calif. Agr. Exp. Sta. Bull. 184:249.

Spiegel-Roy, P., and F. H. Alston. 1979. J. Hort. Sci. 54:115–120.

Steffens G. L., and S. Y. Wang. 1983. Acta Hort. 146:135–142.

Stinchombe, G. R., E. Copas, R. R. Williams, and G. Arnold. 1983. J. Hort. Sci. 59:323–327.

Stokes, P., and K. Verkerk. 1951. Meded. Landb. Hogesch. Wageningen 50:141.

Struckmeyer, B. E., and R. H. Roberts. 1950. Proc. Amer. Soc. Hort. Sci. 56:76–78.

Suranyi, D. 1973. Acta Bot. Hung. 18:179.

Suranyi, D. 1976. Hort. Sci. 11:406.

Swartz, H., and L. Powell. 1981. Acta Hort. 120:173–177.

Swartz, H. J., A. S. Geyer, L. E. Powell, and S. C. Lim. 1984. J. Amer. Soc. Hort. Sci. 109:745–749.

Swietlik, D., and M. Faust. 1984. Hort. Rev. 6:287–338.

Teubner, F. G., and A. E. Murneek. 1955. Missouri Agr. Exp. Sta. Bull. No. 590.

Titus, J., and S. M. Kang. 1982. Hort. Rev. 4:204–246.

Tromp. J. 1982. J. Hort. Sci. 57:277–282.

Tukey, H. B. 1933a. Bot. Gar. 94:433–468.

Tukey, H. B. 1933b. Proc. Amer. Soc. Hort. Sci. 30:209–218.

Tukey, L. D. 1964. Proc. Amer. Soc. Hort. Sci. 84:653–660.

Visser, T., J. J. Verhaegh, and D. P. de Vries. 1976. Acta Hort. 56:205.

Walsch, C. S., H. J. Swartz, and L. J. Edgerton. 1979. Hort. Sci. 14:704–706.

Wang, S. Y., and G. Steffens. 1986. Phytochemistry 24:2185–2190.

Wareing P. F., and W. M. Frydman. 1976. Acta Hort. 56:57.

Wareing, P. F., and P. F. Saunders. 1971. Ann. Rev. Plant. Phys. 22:261–288.

Webster, D. H., and A. D. Crowe. 1969. J. Amer. Soc. Hort. Sci. 94:308–310.

Weinbaum, S. A., C. Giulivo, and A. Ramina. 1977. J. Amer. Soc. Hort. Sci. 102:781–785.

Weinberger, J. H. 1937. Proc. Amer. Soc. Hort. Sci. 37:353–358.

Weinberger, J. H. 1954. Proc. Amer. Soc. Hort. Sci. 63:157–162.

Wertheim, S. Y. 1982. J. Hort. Sci. 57:283–288.

Westwood, M. N. 1962. Proc. Amer. Soc. Hort. Sci. 80:90–96.

Westwood, M. N. 1978. Temperate-zone pomology. San Francisco: Freeman.

Westwood, M. N., and L. P. Batjer. 1958. Prod. Amer. Soc. Hort. Sci. 72:16–29.

Westwood, M. N., and L. P. Batjer. 1960. Prod. Amer. Soc. Hort. Sci. 76:30–40.

Westwood, M. N., L. P. Batjer, and H. D. Billingsley. 1967. Proc. Amer. Soc. Hort. Sci. 91:51–62.

Westwood, M. N., and D. J. Burkhart. 1968. Amer. Fruit Grower 88:26.

Williams, M., and E. A. Stahly. 1968. Hort. Sci. 3:68–69.

Williams, M. W. 1979. Hort. Sci. 15:76–77.

Williams, M. W. 1981. In Reproduction strategies in plants. W. Meudt (ed.). Allanheld Osmum. Granada. London.

Williams, M. W., and L. J. Edgerton. 1981. Fruit thinning of apples and pears with chemicals. U.S. Dept. Agr. Information Bull. 289.

Williams, M. W., and D. S. Letham. 1969. Hort. Sci. 4:215–216.

Williams, M. W., and E. A. Stahly. 1969. J. Amer. Soc. Hort. Sci. 94:17–18.

Williams, R. R. 1970. In Physiology of tree crops. L. C. Luckvill and C. V. Cutting (eds.). London: Academic Press.

Williams, R. R., and D. Wilson. 1970. Towards regulated cropping. London: Grower Books.

Woodbridge, C. G. 1978. J. Amer. Soc. Hort. Sci. 103:278–280.

Yang, S. F. 1980. Hort. Sci. 15:238–243.

Young, E., and L. Edgerton. 1979. Hort. Sci. 14:713–714.

Young, E., and J. Houser. 1980. J. Amer. Soc. Hort. Sci. 105:242–245.

Zeller, O. 1960. Pflanzensuecht 44:175–214.
Zimmerman, R. H. 1971. J. Amer. Soc. Hort. Sci. 96:404.
Zimmerman, R. H. 1972. Acta Hort. 34:139.
Zimmerman, R. H. 1976. Acta Hort. 56:219.
Zimmerman, R. H., M. Faust, and A. W. Shreve. 1970. Plant Phys. 46:839–841.

5

FACTORS DETERMINING SIZE OF FRUIT TREES

The size of the tree plays a central role in orchard management and production of quality fruit. It appears that dwarf trees have many advantages. Light penetrates a dwarf tree better, which favors photosynthesis, and the tree produces more and better quality fruit; the entire tree is closer to the sprayer, and the spraying can be adjusted better, usually resulting in reduced use of chemicals; and dwarf trees are easier to harvest, giving a greater management efficiency to the grower in the most labor-intensive operation of orcharding. Thus interest in dwarf fruit trees has been great for a long time.

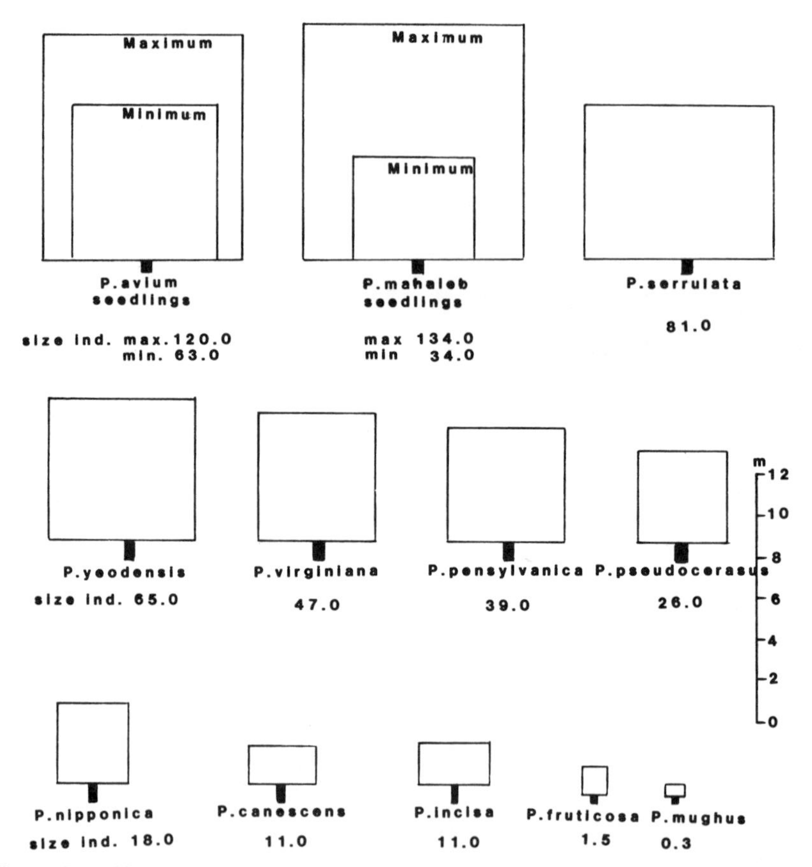

Figure 5.1 Natural variation in size among the *Prunus* species. (Reproduced by permission from Schmidt and Gruppe, 1988.)

If a large group of related species are examined for tree size, the overall differences are great. For developing new and useful types of cherry rootstocks Schmidt and Gruppe (1988) assembled a group of *Prunus* species, belonging to the *Eurocerasus* and *Pseudocerasus* sections, related close enough to cherries to be useful in their breeding program. They examined the size of trees grown for 22 years at the same location. The difference between the largest species, *P. mahaleb*, and the smallest species, *P. mughus*, using the two-dimensional cross section of the canopy (height times width) as an index of size was 446-fold. This enormous size difference illustrates the possibilities that genetic control of tree size can provide. A schematic illustration of size for the species used by Schmidt and Gruppe is provided in Figure 5.1. The overall differences in size are only useful if the components influencing size are determined. There are at least four components that contribute to the size of the tree: internode length, branching angle, location of branching (basitonic, mesotonic, acrotonic), and rate of growth regardless of the other three characters. These components should be treated independently and will be discussed in detail later.

Obviously, in dwarfing, the genetic characteristics of the tree are important. They determine the basic growth habit of the tree, which includes, as already illustrated, the genetic tendency to achieve a certain size. This genetic tendency for a given growth rate (vigor) can be influenced by many horticultural manipulations: grafting the scions on size controlling rootstock; withholding water from the trees by planting them on poor (low-water-capacity) soils, planting them close, or using controlled deficit irrigation; pulling the branches into horizontal position, thus changing the hormonal make up of the tree; or decreasing the GA production sites by repeatedly pruning the young branches or applying GA synthesis inhibitors. Before 1954 only the horticultural methods were used to control vigor and consequently tree size. In 1954 the first so-called 'spur type' 'Delicious' apple tree was discovered, which represented a moderate genetic dwarf with somewhat shortened internodes. The use of genetic dwarfs in fruit production dates from this discovery. However, even today, most of the dwarfing technology employs one or more of the horticultural techniques, and only in warm regions can spur types be used (they are not productive when temperatures are <20°C during the summer.) Thus in considering dwarfing the entire scope of techniques must be discussed. In general, the combination of genetic characters, which determine the growth habit of the tree, and one or more methods of vigor control additively control the size, resulting in semidwarf, dwarf, or extremely dwarf trees. The various dwarfing elements are summarized is Figure 5.2 and will be discussed in detail subsequently.

GENETIC CONTROL OF TREE SIZE

Several independent genetic characters determine tree size. These characters act independently or in combination.

Short Internode

Internode length is variable. In apple the normal internode length during spring flush of growth is about 30 mm. The commercially used spur-type cultivars have an internode length of 22–25 mm. Seedlings in the breeding plots using spur-type cultivars as parents have 15, 10, 5, and even 1 mm internode length types. Usually apple trees with internode lengths above 15 mm are productive, and those with internode lengths shorter

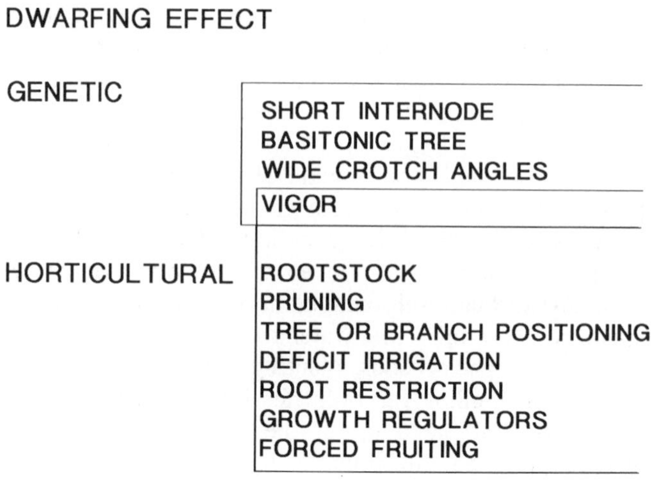

DWARFING EFFECT

GENETIC	SHORT INTERNODE BASITONIC TREE WIDE CROTCH ANGLES VIGOR
HORTICULTURAL	ROOTSTOCK PRUNING TREE OR BRANCH POSITIONING DEFICIT IRRIGATION ROOT RESTRICTION GROWTH REGULATORS FORCED FRUITING

Figure 5.2 Genetic and horticulturally imposable factors determining tree size.

than 15 mm, for unknown reasons, are not productive. In peaches the internode length also varies but with an emphasis on the two extremes, normal length internode (30 mm) and very short internodes (5–7 mm). The short-internode peaches are also productive. Short-internode plants can be found virtually in all tree fruit species. A short-internode pear tree is illustrated in Figure 5.3.

Internode length is not constant during the year. Regardless of the genetically determined length of the internode, spring internodes are longer, and those grown after early June are considerably shorter (Jaumien and Faust, 1984). The commencement of shortening of internode length coincides with the virtual disappearance of gibberellins in the xylem sap (Grochowska et al., 1984). However, there is no experimental evidence that xylem gibberellins determine the length of the internode.

All short-internode trees are characterized by dark green leaves that do not abscise until late in the autumn, much later than long-internode trees.

Wide-Angle Branching

There are many trees that produce branches at a 90° angle. This is a valuable characteristic not only because the tree produces horizontal branches, which is good for fruiting but also because it grows sideways rather than upward. Wide-angle branching can be mimiced by applying cytokinin-type growth regulators such as benzyladenine (BA) to the buds (Williams and Stahly, 1968). The resulting growth usually has a crotch angle close to 90°. Some wide-angle peach trees, commonly called com-

Figure 5.3 Short-internode pear tree, 25 years old, 1 m in height, internode length 1–2 mm.

pact trees and exemplified by 'Compact Red Haven', in addition to producing wide-angle branches also have very limited bud dormancy (apical dominance). All buds of this peach break, which may indicate that the tree is high in cytokinins or low in IAA. There have been no reports on cytokinin content in wide-angle trees.

In contrast, narrow-crotch-angle trees are common. Narrow crotch angle is associated with a relatively high IAA content (Grochowska, personal communication), which is a contrast in tree shape to the wide-angle trees.

Basitonic Tree Types

In certain trees apical dominance is very strong, and budbreak only occurs at the bottom of the branches. This basitonic branching habit

Figure 5.4 Basitonic and wide-angle trees with genetic low- and high-vigor characteristic: (*a*) basitonic; (*b*) wide angle. Left to right: low-, medium-, and high-vigor trees. Trees are 7 years old.

establishes a scaffold system very close to the base of the tree, and consequently the tree will be relatively small (Figure 5.4). The extreme of basitonic types is the apple cultivar 'McIntosh Wijcik,' which branches only to a very low extent (Looney and Lane, 1984). There are differences in branching habit in both major apple types, 'McIntosh' (Looney and Lane, 1984) and 'Delicious' (Walsh and Miller, 1984). Most often the basitonic trees are also short or relatively short-internode trees and usually develop many spurs, although the degree of spurriness depends on the cultivar.

In apple, spur-type cultivars greatly differ in growth and shoot development (Walsh and Miller, 1984). A summary of differences is presented in Table 5.1.

In both types of apple ('McIntosh' and 'Delicious') basitonic-type compactness positively correlates with free IAA levels (Grochowska

Figure 5.4 (*Continued*)

et al., 1984; Looney and Lane, 1984). Treatments with TIBA (an auxin transport inhibitor) indicated accentuated apical dominance in McIntosh Wijcik while it increased branch attachment angles. The inhibitor TIBA also increased the spur number of both compact and standard types (Looney and Lane, 1984). Looney and Lane (1984) were unable to relate basitonic character and spurriness to IAA levels, but they noted the relatively high ratios of free-to-total IAA in this type of tree. Selective removal of buds from 1-year-old shoots of spur-type trees indicates that the fewer buds remain, the more likely they are to develop as shoots rather than spurs (Looney and Lane, 1984). Thus the role of IAA in the development of basitonic trees is unclear.

Gibberellins are difficult to determine in plants in general and especially in fruit trees. In the McIntosh group 'Macspur', 'Morspur Mac,' (both moderately dwarf mutants of 'McIntosh' with 75% internode length,

TABLE 5.1 Growth Types in Spur-Type Apple Trees[a]

	Current Season's Growth		Second-Year Wood	
Cultivar	Short Internodes	Increased Lateral Bud Breaks	Increased Budbreak in Spring	Formation of Long Spurs (5–20 cm)
'Redchief Delicious'[b]	Yes	No	Yes	No
'Macspur'[b]	Yes	No	Yes	No
'Lawspur Rome'[c]	Slight	No	Yes	Possibly
'Granspur Granny Smith'[c]	Slight	Yes	Yes	Possibly
'Greenspur Granny Smith'[c]	Possibly	Yes	Yes	Yes

[a] After Walsh and Miller, 1984.
[b] Denotes basitonic types.
[c] Denotes acrotonic types.

moderately basitonic growth type, and high rate of spurriness), and standard 'McIntosh' all displayed comparable levels of polar gibberellins in their actively growing shoot tips, but McIntosh Wijcik (very dwarf seedling of 'McIntosh' with short internodes, very basitonic growth type, and high rate of spurriness) had only about one-third of the GA content of others (Looney and Lane, 1984). In a 'Delicious' × 'Golden Delicious' second-generation hybrid population GA-like substances measured by bioassay increased in the phloem but virtually disappeared by the end of June from the xylem sap. Gibberellin levels in both phloem and xylem were higher in the basitonic very short internode trees than in the still dwarf but more vigorous trees. As the season advanced, gibberellins shifted toward more polar substances. In general, the more basitonic dwarf trees had more polar GA-like compounds (Grochowska et al., 1984).

Cytokinin levels are comparable between 'Macspur,' 'Morspur Mac,' and standard 'McIntosh' but are four times higher in 'McIntosh Wijcik'. 'McIntosh Wijcik' also tolerates a high level of BA in shoot proliferation cultures, indicating that this cultivar has a way to deal with high cytokinin concentrations (Looney and Lane, 1984). Faust (unpublished) observed that the very spurry basitonic trees and the spur-type, very short internode mesotonic type trees in a seedling population suffered from drought expressed as shoot tip burning during dry periods. This effect could be duplicated in water cultures with seedlings supplied between 1 and 10 μg liter^{-1} BA in the nutrient solution.

Early reports on 'McIntosh Wijcik' indicated that elevated levels of ABA in its shoot tips may explain its growth habit. Later work (Looney and Lane, 1984) determined that ABA levels segregate in hybrid populations of 'McIntosh Wijcik,' and ABA levels are always lower in the more

TABLE 5.2 Characteristics Associated with Vigor in McIntosh Apple Trees[a]

	Trunk Cross-sectional Area (cm^2)		Degree of Spurriness (unit)[b]	Spread (m)
Strain	4 yrs	14 yrs	14 yrs	14 yrs
'McIntosh' standard	22.8	285	0.36	5.1
'Starkspur McIntosh'	16.3	219	0.48	4.6
'Starkspur Ultra Mac'	15.7	159	0.47	4.0
'Macspur'	19.8	222	0.59	3.7
'Morspur Mac'	14.9	156	0.66	3.5

[a] After Looney and Lane, 1984.
[b] The proportion of axillary buds on 13th-year growth to break dormancy and grow in the 14th year.

compact trees on both a per-bud or dry-weight basis. When 'McIntosh' and 'McIntosh Wijcik' trees were compared, the 'McIntosh Wijcik' trees had less free and conjugated ABA on a per-gram basis than the standard McIntosh trees but more on a per-bud basis since the buds of 'McIntosh Wijcik' are much larger than those of standard 'McIntosh.' The generally lower level of ABA in 'McIntosh Wijcik' probably reflects rather than causes the growing habit of this type of tree (Looney and Lane, 1984).

Vigor of Trees

Vigor is the rate of growth. The 'vigorous' tree grows fast; the low-vigor tree grows slowly. Consequently, the low-vigor trees are small compared to vigorous trees. As discussed before, vigor is a characteristic that has a double nature. It can be determined genetically, but it can also be influenced by cultural techniques. Here only the genetic part of vigor control is presented, and the physiological characters of the tree affecting vigor will be discussed later.

Although the growth habit of cultivars is not the same, comparisons of vigor still can be made in reasonable groupings. A comparison of several McIntosh types, all grafted on seedling rootsock, is presented in Table 5.2. It is clear from this type of data that the genetically determined vigor differences are substantial.

The preceding comparison suffers from the difficulty that the various size trees also differ in spurriness and degree of basitony. Changes in basitony and spurriness probably also influence fruiting, and these factors are associated with change in vigor. A much better comparison can be found in apple seedling plots where all the above-described tree types (short internode, wide-angle branching, basitonic) exist as seedlings with various heights. Here comparisons can be made without rootstock, com-

TABLE 5.3 Characterization of Cross Section of Stem at Internodes of 15–16 Stem[a]

	Low-Vigor Tree	High-Vigor Tree
Height of tree, cm	65.2	304.8
Length of terminal, cm	12.4	48.7
Number of internodes	41.1	40.2
Average length of internode, mm	4.1	12.0
C–Ph–X–P ratio, %	22 : 15 : 21 : 42	21 : 15 : 25 : 39

[a] Abbreviations: C, cortex; Ph, phloem; X, xylem; P, pith. After Jaumin and Faust, 1984.

paring trees with the same growing habits. This comparison indicates that vigor is indeed a separate genetic entity (Figures 5.4*a,b*). Because the growing habit of tops were identical with the exception of size, we assume that the character determining vigor resides in the root. More evidence for this supposition will be given when other methods controlling vigor are discussed.

Genetically determined vigor differences are well known in other fruit-producing species; pears, peaches, plums, and apricots. Since the data are very similar, they need not to be discussed here.

Genetically low-vigor trees are different from those developed as size-controlling rootstock. Size-controlling rootstocks usually have enlarged phloem compared to normal trees of the same species. The percentages of cortex, phloem, xylem, and pith remain the same regardless of the vigor of the tree when genetic dwarfs are compared (Jaumin and Faust, 1984). Details of the growth habit and xylem–phloem ratios of genetic dwarfs with low and high vigor are given in Table 5.3.

Occurrence of Genetically Determined Dwarfing Characteristics in Fruit Tree Species

The above-described genetic size-controlling characteristics are known to occur in many of fruit tree species and are summarized in Table 5.4.

It is very possible that some of the characteristics indicated as nonexistent will be discovered when horticulturist become more aware of such characteristics. The four basic genetic characteristics influencing tree size occur generally in woody species and are easily observable in forest or landscape tree material.

Growth Habit and Size of Cultivars

The growth habit and the size of a given cultivar is a mixture of genetically determined characteristics. There are some relatively small and re-

TABLE 5.4 Occurrence of Genetic Dwarfing Characteristics in Fruit Trees

	Short Internode	Wide-Angle Branching	Basitonic Type Tree	Vigor
Apple	x	x	x	x
Pear	x	x	—	x
Cherry (sweet)	x	—	—	x
Cherry (sour)	—	x	x	x
Apricot	x	—	—	x
Plum (domesticated)	—	x	—	x
Plum (oriental)	—	x	x	x
Quince	—	x	x	x
Peach	x	x	—	x

x indicates that specified characteristic has been observed and is present in that species.
Modified from Faust and Zagaja, 1984.

latively large cultivars in the same species. There are upright, spreading types, types with acrotonic or basitonic branching, and types with shorter or longer internodes. Because all the characteristics are mixed and their individual contribution to the growth habit is not clear, efforts in the past were simply to classify growth habits as they appeared to the observer. For example in apple, Brunner (1982) created four categories: weak growing nonupright ('Jonathan'), weak growing upright ('Starkrimson Delicious', 'Goldspur'), strong growing nonupright ('Golden Delicious'), and strong growing upright ('Starking Delicious' and 'Northern Spy'). Lespinasse (1980) developed four categories of apple cultivars based on their growing habits (Figure 5.5) exemplified by 'Starkrimson' (group 1), 'Reine des Reinettes' (group 2), 'Golden Delicious' (group 3), and 'Rome Beauty' or 'Granny Smith' (group 4).

In fruit-producing species other than apple the same principles apply. Scorza (1988) recognized six peach growth types: (1) dwarf, a tree rarely over 2 m tall with extremely short internodes and dense canopy and leaves that are very green and abscise late; (2) compact, trees have shorter internodes than standard trees but longer than dwarf trees, branching occurs with wide-crotch angles, and a large percentage of the vegetative branches develop into lateral branches with leaves that are still very green and abscise late; (3) semidwarf, precise growth habit of semidwarf tree is unclear, it is smaller than the standard tree but otherwise similar in appearance, and it has been obtained by hybridizing dwarf and compact trees; (4) pillar-type tree, the tree is extremely narrow, upright, and columnar and is rarely wider than 2 m; (5) weeping, as its name implies, the branches have no strength to stay in the upright position and they are hanging downward (present use of this type of tree is ornamental); and (6) spur type, fruiting occurs on spurs, which is not

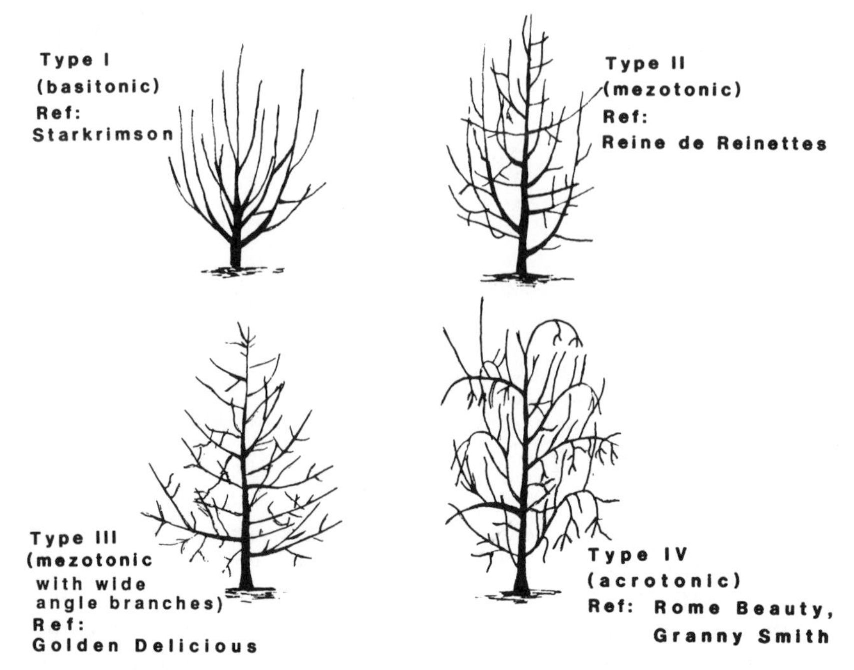

Figure 5.5 Four major growth types of apple cultivars. (Reproduced by permission from Lespinasse, 1980.)

common for peach but exists in almost any other *Prunus* species. Four of these types are illustrated in Figure 1.10.

Recognizing the genetically determined strength and type of the growth of a cultivar is important because other vigor control methods have to be established accordingly.

CONTROL OF VIGOR BY HORTICULTURAL MEANS

As mentioned before, vigor can be controlled by various means. Most of the horticultural techniques described in this section alter the physiology of the tree, resulting in dwarfing. It is very likely that the applied horticultural techniques invoke a different physiological response than those determined genetically. Hence the two effects, genetic dwarfing and horticulturally imposed physiological dwarfing, are usually additive.

Rootstocks

The ability of certain trees to dwarf the one grafted on it has been long recognized. For the early research in this area reviews by Roberts (1949),

TABLE 5.5 Number of Rootstocks of Temperate Zone Fruit Trees listed by Rom and Carlson (1987) and Their Proportion in Various Dwarfing Classes

	Number of rootstocks				
Species	Total	Large	Medium	Semidwarf	Dwarf
Apple	47	13	13	5	16
Pear	54	25	22	3	4
Peach	22	10	6	4	2
Plum	32	13	5	7	7
Cherry[a]	17	7	6	2	2
Total	172	68	52	21	31
Percentage of total		40	30	12	18

[a] Rootstocks introduced in Giessen, Germany.

Rogers and Beakbane (1957), Tukey (1964), Tubbs (1973), and Jones (1976) should be consulted.

In a book on rootstocks edited by Rom and Carlson (1987) authors give long lists of rootstocks developed for the various fruit species. Ferree and Carlson (1987) list 47 apple rootstocks developed at various locations for slightly differing purpose and used in the United States. They include the English (Malling and Malling Merton series), Polish, Russian, Michigan, and Canadian (Ottawa and Kentville) series and various other rootstocks. In apples the interest in dwarf trees dates back at least a half century. Therefore, it is not surprising that 34% of the rootstocks listed for apple are dwarfing. Lombard and Westwood (1987) list 54 pear rootstocks, the majority of which give a tree size ±20% of the standard, and only the related species produce dwarf trees as a rootstock. Quince *Cydonia oblonga* produces a pear tree that is about 50% in size of the standard and two *P. calleryana*, one *Amelanchier* and a *Sorbus* selection, produce pear trees in the 15–25% range of the standard. Among 22 peach rootstocks listed by Layne (1987), the vigor-inducing capacity was evenly distributed among five vigor classes. Of the 17 cherry rootstocks developed in Giessen, Germany, and described by Perry (1987), 2 give tree sizes about 25% of the standard while the others produce relatively large trees. In plums, among the 32 rootstocks listed by Okie (1987), 7 are definitely dwarfing and the others range from semidwarf to large. A summary of the effect of various rootstocks on tree size are given in Table 5.5.

In the nursery the growth of scions grafted on dwarfing rootstocks is almost as vigorous to the observing eye as those grafted on invigorating rootstocks. In contrast, the difference in growth is clearly visible when the trees begin to produce fruit a few years later. Thus one of the first questions raised about dwarfing was whether the rootstock is actually

TABLE 5.6 Length of Shoot (cm) Produced in First Year in Various Apple Rootstock Combinations[a]

	Scion			
Combination On Rootstock	M9a	M26	MM111	Mean Rootstock Effect
M9a	40	55	66	54
M26	69	85	93	82
MM111	74	94	111	93
Mean scion effect	61	78	90	

[a] Similar experiments were carried out with quince, plum, and cherry. Reproduced by permission from Tubbs, 1977.

dwarfing or is it inducing precocious fruiting and through fruiting altering the physiology of the tree, which limits growth. Effect of fruiting on growth is discussed in Chapter 1. Vyvian (1955) found that interactions between scion and rootstock were discernible in the earliest stages of growth of the compound tree. The interactions were apparent when he compared sums of weight of trees with reciprocal unions of 'unlike' combinations such as M9/M12 and M12/M9 with the sums of weights of trees of 'like' combinations such as M9/M9 and M12/M12 (M9 is dwarfing, M12 an invigorating rootstock). The actual growth rates of trees of like combinations were significantly higher than those of unlike combinations (Table 5.6).

Barlow (1971) approached this problem in a different way. He used 'Laxton's Superb' apple cultivar on dwarfing (M9) and vigorous (M16) rootstocks, deblossomed some trees, and allowed others to fruit. He also imposed two levels, low and high, of pruning on the rootstock combinations. His results clearly indicated that the deblossomed trees on dwarfing M9 rootstock were much smaller after 13 years than on the vigorous M16 rootstock, and pruning decreased size by more in the deblossomed trees, 22% for M9/LS (Laxton Superb) and 32% for M16/LS combinations, but 27 and 6%, respectively, in cropping trees (Table 5.7).

On the dwarfing M9 the greater fruitfulness and the precocity of cropping have resulted in the fruit taking over 70% of the total dry-matter accumulation largely to the expense of the structural elements of the tree. In contrast, on the vigorous M16 the crop took 40–50% of the dry matter, and consequently the framework could develop much more. Thus we must conclude that the rootstocks have two interrelated effects on size: one directly on growth, the other indirectly through enhancing cropping. Perhaps the enhanced cropping is dependent on decreased growth itself.

Hard pruning produces the expected results, namely, a smaller tree and a lower level of cropping. The strongest effect of pruning was observ-

TABLE 5.7 Accumulated Weight of 'Laxton's Superb' Apple Cultivar on Two Rootstocks after 13 Years (kg)[a]

Rootstock/Scion Combination	Deblossomed				Cropping				
	Total	Leaf	Wood	Root	Total	Leaf	Wood	Root	Fruit
			Light pruned						
M9/LS	27	5	20	2	20	1	4	0.5	14.5
M16/LS	139	24	90	25	106	9	34	8.0	55.0
			Hard pruned						
M9/LS	21	3	17	1	15	0.5	3	0.5	11.0
M16/LS	84	12	61	11	81	7	35	6.0	33.0

[a] Reproduced by permission from Barlow, 1971.

able on the cropping dwarf tree. Nevertheless, the effect of pruning is additive to the dwarfing imposed by the rootstock.

The root formation of many dwarfing rootstocks of apple is weak. They cannot stand free and require some type of support. A better anchorage can be maintained and dwarfing still can be achieved if a piece of stem of the dwarfing rootstock is grafted between the relatively vigorous rootstock and the scion. This raises the question of whether the rootstock or only a portion of the stem of the rootstock is necessary for dwarfing. A certain length of the 'interstem,' or 'interstock,' is necessary to have a dwarfing effect. The longer the stempiece, the greater the amount of dwarfing. The degree of dwarfing is increased if the entire stempiece is above the ground as opposed to the lower (rootstock–interstem) union being buried under the soil (Ferree and Carlson, 1987). Generally 15–20 cm of interstem of M9 produce trees slightly larger than if they were propagated directly on M9.

Even the bark of the dwarfing rootstock is sufficient to cause dwarfing. Insertion of bark of M26 (dwarfing stock) on a stock scion combination of 'Gravenstein'/MM111 resulted in a dwarfed tree (Lockard and Schneider, 1981). In this case only the bark was transferred, and similar to that of interstem the 20-cm bark graft was more dwarfing than the 10-cm bark graft. It is not necessary to use the bark of dwarfing rootstock to achieve a dwarfing effect. When the bark of a vigorous tree is removed from the stem and regrafted inverted, it produces a stronger dwarfing effect than the interstock or the bark graft of a dwarfing rootstock. Ten centimeter of Gravenstein inverted bark produces an effect similar to that achieved by 20 cm of bark graft of M26, a dwarfing stock, or a dwarfing interstock (Lockard and Schneider, 1981). When a strip of apple bark is inverted and grafted back into the stem, the phloem cells retain their original polarity, and in the reversed position translocation of certain compounds downward is decreased (Antoszewski et al., 1978). The reversed polarity of the cells in the bark strip may persist for more than 60 days (Antoszewski et al., 1978), but the new cells formed under the bark graft will have the usual polarity, and normal translocation will resume after 1 or 2 years (Sax and Dickson, 1956).

Similar to the dwarfing effect of the interstock, which has a stronger effect if it is above the soil line, the height at which the rootstock is grafted has an effect on dwarfing. The tree is considerably smaller if grafting is 25 cm above ground in comparison to a grafted tree when the grafting is only 10 cm above the soil level (Brunner, 1982). Delbard (1961) utilized this principle and grafted the trees at a height determined by the vigor of the scion. He recommends grafting vigorous scion cultivars higher than those of less vigorous scions, which according to him need to be grafted lower.

Throughout the years several aspects of dwarfing have been investigated. Most of these investigations involved apple. Perhaps there are more dwarfing stocks in apple, and clonal (uniform) stock is produced in this species, which makes the investigations much easier. Since altered hormonal control could be a simple answer for the changed tree size, several investigators have studied the hormonal content of or hormone transport in dwarfing rootstocks.

Indoleacetic acid is thought to play a role in the dwarfing of apple. The bark of the more dwarfing apple rootstocks cause a higher rate of auxin destruction (Gur and Samish, 1968), and auxin may accumulate just above the graft union of a dwarfing intermediate stock and in the lower portion of the enlarged tissue of a knotted interstock (Dickinson and Samuels, 1956). In inverted bark, downward translocation of IAA is also decreased (Antoszewski et al., 1978), strengthening the idea that IAA reaching the root is somehow important in creating the vigor of the tree.

A series of investigations focused on phenols of apple bark that may differ from rootstock to rootstock and also may play a role in IAA metabolism. In general, the bark of fruit trees contain many phenolic compounds. Phenolic acids that inhibit growth also enhance oxidative decarboxylation of IAA (Tomaszewski and Thimann, 1966). The monophenols, which contain a single hydroxy group, act as cofactors of IAA oxidase and are also growth inhibitors. In contrast, polyphenols (with two or more hydroxy groups) inhibit IAA oxidase and tend to enhance growth (Grochowska, 1967). Lockard and Schneider (1981) assembled information on phenolic compounds found in apple bark that have a regulatory effect on IAA content in plant tissues (Table 5.8).

Phenols are concentrated mainly in the bark of the tree and in the leaves. The concentration of phloridzin in the bark is very high. On a

TABLE 5.8 Phenolic Compounds Found in Apple Wood and Bark that Have Regulatory Effect on IAA Metabolism

Effect and Compound
IAA SYNERGISTS OR AUXIN–OXIDASE INHIBITORS
Caffeic acid
Chlorogenic acid
Ferulic acid
Protocatechuic acid
Sieboldin
Sinapic acid
IAA ANTAGONISTS OR AUXIN–OXIDASE COFACTORS
Ferulic acid
p-Hydroxybenzoic acid
p-Coumaric acid
Phloretic acid
Phloridzin
PHENOLS WITH NO APPARENT EFFECT ON IAA METABOLISM
Kampferol
Phloretin
Quercitin
Syringic acid
Syringic aldehyde
Vanillin

dry-weight basis concentrations as high as 12% were reported in bark and a fresh-weight basis 1% concentrations were found in leaves. The amount of phloridzin or total phenols is actually lower in the bark of dwarfing compared to the invigorating rootstock. Lockard and Schneider (1981) reported phloridzin in stem bark of 1-year-old trees of MM111 and M26 as 9.2 and 7.7%, respectively. Total phenols were similarly high, 23.8 and 20.4%, respectively, for the two rootstocks. In another set of comparisons between M9 and M16, again the stronger growing rootstock (M16) had more phloridzin in the bark. Martin and Williams (1967) and Hutchinson et al. (1959) reported that the phloridzin content of dormant terminal twigs of apple rootstocks was not related to the vigor they imparted to the scion.

Phloridzin in apple bark is highest just before spring growth resumes, decreases during rapid shoot growth, and increases again when the tree is close to dormancy (Martin and Williams, 1967). Considering the relative proportion of glycolytic pathway–TCA cycle activity, providing energy for growth, and the pentose pathway activity, which is usually high in high-sugar, no-growth conditions, the proportion of phenols in the bark during the growing season makes sense. When no growth occurs, the high pentose pathway activity produces a variety of phenolic compounds, in contrast to growth conditions when the energy-producing pathways are predominant, and less phenolic substances are synthesized.

Free phenols are seldom found in plants. Usually they form glucosides. Because of the high concentration of phenolic glycosides and the substantial quantity of sugar in the molecules, Barlow (1959) considered phloridzin as a storage material in apple. Similarly, Priestley (1960) and Kandiah (1979) listed phloridzin as part of the carbohydrate reserves of the tree.

The types of phenols found in bark are probably genetically determined. Beakbane (1956) found different phenolic substances above and below the graft union in pear–quince combinations. In such grafts phenols are apparently not transported through the graft union (Williams, and Beaklane 1956). Carlson (1974) and Yu and Carlson (1975) found 32 phenols in Mazzard and only 22 in Mahaleb rootstocks and concluded that phenols may be involved in graft incompatibility of these two cherry rootstocks.

Cytokinins are synthesized primarily in the roots and translocated through the xylem to the shoot tip or developing buds where they influence growth. Cytokinins synthesized in the young developing fruits and embryos are probably not important in tree growth (Lockard and Schneider, 1981). Information on cytokinins in grafted trees is relatively limited. Cytokinins have been identified in the xylem sap of apple (Greene, 1975). The cytokinin level in the xylem sap is the highest before budburst and declines prior to cessation of growth (Luckvill and Whyte, 1968). Supraoptimal temperatures, which are known to affect growth adversely, caused a decrease in cytokinin levels in leaves of ungrafted apple rootstocks (Gur et al., 1972). In dwarfing apple rootstocks cytokinin concentration in the xylem sap below the graft union is at least fourfold higher than above the graft union (Jones, 1984). This indicates that the graft union has a yet clearly undefined role to limit the transport of solutes, including cytokinins, in the xylem sap.

Gibberellins are major regulators of growth. Inhibiting the synthetic processes in the gibberellin pathway is a major way to alter tree growth and will be discussed later. Gibberellins are probably also produced by the root in addition to the production by the above-ground part of the tree. Grochowska et al. (1984) identified GA in the xylem sap of the tree in early spring. Robitaille (1970) concluded that there is little evidence to support a role for gibberellins in the rootstock effect. The major GA synthesis inhibitor, paclobutrazol, is only translocated upward, and when the trunk of the tree is treated, very little if any is transported to the root. Therefore, the GA effect on vigor must be assigned to the top of the tree.

TABLE 5.9 Solute Concentration (mg ml^{-1}) and Nutrient Constituents (ppm) in Dwarf Trees Below and Above Graft Union

	Total solute		N		P		K	
Rootstock	B	A	B	A	B	A	B	A
M9	3.4	3.0	230	203	33	31	174	147
M27 (2 times as dwarfing as M9)	3.0	2.1	219	167	26	14	180	123
3426 (4 times as dwarfing as M9)	2.6	1.6	—	—	—	—	—	—

[a] Abbreviations: B, below; A, above the graft union. After Jones, 1984.

Abscisic acid is a hormone that increases markedly in leaves of water-stressed plants and apparently is synthesized in the plastids of the leaves (Milborrow, 1978). Abscisic-acid-like compounds were reported to be higher in dwarf apple trees than in vigorous ones (Robitaille and Carlson, 1978), and ABA was higher in the dwarfing apple stocks themselves (Yadava and Dayton, 1972). In contrast, Feucht et al. (1974) found no correlation between ABA level in *Prunus* species and dwarfing potential. Dwarfing rootstocks often have lower water potential (Olien and Lakso, 1984), which could cause the observed differences in ABA levels between the dwarf and vigorous trees. Therefore, it is unlikely that ABA plays a major direct role in rootstock-induced dwarfing.

The limitation of transport of xylem sap constituents from the rootstock to the scion has been cited as a possible role in dwarfing. In apple the graft union of the dwarfing rootstock or interstock with the scion appears to deplete the solutes of the xylem sap, and this effect increases with the dwarfing effect. The depletion is nonselective, affecting all nutrients (Table 5.9) (Jones, 1984). Studies with interstock indicated that the effect of interstock on sap composition is created by the upper graft, the graft between the dwarfing interstock and the scion. This is analogous to the graft union between the dwarfing rootstock and the scion. Water flow is not affected by the graft union (Jones, 1984).

Modifying Branch Angles as a Means of Vigor Control

Vigor is higher at the top of the tree and much lower close to the base of the canopy. The growth rates of branches in various positions within the tree are illustrated in Figure 6.1. Branches bent at the bottom of the tree will impose more dwarfing than bending to the same degree at the top of the tree. Conversely, to achieve a uniform vigor control throughout the tree, the lower branches should have a more upright angle and the upper branches a more horizontal angle. Applying the principles of the reaction of the tree to bending, an infinite number of production systems can be developed.

Bending changes the hormonal makeup of the branch, and this creates a desirable effect but also causes some problems. The closer the branch is pulled to the horizontal, the more the apical dominance is eliminated, and it is more likely that the buds located at the base of the shoot will break and produce a strong upright shoot. There are two ways to deal with such shoots: prune them out, which is labor consuming, or fit them into the tree design. Pulling the branches down has been utilized in many canopy designs, and the possibilities of this technique are almost endless. In Figures 5.6 and 5.7 a number of canopy designs are illustrated. The traditional palmette (Figure 5.6B) uses branches forced into 45° angles. This keeps the branches partially vigorous and minimizes pruning, but the vigor usually concentrates at the top of the tree, and it is virtually impossible to grow fruiting lateral branches in the lower empty triangle of the tree. In contrast, the Haag trellis (Figure 5.6A) uses all horizontal branching, which decreases vigor more, but because of the high rate of budbreak, there is increased pruning. The modified palmette (Figure 5.6C) uses a combination of the two designs. The lower scaffolds of the modified palmette are developed in a 45° angle, but the higher branches are closer and closer to the horizontal. This design distributes the growing strength of the tree. The lower branches, in which the vigor is less, are given a chance to grow more by the steeper angle, which allows more vigor. In comparison the higher branches, which are more vigorous and need more control, are pulled closer to the horizontal to control vigor. The size control imposed by the palmette method can be further augmented by twisting the secondary branches and bending them downward, as is illustrated in the palmette with bent branches (Figure 5.6D). A new version of the palmette developed for mechanical harvesting is the Tatura trellis primarily used for peaches (Figure 5.6H).

The strong upright growth developing at the base of the branch after it is pulled into a horizontal direction can be dealt with by developing a second tier of branches using this upright shoot. Often a third tier of scaffolds is developed from the shoots of the second tier. Such systems are the Vincent trellis (Figure 5.6E), which employs this method in both directions, and the Lepage trellis (Figure 5.6F), which applies this technique only in one direction. A modern version of the horizontal trellis is the Lincoln trellis developed for apples (Figure 5.6G). The horizontal scaffold systems are more successful in controlling vigor in cool climates. If the day temperature does not exceed 20° and the night temperature is around 16°C, the upright regrowth is relatively limited. If the day temperature is in the 30° range and the night temperature exceeds 25°, the

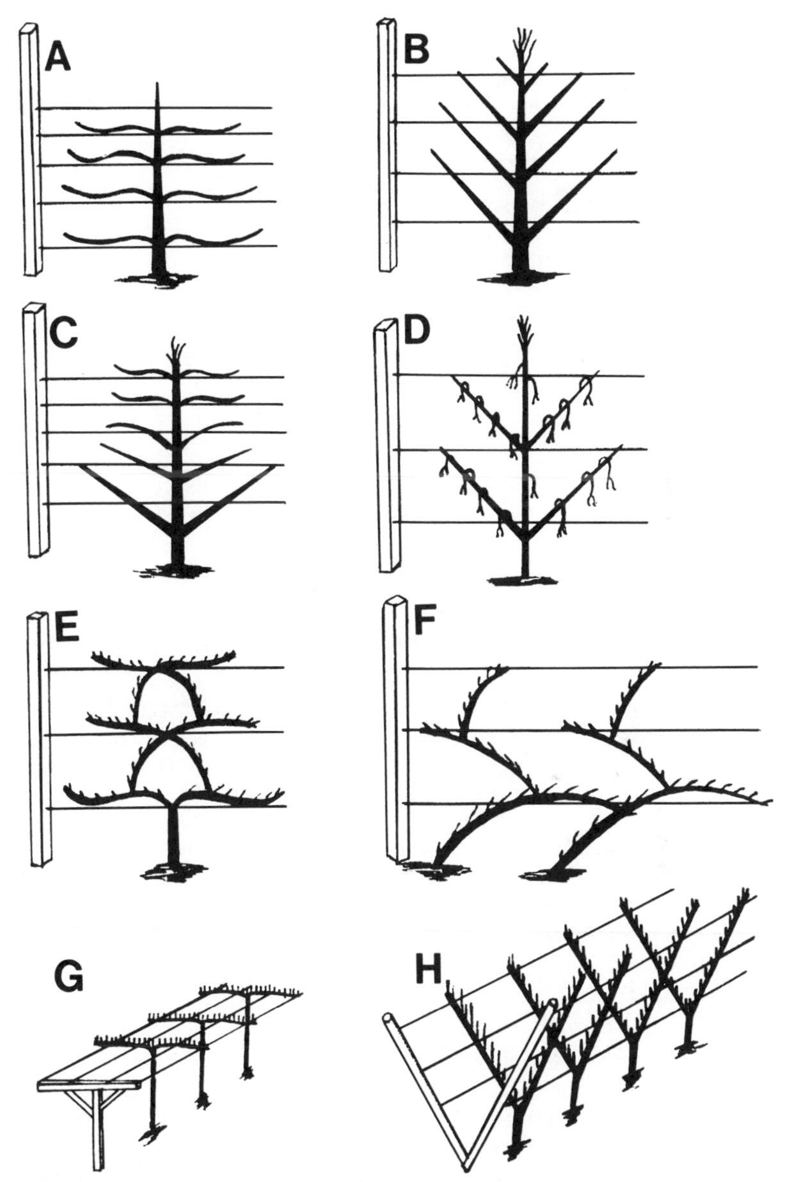

Figure 5.6 Tree designs utilizing bending the branches into desired position to dwarf the tree: A, Haag trellis (Schmitz-Hübsch and Furst, 1959); B, palmette (Stark, 1974); C, modified palmette (Stark, 1974); D, palmette with bent branches (Fejes et al., 1972); E, Vincent trellis (Fejes et al., 1972); F, Lepage trellis (Coutanceau, 1962); G, Lincols trellis (Dunn and Stolp, 1987); H, Tatura trellis (Chalmers and van den Ende, 1975).

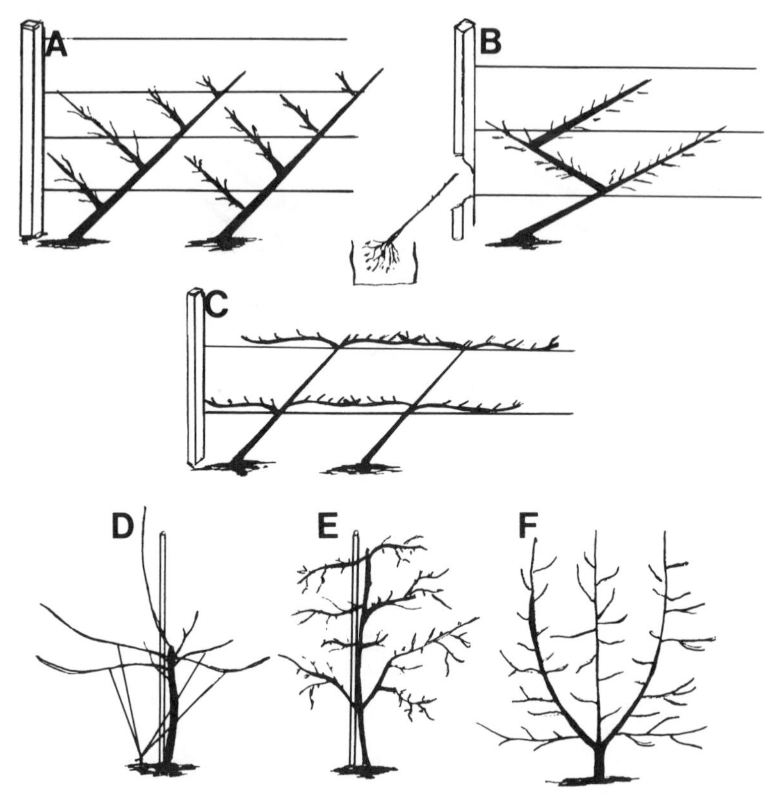

Figure 5.7 Tree design utilizing slanted tree planting: A, Marchand trellis (Soubeyrand, 1965); B, Bouche-Thomas trellis (Bouche-Thomas, 1953); C, seabrook trellis (Brunner, 1982); D, young tree with its branches pulled down to force the development of wide-crotch angles; E, slender spindle (Wertheim, 1978) utilizing dwarfing rootstock, wide-angle branching, and pruning; F, triple axis palmette (Lespinasse, 1980).

regrowth is too strong and a system such as the Vincent or Lepage trellis is inevitable.

In some cases, especially with stone fruits, bending the branches and controlling the subsequent growth is not easy. Thus it is desirable that the entire tree is started in a slanted position. Consequently, the trees are planted in an angle, as illustrated by the inset of Figure 5.7A. There are slight variations on how to handle the tree after it is planted, and the Marchand trellis (Figure 5.7A), Bouche-Thomas trellis (Figure 5.7B), and Seabrook trellis (Figure 5.7C) only differ in detail techniques, not in the principles.

The most often used technique is the so-called slender spindle, which

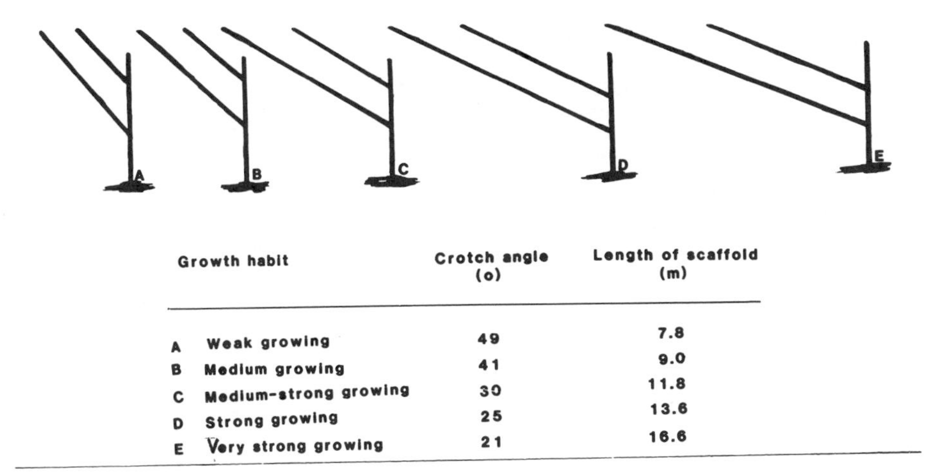

	Growth habit	Crotch angle (o)	Length of scaffold (m)
A	Weak growing	49	7.8
B	Medium growing	41	9.0
C	Medium-strong growing	30	11.8
D	Strong growing	25	13.6
E	Very strong growing	21	16.6

Figure 5.8 Relationship between bending of branches and tree vigor. (Reproduced by permission from Delbard, 1961.)

starts by tying down the scaffold to the horizontal position and follows up with pruning that accentuates the horizontal branching. This technique is well suited for apples and pears (Figures 5.7D,E). When the tree is strong growing and there is little possibility to control its growth by pulling the branches down or pruning them, the solution is often to let the tree grow and flatten the secondary growth. This principle is applied in the triple-axis palmette (Figure 5.7F).

Delbard (1962) realized that the success of bending greatly depends on the genetic vigor of the tree, and bending needs to be adjusted accordingly. If the genetically determined vigor of the cultivar of rootstock–scion combination is low, then the branches are developed to a higher angle (49°), and as the vigor increases, the branch angle is decreased to 21° (Figure 5.8). Delbart combined the changing of the branch angle with varying the planting distance of the trees. The stronger the tree is, the further apart Delbart places the trees. Thus he allows space for the scaffolds to develop to the maximum length.

Effect of Fruiting on Tree Size

Increased cropping is usually accompanied by decreased vegetative growth in the same year with reduced dry-weight increment to the leaves, stems, trunks, and roots. The treshold crop below which no such effect occurs is very low (Jackson, 1984). Although cropping increases photosynthetic efficiency, the production of additional photosynthates is not great enough in heavily cropping trees to prevent the decrease in vegeta-

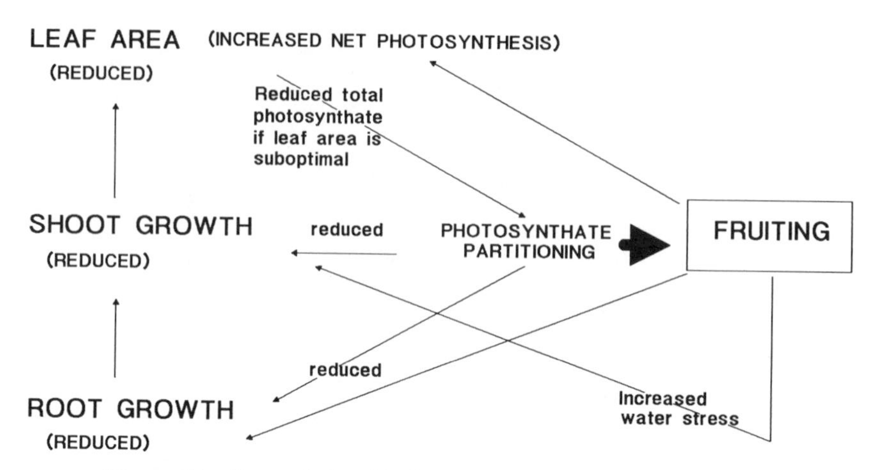

Figure 5.9 Interrelationship between cropping and tree growth.

tive growth. The effect on shoot growth appears to be a function of total fruit yield, not the number of fruit per se. Root growth is very sensitive to effects of cropping and can be almost totally stopped (Jackson, 1984).

In some fruits the total length of shoots growing in the year following the heavy crop is reduced to about the same extent as is shoot growth in the cropping year. The cumulative effect of cropping is such that it leads to great size differences between trees deblossomed from planting and those allowed to crop normally. Details of cropping on distribution of carbohydrates are discussed in Chapter 1 and previously in this chapter under rootstock effect.

In addition to utilizing most of the carbohydrates, cropping also has an effect in lowering the water potential of the tree, which has a secondary effect in decreasing shoot growth. Obviously the two effects act synergistically in reducing tree growth. Which of the two effects is more important depends on the environment to which the trees are exposed. Maggs (1963), Avery (1970), Head (1969), Hansen (1980), and Barlow (1964, 1975), all working in cool, low-light environments (England and Denmark) with apples, place the emphasis on carbohydrate partitioning. Chalmers et al. (1984) working in a high-energy environment (Australia) with peaches has shown that the photosynthate component is less important than the water status of the tree, and in peaches the crop itself cannot decrease tree size considerably. A schematic diagram of the relationship between fruiting and tree growth is presented in Figure 5.9.

The residual effect of cropping extending for at least 1 year after the heavy crop has been explained by depleting reserves (Rogers and Booth, 1964). Probably the effect of fruiting on root growth is more important in the residual effect than generally considered. Heavy cropping effectively stopped root growth in apples on dwarfing rootstock (Avery, 1970). Maggs (1964), Head (1969), and Hansen (1980) all reported that cropping essentially stopped root growth in apple. The magnitude of the residual root effect may be more severe in nonirrigated conditions where the size of the root system is more critical for water and nutrient uptake.

Effect of Water Stress on Tree Size

Several means used to manipulate plant growth involve, in principle, the manipulation of plant water potential. Naturally this is true when reduced irrigation is used or the water supply of the tree is limited by close planting or planting on low-water-capacity soils. The effect of reduced xylem water potential exerts its effect on growth in two possible ways: through (1) a phenologically based mechanism and (2) organ- or tissue-specific responses to xylem water potential.

The phenologically based mechanism requires that on a whole-tree basis the predominantly vegetative growth is followed by predominantly fruit growth in each season or vice versa. This allows an irrigation strategy that suppresses one form of growth without suppressing the other. Chalmers et al. (1984) developed such an irrigation system for 'Golden Queen' peach and pears. The system is discussed in detail in Chapter 3. In both cases only 25–30% of fruit growth coincides with shoot growth. In other fruit cultivars the overlap is considerably more, and deficit irrigation cannot be applied to decrease shoot growth.

Different plant processes respond differently to reduced water potential. Cell expansion depends directly on turgor pressure. Photosynthesis and translocation occur throughout the day, at least in peach, even at water potentials low enough to make the plant shrink. The distribution of photosynthate that occurs during the low-water-potential period has an important bearing on the subsequent distribution of water (and ensuing growth) during the period of high water potential that occurs during the night (Chalmers et al., 1984). Irrigated peach trees carrying a full crop load rarely have a sufficiently high water potential to permit growth of vegetative tissues during the day (Chalmers and Wilson, 1981). Within a normal irrigation cycle tree water potential recovers fully at night but tree growth occurs only when fruits are not growing rapidly. Net growth of peach fruit, however, occurs each night, indicating that the fruit will grow first when the tree's water potential is recovered. Fruits are sinks for photosynthates translocated from the leaves throughout the day (Chalmers et al., 1975). An osmotic potential accumulated in the fruit will attract water first from the apoplast when xylem potential is low during the day due to deficit irrigation. Although the theory that specific organs respond to water stress differentially is attractive (Chalmers et al., 1984), it has not yet been established experimentally.

Effect of Plant Density on Tree Size

The growth of vegetative shoots of peaches is strongly inhibited by severe root restriction (Richards and Rowe, 1977a). It does not make much difference how the restriction is created. In Richards and Rowe's experiments the restriction was created in a limited soil layer over a nonpermeable clay. If trees are planted close together, the peach tree root develops under the tree only, and its horizontal development is restricted by neighboring roots. Tree vigor is strongly reduced by close within-the-row spacing and further exacerbated when between-the-row spacing was reduced (Chalmers et al., 1984). Close planting needs to be used with care. If the proper framework design is used, such as the Tatura trellis for peach, the trees receive enough light to stay productive. However, the desire to reduce tree size by crowding often leads to overcrowding and highly shaded conditions, which defeats the purpose of using reduced growth for smaller tree size coupled with increased fruiting.

Chalmers et al. (1984) reported that fruit was larger on closely planted trees receiving reduced irrigation in comparison to widely planted trees receiving the same water regimes. They explained this by the fact that the widely spaced trees grew more and were unable to respond with increased fruit size to the deficit irrigation because shoot growth was still too much and competed successfully with fruit enlargement. Thus they concluded that close spacing and deficit irrigation interact in creating a small tree.

Role of Gibberellins in Controlling Tree Size

Through the use of GA biosynthesis inhibitors, it has become clear that by regulating GA levels in trees, tree size can be also regulated. The success of GA biosynthesis inhibitors in controlling growth compels us to review the limited knowledge of gibberellins in fruit trees.

All GAs have the same chemical structure in which the carbon atoms are numbered as shown in Figure 5.10. The GAs are subdivided into C_{20}–GAs, containing all 20 carbon atoms of their diterpenoid precursors, and C_{19}–GAs, which have lost carbon 20 (Jones and MacMillan, 1984). In general, three phases of GA synthesis can be distinguished. The first phase, from mevalonate to *ent*-kaurene, involves nine steps that are catalyzed by soluble enzymes; the second phase, from *ent*-kaurene to GA_{12}–7-aldehyde, involves five steps that are catalyzed by microsomal enzymes; and the third phase, which involves hydroxylation of C_{20}–GAs and later the C_{19}–GAs. Hydroxylation at carbon 13 occurs early in the pathway, and it is catalyzed by microsomal enzymes whereas 3B-hydroxylation occurs late in the pathway and is catalyzed by soluble enzymes (Graebe, 1986).

The various GAs reported in apples are shown in Figure 5.10. A considerable literature exists on the various GAs diffused from apple

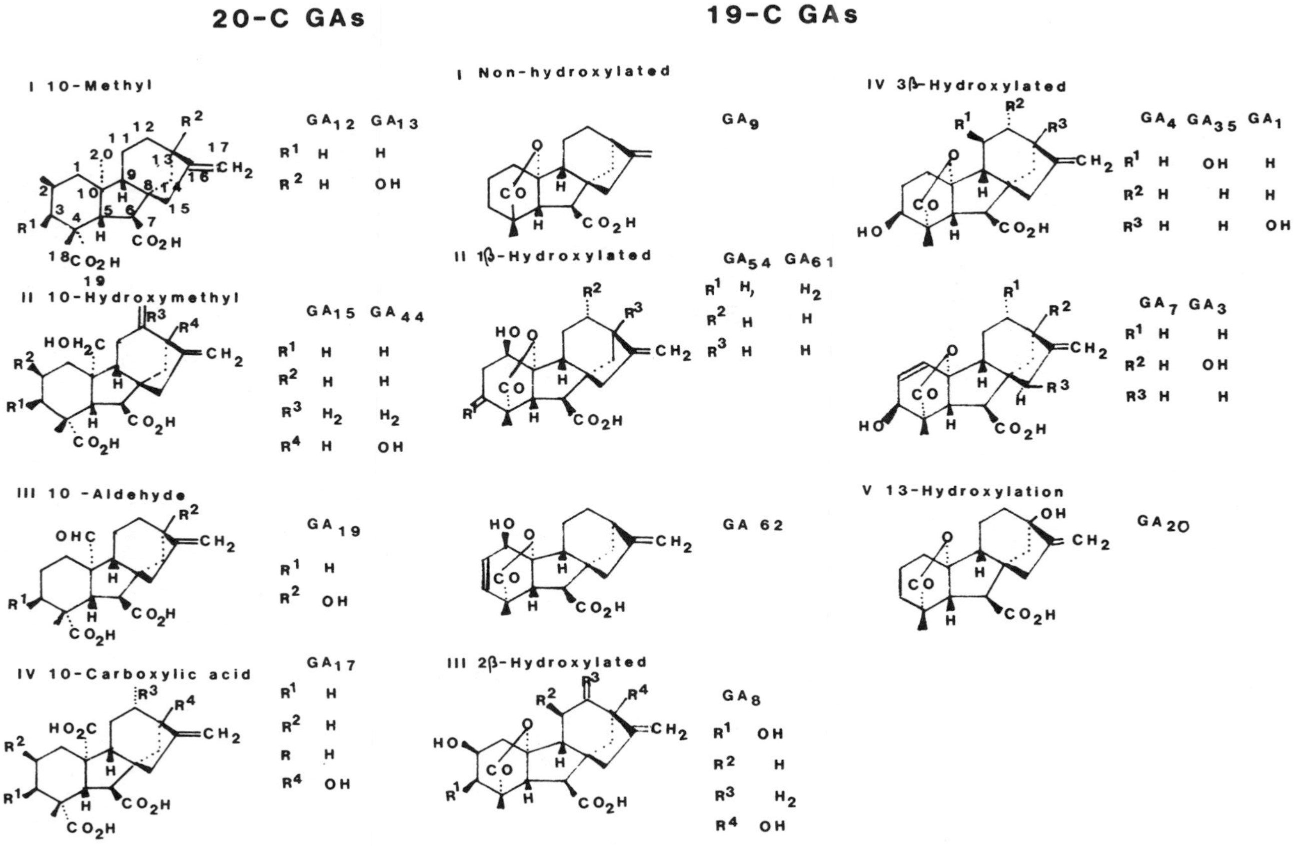

20-C GAs

I 10-Methyl

	GA_{12}	GA_{13}
R^1	H	H
R^2	H	OH

II 10-Hydroxymethyl

	GA_{15}	GA_{44}
R^1	H	H
R^2	H	H
R^3	H_2	H_2
R^4	H	OH

III 10-Aldehyde

	GA_{19}
R^1	H
R^2	OH

IV 10-Carboxylic acid

	GA_{17}
R^1	H
R^2	H
R	H
R^4	OH

19-C GAs

I Non-hydroxylated

GA_9

II 1β-Hydroxylated

	GA_{54}	GA_{61}
R^1	H,	H_2
R^2	H	H
R^3	H	H

GA 62

III 2β-Hydroxylated

	GA_8
R^1	OH
R^2	H
R^3	H_2
R^4	OH

IV 3β-Hydroxylated

	GA_4	GA_{35}	GA_1
R^1	H	OH	H
R^2	H	H	H
R^3	H	H	OH

	GA_7	GA_3
R^1	H	H
R^2	H	OH
R^3	H	H

V 13-Hydroxylation

GA_{20}

Figure 5.10 Structure of various gibberellins reported in apples.

seeds and their effect on flower bud formation (Buban and Faust, 1982; Dennis 1976; Hoad 1978). Pharis and King (1985) summarized the seed GAs in apples. The major GAs identified were GA_4 and GA_7 with 12 other GAs also characterized by gas chromatography–mass spectrometry (GCMS). In very immature seeds large amounts of GA_3 and lesser amount of GA_1 were also found. To date, in immature apple seeds $GA_{1,3,4,7,8,9,12,15,17,20,35,44,53,54,61,62,63,68}$, dehydro GA_4, 16,17-dihydroxy-GA_4 and 16,17-dihidroxy-GA_9, iso-GA_7, and C-3 epi-GA_{54} were identified. The major GA in the vegetative tissue of apple, GA_{19}, was not identified in the seed, which points to the fact that by determining seed GAs, one cannot generalize to the GAs of the vegetative tissues.

In fruit trees there are two aspects of gibberellins that concern horticulturists. They are the GA content and function in genetic dwarf fruit trees and the use of GA synthesis inhibitors to influence growth of fruit trees. Dwarf plants of peas, rice, wheat, maize, beans, and *Cucurbita maxima* were examined for GA content and possible metabolic blocks in the biosynthetic pathway of particular GAs. These plants also served to draw conclusions about the active GAs in each of these plants (Graebe, 1986; Jones and MacMillan, 1984; MacMillan, 1987). In dwarf types of these plants the biosynthetic pathway is blocked either in the early steps of the pathway soon after GA_7–12-aldehyde or just before GA_1 (GA_1 is considered to be the active GA in these plants). A composite of metabolic steps in the GA biosynthesis is presented in Figure 5.11 showing in which plant the various steps were identified. Unfortunately, this type of work has not been done in fruit trees. Therefore, we can only speculate what may happen in the trees. Basically there are three types of trees among the fruit-bearing, commercially important species: (1) normal size trees; (2) moderately dwarf, which are anywhere from 75 to 50% of normal size; and (3) extreme dwarfs, which have an internode length of 50% to as low as 5% of normal. In all three types internode length greatly decreases during the hot days of summer even though the trees are irrigated and photosynthesis and turgor are satisfactory. In the extremely dwarf apples, in addition to shortening the internodes, the leaves are also small and narrow, displaying the so-called 'June effect' (Figure 5.12). The possibility exists that the extreme dwarfs have a metabolic block in their GA biosynthesis, and this is the reason for the short internodes. Since GA_1 has been identified in apple and GA_4 can be converted to GA_1 (not shown in apple), GA_1 could be the central active GA in apple also.

Apples accumulate 10 times the level of various GAs during the time growth slows and internodes become shorter (Grochowska et al., 1984). They do not respond to GA sprays. In general, this supports the theory that there is a block in the synthesis of gibberellins close to the active GA; perhaps GA_1 and other inactive GAs accumulate. The internode

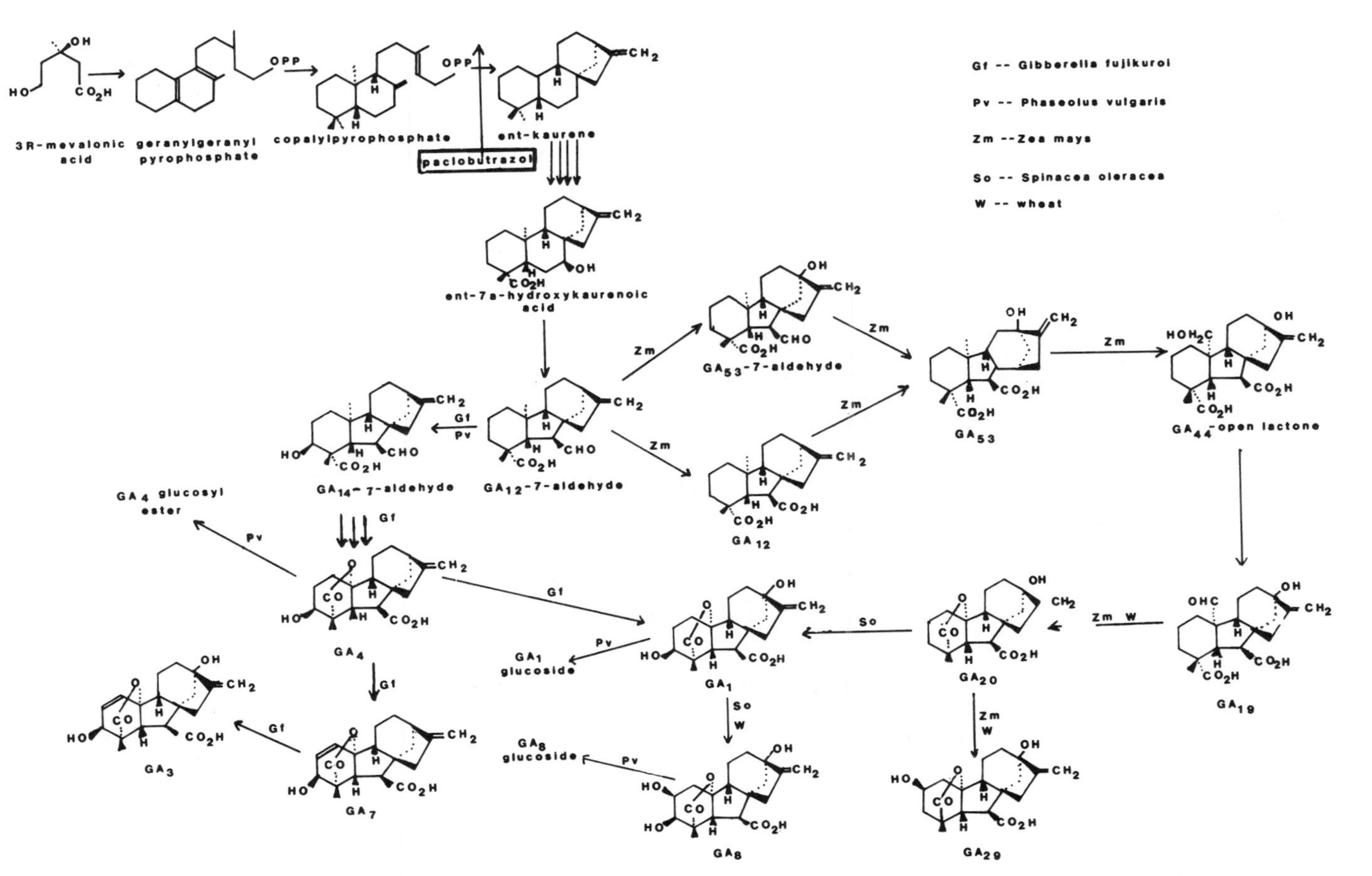

Figure 5.11 Theoretical scheme showing interrelationship between various gibberellins.

Figure 5.12 Apple shoot showing the 'June effect'. Note the size of leaves below the arrow (spring leaves) and above the arrow (developed after early June).

length is highly variable in apple, in contrast to pear and peach, where the internodes are either short or long. Thus the apple may have 'leaky' genes similar to the Le gene of dwarf pea (Ingram et al., 1986). Neither of the preceding speculations explains the shortening of internodes in June (Jaumien and Faust, 1984) when hot temperature commences. During the summer GA concentration is very high in the phloem even in normal apples (Grochowska et al., 1984). The accumulation in dwarfs could be explained by metabolic blocks, but the accumulation in normal trees cannot happen unless a metabolic block is sensitive to temperature. Although GA was determined by bioassays and the specific GAs are not known, the possibility exists that another mechanism involving receptor sites, proposed by Trewavas (1987), operates when environmental conditions change. Yet another possibility exists in trees. The xylem GA, supposedly produced by the roots, decreases to a very low level about the time when internodes shorten and the June effect is observed (Grochowska et al., 1984), which raises the likelihood that root GAs are involved in internode elongation in fruit trees. Grafting experiments (dwarf peach on normal root produce longer internodes than dwarf peach on self-root; dwarf apple on normal root also produces longer internodes than dwarf apple on root of dwarf apple) also indicate that the root has some involvement in determining internode length, but it cannot completely overcome the genetically determined length (Faust, unpublished).

An alternative approach to evaluate the involvement of GA in tree growth is to apply specific inhibitors of GA biosynthesis to reduce GA concentration and decrease growth. Among the growth regulators, paclobutrazol is the most useful for this purpose. Paclobutrazol, (2RS,3RS)-1, 4-(4-chlorophenyl)-4,4-dimethyl-2-(1,2,4-triazol-l-yl)pentan-3-01, is a potent plant growth retardant that is active on a broad range of plant species. It inhibits specifically the three steps in the oxidation of the GA precursor *ent*-kaurene to *ent*-kaurenoic acid in cell-free extract of plants including apple (Hedden and Graebe, 1985). Structurally, paclobutrazol is a substituted triaziole with two assymetric carbon atoms, and it is produced as a mixture of the 2R,3R and 2S,3S enantiomers. In a cell-free system of *C. maxima*, the 2S,3S enantiomer inhibited *ent*-kaurene oxidation more efficiently than did the 2R,3R (Hedden and Graebe, 1985). The 2R,3R enantiomer, however, effectively inhibits the C_{14} demethylation of fungi sterols (Baldwin and Higgins, 1984). Thus the mixture of two enantiomers in paclobutrazol may have some effect on sterols in addition to inhibiting GA biosynthesis.

The regulator can be applied to trees as a spray, a soil drench, or as a paint on the trunk. There is no translocation of this compound from mature leaves. The uptake of the effective dose occurs very likely through the apical bud or the soft stem immediately behind it (Lever, 1986). For this reason multiple, low-rate spray applications are more effective than a single spray (Quinlan and Richardson, 1986). Paclobutrazol binds reversibly to the soil and after uptake to the xylem elements in the stem. The chemical reversibly bound to the vascular tissue constitutes a reservoir-releasing chemical to give extended control often for several years (Lever, 1986). Because of the slow movement through the vascular elements, there is usually a delay between application of soil treatment and the exhibition of growth effects. The effect is often only visible in the year after the treatment.

Paclobutrazol treatments reduce shoot elongation and leaf expansion more than leaf or node number in apple. Foliar applications of GA_3 are able to restore leaf area expansion more effectively than shoot growth (Steffens and Wang, 1986). The small amount of GA applied at bloom to overcome reduction in size does not affect shoot growth (Williams et al., 1986). Occasionally shoot number is inhibited more than shoot elongation (Elfving and Proctor, 1986). In apple, fruit weight is also decreased when shoot length is reduced by 40 (Miller and Swietlik, 1986) to 50% (Elfving and Proctor, 1986). Paclobutrazol is similarly effective in reducing shoot growth in other fruits. It reduces shoot elongation in apricot (Gaash, 1986), cherry (Bargioni et al., 1986; Webster and Quinlan, 1986). nectarine (Costa et al., 1986; Erez, 1986), peach (Coston, 1986; Shearing and Jones, 1986), pear (Stan et al., 1986), and plum (Gaash, 1986). When high rates of the chemical are applied, leaf size is also reduced in pear and some stone fruits (Williams et al., 1986). Fruit size is not affected in stone fruits. With 40–50% shoot reduction, fruit size was the same in peach (Shearing and Jones, 1986) and in nectarines (Costa et al., 1986; Erez, 1986). It is possible that a larger dose is needed in apple to achieve the same degree of reduction than in stone fruits. Examining the growth curves of fruit trees treated with paclobutrazol, it is evident that there is a double effect of this chemical on shoot growth. The early spring growth is slowed as a result of inhibiting GA synthesis. This is perhaps the result of shortening the internodes. In addition to the spring growth effects, the ceasing of summer growth commences much earlier on treated trees than on controls (Sansavini et al., 1986; Sharing and Jones, 1986). Previously we discussed the slowing of apple growth, which may include the cessation of all growth. The term used for this phenomenon is the *June effect.* It appears that paclobutrazol treatments trigger the June effect and carry it to the cessation of growth. We have speculated that the June effect is associated with xylem GA concentrations. There is no record that paclobutrazol influences xylem GA concentrations. Until such data exist, conclusions cannot be made about forcing the cessation of growth when the GA content of the plant is reduced.

Concomitant to the reduction in growth, a number of physiological changes can be observed in the tree. Some of these changes are likely to be direct whereas others are indirect, a result of decreases in GA content. A short discussion of the observed changes follows.

1. The Root growth of apple is altered. The number of short roots with an enlarged diameter is increased in paclobutrazol-treated apple seedling, an effect that could be reversed by foliar-applied GA_3 (Steffens and Wang, 1986).

2. A higher portion of lateral buds tend to become flower buds, and consequently the number of flower buds and flowers per tree increases (Williams, 1984). The result is more but smaller fruit and the potential of biennal bearing. The overall crop on paclobutrazol-treated trees is usually higher, the small reduction in weight of individual fruits more than compensated for by the larger number of fruits (Steffens and Wang, 1986; Williams et al., 1986). Such trees may need extra care in thinning.

3. In trees with lowered GA content assimilate partitioning is shifted from leaves to roots and carbohydrate concentration is increased in all parts of the tree (Steffens and Wang, 1986) with the exception of fruits. The carbohydrate content increases in wood grown before the treatment

as well as in wood whose growth was reduced by the treatment (Steffens and Wang, 1986). The change in partitioning is obviously a secondary effect, a reaction to the elimination of the shoot as a sink.

4. The chlorophyll content on a leaf area basis is higher than in the normal plants. This is especially true for leaves developed after treatment (Steffens and Wang, 1986). The higher chlorophyll content in paclobutrazol-treated plants is very similar to the higher chlorophyll in leaves of genetic dwarfs that are also possibly low in GA.

5. Dark respiration of shoots of paclobutrazol-treated apple seedling is lower than untreated seedlings (Steffens and Wang, 1986). This is in contrast with the higher rate of respiration of young roots that developed as the result of paclobutrazol treatment (Steffens and Wang, 1986).

6. Under a non–water stress situation paclobutrazol-treated plants use less water than control plants (Steffens and Wang, 1986). In experimental conditions detached leaves of control plants lose 15% of their water content within 1 hr, whereas paclobutrazol-treated plants do so in 2 hr. The components of the water potential were not determined concomitantly with the water loss. Since leaves of paclobutrazol-treated plants are higher in carbohydrates in general, it is possible that the osmotic potential of these leaves is higher and they retain water better.

7. The polyamine content of paclobutrazol-treated plants is lower (Steffens and Wang, 1986).

8. Russeting the fruit on cultivars of apple such as 'Golden Delicious' is usually increased by paclobutrazol if conditions are conducive for russeting (Williams et al., 1986). As discussed in Chapter 5, russeting is related to the growth of fruit, and through regulating growth it is influenced by GA. Russeting can be reversed by GA sprays.

9. In apple the length–diameter ratio of the fruit changes. Fruit length is decreased when GA is limited. Thus it is not surprising that paclobutrazol-treated trees produce shorter fruit.

Theoretical Mechanisms Determining Tree Size

Almost 50 years of research have not totally explained the mechanism of dwarfing and the biochemical or physiological processes involved in determining tree size. At the same time the horticultural manipulation of tree size is much more advanced and probably remains ahead of an understanding of the phenomenon itself. From horticultural, genetic, and physiological data one can conclude that more than one mechanism operating in the tree determines its size.

It is unquestionable that GA synthesis inhibitors greatly decrease growth and reduce tree size. Thus, in case of reduced growth, especially when internodes are shortened, the theory that tree growth is greatly dependent on the availability of GA must be accepted. Some data indicate that a similar but not well-understood mechanism operates in genetic dwarfs with short internodes. The fact that in short-internode plants xylem GAs decrease to very low levels and phloem GAs, although with different polarity, increase when internodes shorten during the summer cannot be explained simply by insufficient GA levels for growth. Thus reduced-stature tree habit based on short internodes only can be tentatively assigned to GA levels in the growing tissues.

In general, it is agreed that controlled water stress decreases tree size. This has been discussed previously. One could speculate then that the various influences supposedly affecting the root function also include uptake of water. However, water loss measured from severed scions indicates that the transpiration rates of composite apple trees are similar irrespective of the type of rootstock on which they were grown (Jones, 1984). Thus at least the root effects do not include the uptake of water. Roots take up water according to its availability and the water potential of the tree. This gives a practical possibility to control tree size separate from other mechanisms.

Vigor is a much more complicated mechanism because it can be affected by many factors. Lockard and Schneider (1981) proposed that the level of auxin reaching the root in active form influences root growth and metabolism, including the synthesis of cytokinins, which in turn should promote growth on the top. They tried to provide evidence that in dwarf trees IAA is metabolized by the bark and does not reach the root. The proposed mechanism should be counteracted by either applying IAA to the root or cytokinins (CK) to the growing point. They admitted that in their experiments neither of these compounds induced extension growth in an intact plant, dwarfed or not.

The hypothesis of Lockard and Schneider should be looked upon in a larger context. They are probably correct that the root is the controlling organ in the rate of growth of the tree, but IAA is perhaps only one factor in the matrix determining vigor. It has been discussed before that in all likelihood the genetic control of vigor locates in the root. Jones (1984) indicated that the xylem sap contained much lower concentrations of nutrients in a dwarf than in a vigorous tree. He thought that this could determine vigor. There is no indication that the leaf content of nutrient elements in dwarf trees is different than in vigorous trees. Thus the xylem sap concentration of nutrients is more an indication of the root function than a factor determining the rate of growth. Fruiting has a profound effect on tree size. Fruiting also greatly decreases carbohydrate partitioning to the roots. Thus fruiting can and in all probability does influence the functioning of roots. The bark of the rootstock or interstock and the reversed polarity bark all decrease the transport downward. This can be seen from the swelling above the graft union in many dwarfing graft combinations and the data presented on IAA transport by Lockard and Schneider (1981).

Decreasing the size of the root by pruning also decreases the size of the tree. This is additional evidence of the leading role of the root in determining vigor. Transplanting is an unavoidable operation that is associated with root pruning. Total shoot growth was 32% less than similar trees propagated in the field and not transplanted (Geisler and Ferree, 1984). Root pruning was practiced in the 1910s to dwarf apple trees grown on dwarfing rootstock. After 15 years the total weight of trees pruned annually was only 3%, those pruned biennially 7%, and those pruned every 4 years 43% of trees receiving no root pruning (Bedford and Pickering, 1919). Root pruning is effective for dwarfing apple trees when done in July but not when done in late summer (Maggs, 1964, 1965). Perhaps the regrowth of roots was more difficult if pruned at a time when root growth was normally limited.

Several theories were advanced for the dwarfing effect of root pruning: (1) Root-pruned trees may suffer from water stress (Geisler and Ferree, 1984) and restoration of functional equilibrium corresponds to the recovery of root volume in peach (Richards and Rowe, 1977b). (2) Retarded shoot growth may be caused by a limited supply of cytokinins produced in the root. Root pruning could be substantially reduced if part of the root volume is removed (Richards and Rowe, 1977a). (3) Regrowth of root requires carbohydrates, and the redistributed carbohydrates limit shoot growth. In peach the change in redistribution of carbohydrates lasts only to 25 days, after which the equilibrium is restored (Richard and Rowe, 1977a,b). In apple the restoration of equilibrium may be as long as 1 year (Harris et al., 1971). The short restoration period of equilibrium between root and shoot growth points to the fact that if root pruning is done to control tree size, then root pruning needs to be repeated perhaps several times during the season or during consecutive years.

Bending of branches also decreases translocation to the roots and so does summer pruning, which decreases the available carbohydrates to the roots. At this time it is not possible to pinpoint which aspect of the root determines the rate of growth at the top. Nevertheless, it is visualized, at least by this author, that vigor control is achieved by a matrix of factors rather than one single biochemical action.

The effect of various horticultural techniques in the dwarfing of fruit trees is additive because these procedures influence various mechanisms. For example, spur-type tree (short internode), dwarfing rootstock, and reduced irrigation all produce reduced size trees. All three combined produce an extremely dwarf tree.

Some facts that are not easy to fit into a unified theory or they only can be accommodated in a highly speculative manner. Jones (1984) reported that cherry clone 15 (*Prunus avium* × *P. pseudocerasus*) produces very large trees as a rootstock but grows slowly as a scion. As an interstock it has no effect on tree size unless small twigs are allowed to develop on the interstock, which results in a dwarfing effect. This suggests that the leaves

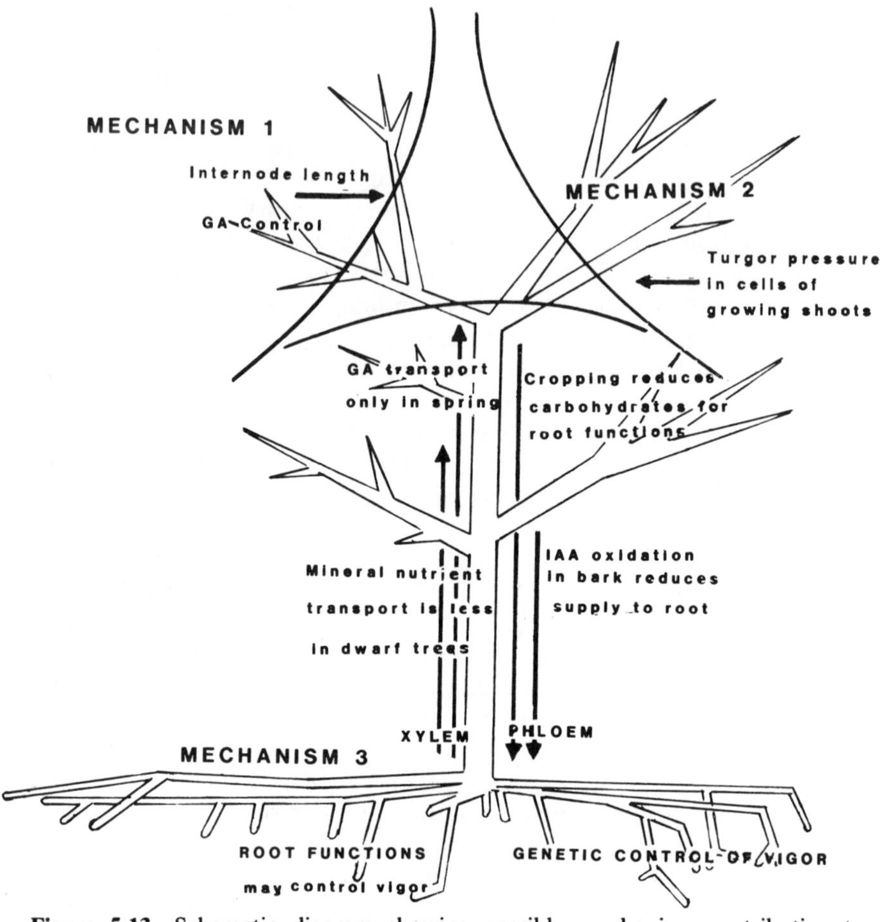

Figure 5.13 Schematic diagram showing possible mechanism contributing to decreased tree size.

produce some kind of inhibitor, and the dwarfing effect is manifested only when this leaf-produced inhibitory substance is translocated to the growing points.

From the preceding discussion it is obvious that three mechanisms operate in trees to determine tree size: a hormonal mechanism centered on gibberellin availability, turgor pressure in the cells of young shoots, and vigor.

1. Gibberellic acid is the most important hormone in the hormonal mechanism. This mechanism operates in the above-ground portion of the tree. Giberrelic acid added or taken away for the above-ground part of the tree has a significant effect. It is likely that cytokinins and perhaps inhibitors are also part of this mechanism.

2. The turgor pressure of the cells in the extending shoot control the second mechanism, and it is determined by water availability to the plant. Although water is taken up by the roots, this mechanism also involves the above-ground portion of the tree.

3. Vigor is determined by the functioning of the root. In turn root functions are affected by a great variety of influences ranging from IAA transport to the root to carbohydrate supply reaching the root, confinement of the root system, rootstock, and so on. It is possible that when the tree is dwarfed by mechanism 1 or 2, the dwarfing itself has a lasting effect by influencing the root system. However, presently it appears that the three mechanisms are distinct and separate. A schematic illustration of the dwarfing mechanism is presented in Figure 5.13.

REFERENCES

Antoszewski, R., U. Dzisciol, A. Mika, and A. Czynczyk. 1978. Acta Physiol. Plant. 1:35–44.

Avery, D. J. 1970. New Phytologist 69:19–30.

Baldwin, B. C., and T. E. Higgins. 1984. Pestic. Sci. 15:156–166.

Bargioni, C., C. Madinelli, A. Ramina, and P. Tonutti. 1986. Acta Hort. 179:581–582.

Barlow, H. W. G. 1959. Scientia Hort. 14:35–41.

Barlow, H. W. G. 1964. Ann. Rep. E. Malling Res. Sta. 1963:104–107.

Barlow, H. W. G. 1971. Ann. Rep. E. Malling Res. Sta. 1970:52–53.

Barlow, H. W. G. 1975. Climate and the orchard. Slough, England: Farnhar Royal, pp. 98–102.

Beakbane, A. B. 1956. Ann. Appl. Biol. 44:517–521.

Beakbane, A. B., and E. C. Thompson. 1939. J. Pomol. Hort. Sci. 17:141–149.

Bedford, H. A. R., and S. U. Pickering. 1919. Science and fruit growing. London: Macmillan, pp. 45–80.

Bouche-Thomas, E. 1953. La methode Bouche-Thomas.

Brunner, T. 1982. Torpegyumolcsfa-neveles (Development of dwarf trees). Budapest: Mezogazdasagi Kiado.

Buban, T., and M. Faust. 1982. Hort. Rev. 4:174–203.

Carlson, R. F. 1974. Proc. Xixth, Jutern. Hort. Congr. Warszawa. pp. 294–302.

Chalmers, D. J., R. L. Canterfold, P. H. Jerie, T. R. Jones, and T. D. Ugalde. 1975. Austral. J. Plant Physiol. 2:635–645.

Chalmers, D. J., P. D. Mitchell, and P. D. Jerie. 1984. Acta Hort. 146:143–149.

Chalmers, D. J., and B. van den Ende. 1975. J. Agr. Victoria 73:473–476.

Costa, C., R. Biasi, A. Ramina, and P. Tonutti. 1986. Acta Hort. 179:567–570.

Coston, D. C. 1986. Acta Hort. 179:575.

Coutanceau, M. 1962. Arboriculture fruitiere. Paris: Baillere.

Delbard, G. 1961. Arboriculture Fruitiere 12:18–25.

Delbard. G. 1962. La production fruitiere. Arboriculteur exp. Paris.

Dennis, F. G. 1976. J. Amer. Soc. Hort. Sci. 101:629–633.

Dickinson, A. G., and E. W. Samuels. 1956. J. Arnold Arboretum 37:307–313.

Dunn, J. S., and M. Stolp. 1987. Hort. Sci. 22:568–572.

Elfving, D. C., and J. T. A. Proctor. 1986. Acta Hort. 179:473–480.

Erez, A. 1986. Acta Hort. 179:513–520.

Faust, M., and S. W. Zagaja. 1984. Acta Hort. 146:21–27.

Fejes, S., E. Horn, and T. Brunner. 1972. Gyumolcssoveny (Fruit trellis). Budapest: Mezogazdasagi Kiado.

Ferree, D. C., and R. F. Carlson. 1987. In Rootstocks for fruit crops. R. C. Rom and R. F. Carlson (eds.). New York: Wiley, pp. 107–144.

Feucht, W. M. M. Khan, and J. P. Daniel. 1974. Phys. Plant. 32:247–252.

Fidighelli, C., G. D. Strada, and R. Quarta. 1984. Acta Hort. 146:47–57.

Gaash, D. 1986. Acta Hort. 179:559–562.

Geisler, D., and D. C. Ferree. 1984. Hort. Rev. 6:155–188.

Graebe, J. E. 1986. In Plant growth substances 1985. M. Bopp (ed.). Berlin: Springer-Verlag, pp. 74–81.

Greene, D. W. 1975. Hort. Sci. 10:73–74.

Grochowska, M. J. 1967. Bul. l'Academie Polish des Sciences 15:455–459.

Grochowska, M. J., G. J. Buta, G. L. Steffens, and M. Faust. 1984. Acta Hort. 146:125–134.

Gur, A., B. Bravdo, and Y. Mizrahi. 1972. Physiol. Plant. 27:130–138.

Gur, A., and R. M. Samish. 1968. Beitr. Biol. Pflanzen 45:91–111.

Hansen, P. 1980. In Mineral nutrition of fruit trees. D. Atkinson, F. E. Jackson, R. O. Sharples, and M. Waller (eds.). London: Butterworths, pp. 201–212.

Harris, R. W., W. B. Davis, N. W. Stice, and D. Long. 1971. J. Amer. Soc. Hort. Sci. 96:109–111.

Head, G. C. 1969. J. Hort. Sci. 44:175–181.

Hedden, P., and J. E. Graebe. 1985. J. Plant Growth Regul. 4:611–617.

Hoad, G. V. 1978. Acta Hort. 80:93–103.

Hutchinson, A., C. D. Taper, and G. H. N. Towers. 1959. Canad. J. Biochem. Physiol. 37:901–910.

Imgram, T. J., J. B. Reid, and J. Macmillan. 1986. Planta 168:414–420.

Jackson, J. E. 1984. Acta Hort. 146:83–88.

Jaumien, F., and M. Faust. 1984. Acta Hort. 146:69–79.

Jones, O. P. 1976. Ann. Bot. 40:1231–1235.

Jones, R. L., and J. MacMillan. 1984. In Advanced plant physiology. M. B. Wilkins (ed.). Melbourne: Pittman, pp. 21–52.

Jones, O. P. 1984. Acta Hort. 146:173–182.

Kandiah, S. 1979. Ann. Bot. 44:175–183.

Layne, R. C. 1987. In Rootstocks for fruit crops. R. C. Rom and R. F. Carlson (eds.). New York: Wiley, pp. 185–216.

Lespinasse, J. M. 1980. Perspectives de l'horticulture. no. 15:27–102. Ministere de l'Agr. Quebec, Canada.

Lever, B. G. 1986. Acta Hort. 179:459–466.

Lockard, R. G., and G. W. Schneider. 1981. Hort. Rev. 3:315–371.

Lombard, P. B., and M. N. Westwood. 1984. Acta Hort. 146:197–202.

Lombard, P. B., and M. N. Westwood. 1987. In Rootstocks for fruit crops. R. C. Rom and R. F. Carlson (eds.). New York: Wiley, pp. 145–184.

Looney, N. E., and W. D. Lane. 1984. Acta Hort. 146:31–46.

Luckvill, L. C., and P. Whyte. 1968. Plant Growth Regulators. Scientific Monograph 31:87–101.

MacMillan, J. 1987. In Hormone action in plant development. A critical appraisal. G. V. Hoad, J. R. Lenton, M. B. Jackson, and R. K. Atkin (eds.). London: Butterworths, pp. 73–86.

Maggs, D. H. 1963. J. Hort. Sci. 38:119–128.

Maggs, D. H. 1964. J. Exp. Bot. 15:574–583.

Maggs, D. H. 1965. J. Exp. Bot. 16:387–404.

Martin, G. C., and M. W. Williams. 1967. Hort. Sci. 2:154.

Milborrow, B. V. 1978. In The biochemistry of phytohormones and related compounds, Vol. 1, D. S. Letham, P. B. Goodvin, and T. Higgins (eds.). pp. 295–348.

Miller, S. S., and D. Swietlik. 1986. Acta Hort. 179:563–566.

Okie, W. R. 1987. In Rootstocks for fruit crops. R. C. Rom and R. F. Carlson (eds.). New York: Wiley, pp. 321–360.

Olien, W. C., and A. N. Lakso. 1984. Acta Hort. 146:151–158.

Perry, R. L. 1987. In Rootstocks for fruit crops. R. C. Rom and R. F. Carlson (eds.). New York: Wiley, pp. 217–264.

Pharis, R. P., and R. W. King. 1985. Ann. Rev. Plant. Physiol. 36:517–568.

Priestley, C. A. 1960. Ann. Rep. E. Malling Res. Sta. for 1959:70–77.

Quinlan, J. D., and P. J. Richardson. 1986. Acta Hort. 179:443–451.

Richards, D., and R. N. Rowe. 1977a. Ann. Bot. 41:729–740.

Richards, D., and R. N. Rowe. 1977b. Ann. Bot. 41:1211–1216.

Roberts, R. H. 1949. Bot. Rev. 15:423–463.

Robitaille, H. 1970. The relationship of endogenous hormones to growth characteristics and dwarfing in Malus. Ph.D. Dissertation. Michigan State University.

Robitaille, H., and R. F. Carlson. 1971. Hort. Sci. 6:639–640.

Robitaille, H. A., and R. F. Carlson. 1976. J. Amer. Soc. Hort. Sci. 101:388–392.

Rogers, W. S., and A. B. Beakbane. 1957. Ann. Rev. Plant Physiol. 8:217–236.

Rogers, W. S., and Booth, G. A. 1964. J. Hort. Sci. 39:61–65.

Rom, R. C., R. F. Carlson. 1987. Rootstocks for fruit crops. New York: Wiley.

Sansavini, S. R., Bonomo, A. Finotti, and U. Palara. 1986. Acta Hort. 179:489–496.

Sax, K., and H. Q. Dickson. 1956. J. Arnold Arboretum 37:173–179.

Schmidt, H., and W. Gruppe. 1988. Hort. Sci. 23:112–113.

Schmitz-Hübsch, H., and Furst, L. 1959. Intensiv Obstbau in Heckenform. Stuttgart.

Scorza, R. 1988. Compact fruit. 21:92–98.

Shearing, S. J., and T. Jones. 1986. Acta Hort. 179:505–512.

Soubeyrand, G. 1965. Arboriculture Fruitiere 136:47–54.

Stan, S., I. Popescu, M. Cotorobai, and M. Radulescu. 1986. Acta Hort. 179:555–558.

Stark, P., Jr. 1974. Amer. Fruit Grow. 94(11):12–14.

Steffens, G. L., and S. Y. Wang. 1986. Acta Hort. 179:433–442.

Tomaszewski, M., and K. V. Thimann. 1966. Plant Physiol. 41:1443–1454.

Trewavas, A. 1987. In Hormone action in plant development. A critical appraisal. G. V. Hoad, J. R. Lenton, M. B. Jackson, R. K. Atkin (eds.). London: Butterworths, pp. 19–38.

Tubbs, F. R. 1973. Hort. Abstr. 43:247–253; 43:325–334.

Tubbs, F. R. 1977. J. Hort. Sci. 52:37–48.

Tukey, H. B. 1964. Dwarfed Fruit Trees. New York: Macmillan.

Vyvian, M. C. 1955. Ann. Bot. 19:401–423.

Walsh, C. S., and A. N. Miller. 1984. Acta Hort. 146:211–214.

Webster, A. D., and J. D. Quinlan. 1986. Acta Hort. 179:577–580.

Wertheim, S. J. 1978. Acta Hort. 65:173–179.

Williams, A. H., and B. B. Beakbane. 1956. Ann. Appl. Biol. 44:517–521.

Williams, M. W. 1984. Acta Hort. 146:97–104.

Williams, M. W., E. A. Curry, and G. M. Greene. 1986. Acta Hort. 179: 453–458.

Williams, M. W., and E. A. Stahly. 1968. Hort. Sci. 3:68–69.

Yadava, U. L., and D. F. Dayton. 1972. J. Amer. Soc. Hort. Sci. 97:701–705.

Yu, K. S., and R. F. Carlson. 1975. Hort. Sci. 10:401–403.

6

PRUNING AND RELATED MANIPULATIONS: PHYSIOLOGICAL EFFECTS

One of the most basic production practices in fruit growing is cutting away shoots or roots. Trees have parts cut away for the purpose of changing some of their physiological functions. On young trees cutting away shoots establishes a functional framework. On older trees cutting away branches, if practiced correctly, allows more light penetration into the canopy of the tree, influences photosynthate translocation to the fruit and to the root, and regulates flower bud formation.

In general, cutting away parts of the tree is called pruning. Pruning is a process that may influence the physiology of the tree for several years, if not for the entire life of the tree. It takes both a careful consideration and an understanding of the entire physiology of the tree to successfully accomplish the goals set out for the pruning process. Pruning should be looked upon not as a one-time operation but as a series of operations started on a young tree in the year of planting and continued for years with slightly changing objectives as the tree matures. In contrast, pruning can be an operation that lasts only for a short duration when corrective action is needed to set the tree on a certain course of response that is best achieved by pruning. Because pruning is only one (perhaps the most drastic) manipulation that influences the physiology of the tree, it always has to be considered together with other manipulations planned concurrently.

Before performing the operation of pruning, one has to decide when the pruning should be done and how heavily the tree should be pruned. Trees can be pruned either during the dormant season or during the vegetative season. Results differ according to the time the tree is pruned. Trees also can be pruned lightly or heavily depending on the amount of wood cut away. Pruning produces different results corresponding to the location of the cut that has been made and the type of tissue removed. If the cut takes away the terminal bud or young tissues at the tip of the shoot, the tree responds differently than if the entire shoot is removed. Various species, and often different cultivars within the species, react differently to the same pruning procedure. It is impossible, therefore, to describe all the variations of responses to pruning of all deciduous orchard species. Thus only major generalities are presented here.

In connection with pruning, there are several other manipulations that are applied by the practical orchardist. These usually include the bracing of branches to bring them closer, if not entirely, to the horizontal position, looping the branches to decrease their rate of growth, and positioning of certain branches in locations that are convenient for future operations in the orchard and to influence crotch angle in general. Because these operations are part of the tree shaping or pruning process, they must be considered concurrently with pruning.

Pruning as a procedure to dwarf trees was widely employed in China during the Sixth Dynasty in the first and second century. The procedure was designed for ornamental plants. The technique, panjing, reached its height during the Tung Dynasty from the eleventh to thirteenth century, and it included apricot and apple among the species cultivated. The Japanese version of the technique, bonsai, is more recent, but records indicate that apricots were grown as bonsai around 1700 A.D. by the samurai.

Pruning fruit trees to shape them has its origin in the fourteenth or fifteenth century. Pierre Belon of Mons, France, mentions that hedgerows of fruit trees on both sides of garden pathways existed in 1558. The Italians, he said, called these hedgerows *spalieres* using a version of the French word *espaulière*, meaning 'palisade,' a living wall of trees shaped in various forms. Although the first references in the literature to the term 'espalier' are from the sixteenth century, there is evidence in the form of paintings that this form of pruning had been practiced at least one or two centuries before (Tukey, 1966).

Around the turn of the twentieth century Lorette (1926) at Wagnonville, France, developed a system of summer pruning of apples for promoting flower bud development. His system was built on summer pinching that was practiced in the mid-nineteenth century particularly for pears and peaches. From then on, many practical developments have followed. Since about 1915, a great deal of attention has been focused on the physiological basis of tree response to pruning. These studies, summarized by Mika (1986), are the subject of this chapter.

TYPES OF PRUNING

In general, two types of cuts are applied to a tree: shoot shortening or thinning. Shoot shortening is called pinching, tipping, heading, or stubbing, reflecting the 'severity' of the cut, which is directly related to the amount of growth removed. During the summer, shoots are normally pinched, tipped, or headed, whereas during the winter they are headed or stubbed. It is important to consider that heading, regardless of its severity, removes the growing point and developing leaves if applied during the summer and the terminal bud if applied during the winter. This operation severely changes the hormonal balance of the shoot and forces the plant to react accordingly. The contrasting operation, removal of entire shoots or branches at their base, is called thinning. Thinning does not change the relationship of various parts of the shoot or branch to each other because either the entire shoot or branch is removed or they are not touched at all.

Some tree species such as apple or pear are headed only during the early part of their life (1–4 years) mostly for developing the desired form of the canopy. Other species such as peaches, in addition to early heading, are regularly headed yearly throughout their life to promote growth. The important difference is that pome fruits produce flowers and fruit mostly on multi-year-old structures, whereas peach flower buds and fruit are produced on 1-year-old wood. Thus, it is essential that a peach tree grow sufficiently every year to produce structures for flowering and fruit.

Gyuro (1980) classified the degrees of pruning inflicted upon a tree. He considered the pruning rate of individual shoots and also the rate of pruning applied to the entire tree. Pruning of shoot, considering the portion retained, can be long, medium, short, or very short. The amount of wood removed is one-third, one-half, or two-thirds of the shoots or more in the case of very short pruning when branches are cut to an undeveloped bud located at the basal portion of the shoot or to the

TABLE 6.1 pruning Rate of Shoots[a]

Pruning Rate of Shoots	Pruning Method
Long	Terminal bud and one-third of shoot removed
Medium	Terminal bud and one-half of shoot removed
Short	Terminal bud and two-thirds of shoot removed
Very short	Cut to underdeveloped bud or ring

[a] Reprinted by permission from Gyuro, 1980.

TABLE 6.2 Pruning Rate of Entire Trees[a]

Pruning Rate of Tree	Pruning Method
Strong	Large branches are removed, all shoots are cut short
Medium	Large branches are removed, all shoots are cut long
Light	Thin branches are removed, no shoots are cut

[a] Reprinted by permission from Gyuro, 1980.

growth ring at the base of the shoot (Table 6.1). Pruning rate of trees is a combination of severity of thinning and heading (Table 6.2).

Gyuro (1980) also considered pruning systems that include the degree, character, time, and purpose of pruning with additional complementary methods (Table 6.3). Multiple combinations between entries in columns (Table 6.3) is possible. It is obvious that the effect of pruning on the physiology of the tree is dependent on the type or combination of operations listed.

EFFECT OF PRUNING ON APICAL DOMINANCE

Pruning performed on growing or dormant shoots removes apical dominance, releases buds from correlative inhibition, and the resulting growth changes tree construction (Mika, 1986). This reaction of the tree is widely used for positioning scaffold branches on the main stem by heading young trees, called whips, at the right position or heading back a scaffold branch to produce secondary branching.

Barlow and Hancock (1960, 1962) showed experimentally that removing either the shoot tip of the young growing leaves stimulated the growth of axillary buds into laterals. They proposed the concept that the axillary meristems are inhibited in growth by the main axis, but the inhibition is caused by the young apex. The fact that the removal of unexpanded leaves is sufficient to induce bud growth was confirmed by Mika (1971). Removal of fully expanded leaves cannot stimulate axillary bud growth

TABLE 6.3 Classification of Pruning Systems by Purpose

Factors Considered	Training	Canopy Thinning	Forcing Flowering	Rejuvenation	Hedging
Character of pruning	Complete	Partial	Complete	Partial	Complete
Frequency of pruning	Regular during training	Irregular after training	Regular or irregular	Irregular	Regular
Time of pruning	Dormant	Dormant or summer	Dormant or summer	Dormant	Dormant or summer
Complementary methods used	Pulling branches into position	None	Pulling branches into position	Chemicals after pruning	None
Degree of pruning the tree	Medium to weak	Medium to weak	Weak	Strong	Strong to weak
Degree of pruning the shoots	Long to medium	None	Medium to short	Short to very short	Long to short

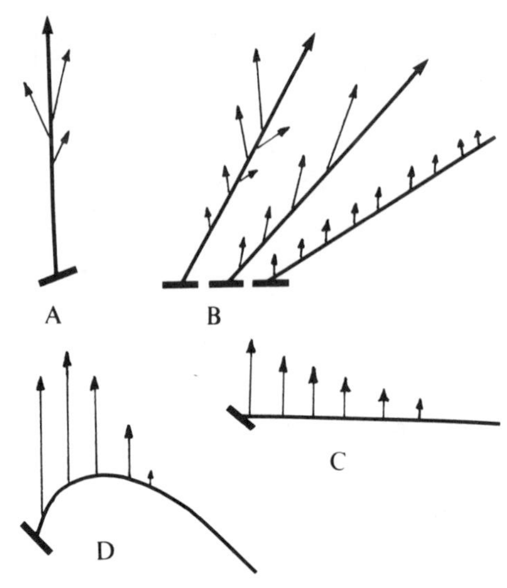

Figure 6.1 Effect of branch position on growth: A, vertical; B, angles between 30 and 55°; C, horizontal; D, arched. (Reproduced by permission from Gyuro, 1980.)

(Mika, 1971). Heading the shoot near its top induces the top five to seven buds to grow. Pinching done at the middle or at the base of the growing shoot very often induces only the uppermost bud to grow (Mika, 1986).

Apple has much stronger apical dominance than peach. Within each species, there are well-known differences. In apples, the so-called spur-type cultivars have much stronger apical dominance than the standard types. In peach, the standard cultivars have more apical dominance in comparison to types such as the cultivar 'Compact Redhaven,' which has practically no apical dominance. Consequently, pruning-induced removal of apical dominance greatly depends on the type of tree pruned. Heading shoots of spur-type apples close to the apex produces budbreak similar to heading shoots of standard trees close to their base. Heading a 'Compact Redhaven' peach does not produce any additional budbreak because the buds would break anyway.

Bending the shoots also changes the influence of apical dominance on individual buds. Bending the branch below the horizontal position effectively removes apical dominance from the uppermost bud. The effect is similar to that of pruning, with some exceptions (Figure 6.1). Crabbe (1984) has clearly shown that bending has a reversible effect on apical dominance. Budbreak occurred on bent and straightened shoots of apples according to the length of time the shoot spent in the bent position (Figure 6.2).

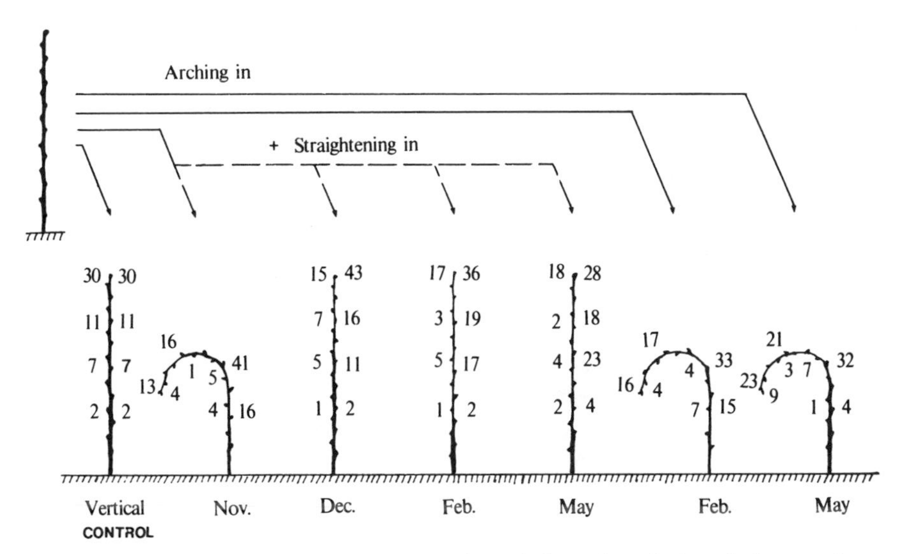

Figure 6.2 Percentage of distribution of total shoot length in relation to time of arching and duration of arched position. (Reproduced by permission from Crabbe, 1984.)

In spur-type apple trees, a second-year shoot produces only one to three strong laterals from its uppermost buds. The other buds develop into spurs (Figure 6.3) or remain dormant. If the shoot is headed during the early part of the growing season, the lower buds develop into strong laterals, but the overall growth is not changed significantly (Mika et al., 1977). If the shoot is headed during the dormant season, the number of strong laterals increases significantly, and the proportion of long shoots to spurs is also altered markedly in favor of shoots (Mika, 1975). Heading the dormant shoot of apple also changes the pattern of branching. On an unheaded shoot, the first bud produces upright growth, and the lower buds develop into laterals with wide crotch angles. On headed shoots, the majority of crotch angles are narrow.

The results of heading a shoot according to Mika (1986) are also influenced by the 'condition' of the bud that becomes the uppermost after heading. Strong apple shoots have well-developed buds on the upper three-fourths of the shoot length. The shoot base has weaker buds. The strongest regrowth occurs on the upper three-fourths of the shoot (Mika, 1986). A growth potential gradient similar to that observed on shoots also exists within the whole tree. The growth potential on the top of the tree is much greater than at the base of the canopy. Pruning at the top of the tree produces stronger growth response than at the base of the tree (Figure 6.4). Pulling branches down in a horizontal position decreases the

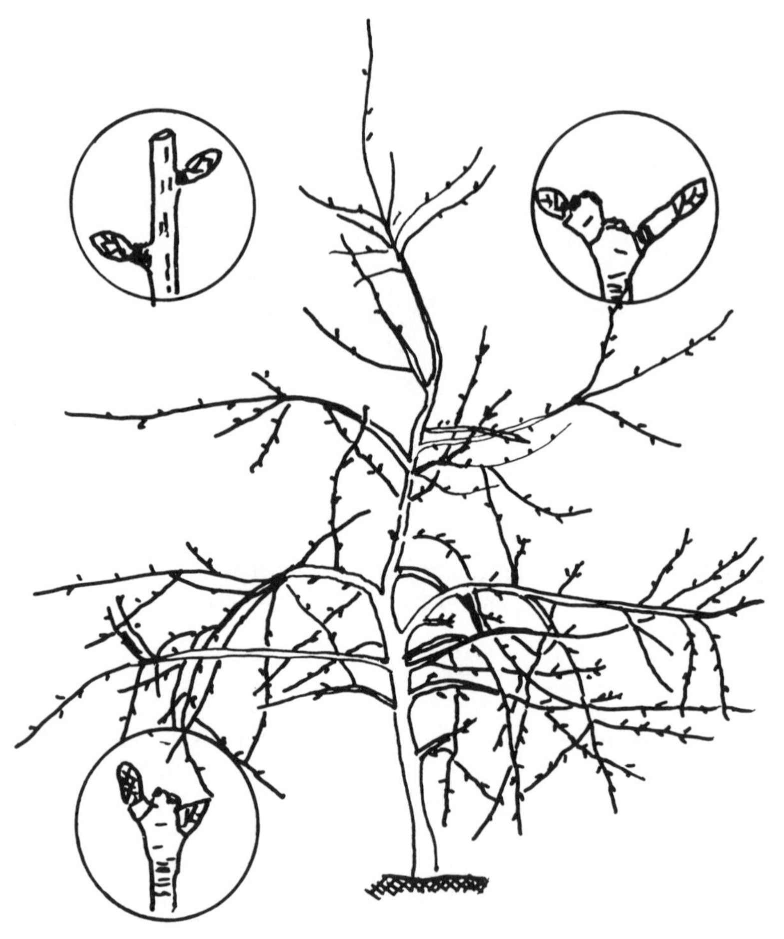

Figure 6.3 Spur development of apple.

rate of growth. If pruning and positioning of branches are used concurrently, they can produce the desired rate of growth regardless of the position of the cut within the tree. This is well illustrated in the development of the so-called Italian palmette form of trees. The growth angles in the basal region of the tree are allowed to be narrow as they develop naturally. With each successive tier of scaffolds the growth angles are widened gradually. At the top of the tree the growth angles are almost horizontal. This balances growth. If growth angles along the length of the tree were equal, yearly pruning of peach growth would occur solely at the top. This growth would shade the rest of the tree and result in decreased productivity. At the same time growth at the base of the canopy would diminish. By balancing the two procedures, pruning and branch positioning, the growth response to pruning can be effectively regulated. In the

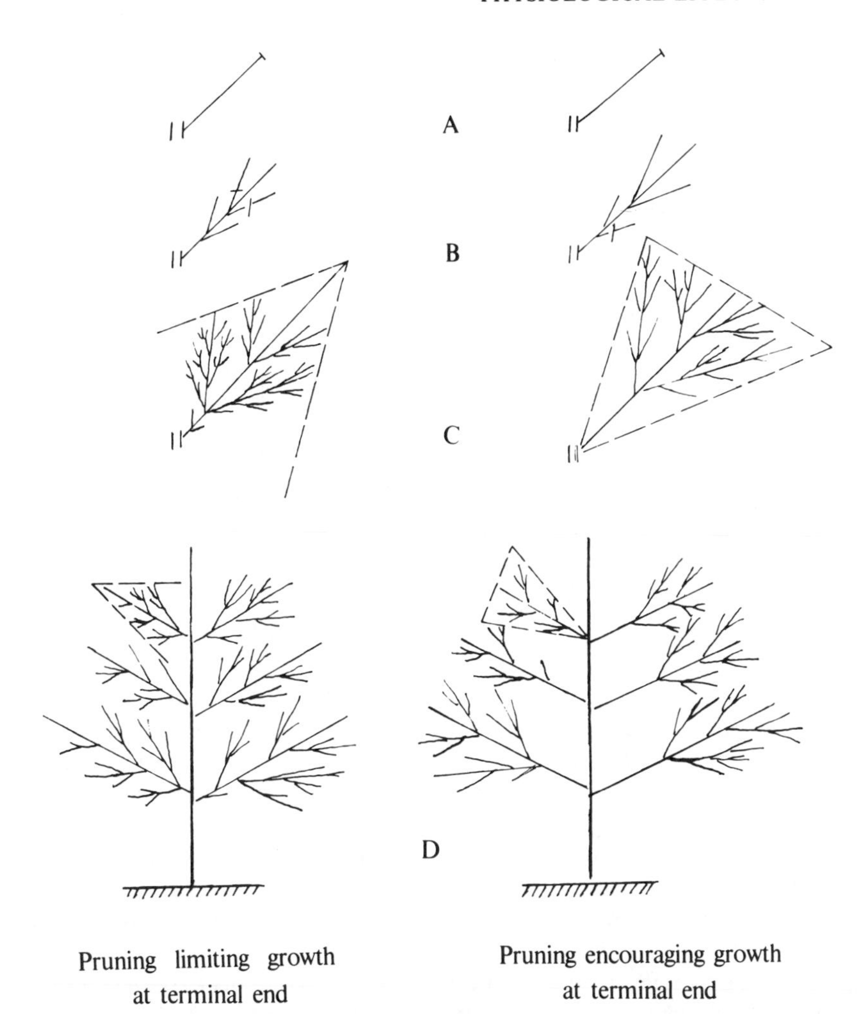

Figure 6.4 Pruning type limiting or encouraging growth at terminal end of apple: A, first cut; B, regrowth and second-season cut; C, resulting growth after cuts for B; D, entire tree. (Reproduced by permission from Gyuro, 1980.)

final analysis, the effect of pruning on apical dominance can be tempered by shoot angle and shoot position within the tree (Figure 6.5).

EFFECT OF PRUNING ON GROWTH OF THE TREE

In considering the effect of pruning on tree growth, it is necessary to consider the time of pruning. Dormant pruning is generally invigorating, whereas summer pruning is believed to weaken growth. Early ex-

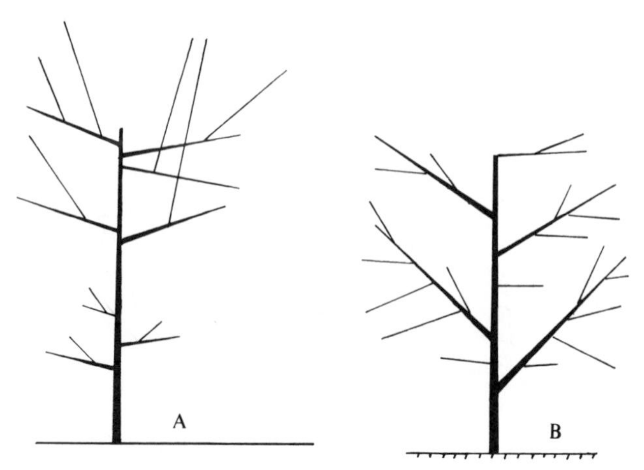

Figure 6.5 Effect of shoot positioning on balance of growth within the tree. (A) Lower scaffolds have wide angles; growth is on the top. (B) Lower scaffolds have narrow angles, upper scaffolds wide angles; growth is evenly distributed. (Reproduced by permission from Gyuro, 1980.)

periments on pruning (Alderman and Auchter, 1916; Gardner et al., 1922; Knight, 1934) demonstrated that pruning stimulates new growth mostly by releasing apical dominance. This usually produces more visible growth, and because of this, dormant pruning is considered invigorating. Grubb (1937) demonstrated that despite the existence of longer shoots on pruned trees, the pruned trees in their entirety were smaller than their unpruned counterparts. Apparently growth of long shoots is not enough to equal the growth of the unpruned tree plus the amount of wood that has been cut away. Thus pruning to a certain extent effectively controls growth. Maggs (1965) supported this opinion, basing his experience on a determination of tree parts of 1-year-old apple trees. With increasing severity of pruning the weight of the new shoot growth increased, whereas the circumference of trunk growth decreased. Root growth also decreased with increased severity of pruning. The change in proportion of the tree parts after 1 year is very little. However, reduction of trunk circumference in every year may lead to significant dwarfing after continued pruning for several years (Preston, 1968).

In production practices this is clearly reflected in the recommendations for establishing the allotted space for each tree in the orchard. The consideration for establishing the needed space for apple trees includes factors for soil; genetic ability of the cultivar to grow and type of growth; (e.g., upright vs. spreading); rootstock, which influences size; and most importantly for the present consideration, 'orchard management.' Orchard management in this context means the amount of pruning plan-

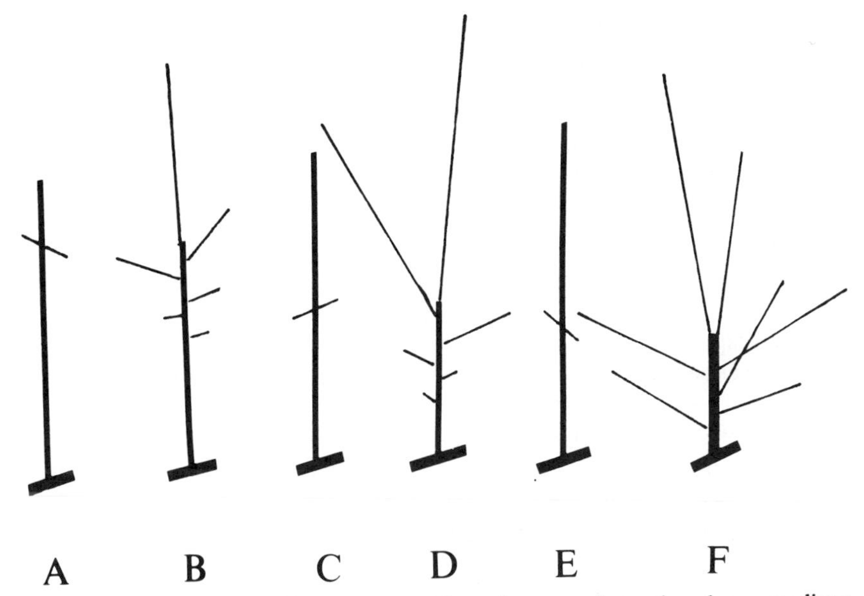

Figure 6.6 Regrowth resulting from various degrees of pruning: long, medium, and short. (A, C, E) Location of cut long, medium, short, respectively. (B, D, F) Corresponding regrowth. (Reproduced by permission from Gyuro, 1980.)

ned yearly. More pruning has to be planned the smaller the space the tree is given to fill.

Koopmann (1896) determined that new shoot growth increases with severity of pruning. This was later confirmed by Knight (1934) and by Jonkers (1962, 1982). When stems of 'Golden Delicious' and 'Cox's Orange Pippin' apple cultivars were shortened by 0, 20, 40, 60, or 80%, the new terminals became larger with increasing pruning severity. A similar growth rate was experienced by the author in the case of peaches and especially in apricots. Regrowth of shoots is illustrated in Figure 6.6.

A series of experiments on dormant pruning of apples was carried out by Maggs (1959, 1965) in England. Based on this experience, he summarized the reaction of apple trees to dormant pruning as follows:

1. The individual shoots arising from a pruned branch are longer than those from an unpruned branch.
2. The final size of the pruned tree does not equal the unpruned tree despite the faster growth of its shoots.
3. Regrowth of shoots is positively correlated with the length of stem before pruning.

TABLE 6.4 Effect of Shoot Angle on Regrowth in Apples[a]

Effect	Pruned Shoot Horizontal	Pruned Shoot at 30° Angle	Pruned Shoot Vertical
Growth of terminal shoots, cm	26	29	51
Growth of lateral shoot, cm	42	16	57
Number of branches	5.7	4.9	5.4

[a] Reproduced by permission from Gyuro, 1980.

Tree response to pruning is also influenced by the size and type of cuts. Numerous small cuts stimulate growth more than a few larger cuts (Mika, 1982), and heading cuts stimulate more growth than thinning cuts (Mika, 1982) when a comparative amount of wood is removed in both cases.

The position of shoot angles strongly influences the rate of regrowth. Gyuro (1980) reported that the strongest regrowth occurred on vertical branches and the weakest on branches positioned at a 30° angle (Table 6.4).

Since strong growth occurs at the top of the tree, the training system compensates for this by positioning lower scaffold branches in the canopy with more vertical angles and higher scaffold branches with more horizontal angles. This compensates for the difference in growth between lesser basal growth and more vertical angles and more top growth but more horizontal angles (Figure 5.6C).

It is well established, based on several experiments, that summer pruning decreases growth of young trees even more than dormant pruning or no pruning (Alderman and Auchter, 1916; Aselage and Carlson, 1977; Ferree and Stang, 1980; Maggs, 1965; Mika et al., 1983). In older bearing trees summer pruning retards root growth and restricts canopy dimensions but is ineffective in decreasing shoot growth (Beakbane and Preston, 1962; Engel, 1974; Parnia et al., 1977; Taylor and Ferree, 1981; Utermark, 1977). Maggs (1965) calculated that summer pruning produced less growth, equivalent to a shortened growing season. Reduction of tree growth was equal to growth that has taken place during about one-third of the growing season, which is about 50 days under English conditions. Maggs (1965) pointed out that following winter pruning buds develop rapidly, whereas after summer pruning buds need time to complete differentiation, and thus the development of shoots is retarded.

In a different set of experiments Mika et al. (1983) compared summer pruning and thinning of young trees with similar dormant treatments. Both summer pruning treatments decreased growth by one-fifth compared to dormant pruning. Summer pruning produced short shoots and spurs, whereas winter pruning produced fewer longer shoots. Engel (1974) and Heinicke (1975) also noted that summer pruning produces more spurs.

TABLE 6.5 Effect of Pruning on Shoot Growth (cm)

Time	Short Pruning	Medium Pruning	Long Pruning
	TERMINAL		
June	40	37	34
July	43	47	44
October	58	40	34
March	47	39	39
Average	47	41	38
	LATERAL		
June	22	25	49
July	25	58	52
October	34	41	40
March	43	38	42
Average	31	40	46

[a] Reproduced by permission from Gyuro, 1980.

Several other investigators have measured the effect of summer pruning compared to corresponding dormant pruning. It is unquestionable that summer pruning decreases growth. In a detailed study Gyuro (1980) compared the combined effect of pruning severity and pruning time on shoot growth of apple. His data clearly indicate that on terminal shoots pruning severity increases regrowth, but on lateral shoots severity of pruning does not elicit strong regrowth (Table 6.5).

Some investigators used summer thinning done by hand pruning in an orchard. Others stimulated the effect of hedging machines, which obviously inflict heading-type cuts on the trees. Accordingly, results differ. Generally, the earlier the summer pruning is performed during the growing season, the greater the amount of regrowth that occurs (Ferree and Stang, 1980; Myers and Ferree, 1983b; Rom, 1982). Detailed results of summer pruning have been summarized by Ferree et al. (1984) and by Mika (1986).

Peach also responds to summer pruning. Blake (1917) recommended summer pruning for peaches to improve form and fruitfulness. Marini (1985) compared dormant pruning, summer pruning, and summer topping of peaches in June or July (60 or 90 days after bloom). Summer pruning and topping stimulated growth of 'Cresthaven' but not 'Sungreen' or 'Loring' peach cultivars. Thus a cultivar effect should be considered when summer pruning of peaches is planned.

At present high-density systems utilize mowing, which is summer heading of shoots. Although these studies examine the entire orchard system rather than pruning treatments, summer mowing is an increasing practice in high-density peach plantings. Hedging is distinguished from summer pruning by the fact that hedging is done by machines, which make indiscriminate heading cuts, whereas summer pruning is done by hand with the desired combination of heading or thinning cuts. In many ways the advantages offered by summer pruning in decreasing tree size are theoretical only because there is no practical way to carry them out. Hedging has the advantage that machine hedging can be practiced at times when hand labor is not available for summer pruning. Thus summer hedging is a form of summer pruning that has to be considered. Emmerson and Hayden (1985) used a complete program of hedging that starts the first 2 years of the tree's life by tipping the terminals by hand. From the third year on hedging is practiced relatively early in the season when terminals are 20–25 cm long. Hedging is repeated again 30 days later. Late summer hedging in peach, compared to the technique described in the preceding, has the disadvantage that regrowth is too weak for flower bud development and growth continues late in the season, causing insufficient hardiness. Hedging in apples is most advantageous on semidwarf to strong growing trees grafted on M7 or MM111 rootstock to control tree size.

EFFECT OF PRUNING ON PHOTOSYNTHESIS

Both dormant and summer pruning increase light penetration into the tree, which influences leaf structure and photosynthesis. Both dormant pruning (Mika, 1975; Mika et al., 1977) and early summer pruning (Taylor and Ferree, 1981) decrease leaf area of apple trees, but regrowth could compensate for the loss. Summer pruning, done in the middle of the season or later on apples, usually causes no regrowth. Thus pruning at this time permanently decreases the leaf area of the trees. If an excessive number of heading cuts are made during dormant pruning, excessive regrowth of new shoots the following season forms an outer mantle of the tree canopy, which hinders light penetration.

Dormant pruning can enlarge leaf size and mesophyll cell size and increase chlorophyll content per leaf area (Christiev, 1970; Polikarpov and Adaskalieij, 1973; Zelev, 1972, 1977). It also can extend the period when stomata are open daily by increasing the leaf water content (Aldrich, 1935). Heading cuts also influence photosynthesis by creating new sinks in the form of numerous growing points (Mika and Antoszewski, 1973). Shoot heading also appears to delay shoot senescence. During the growing season the photosynthetic activity of leaves slowly declines with age. Heading cuts seem to be able to restore photosynthetic activity to its previous high and decrease the rate of decline thereafter. Taylor and Ferree (1981) reported that summer heading of container-grown 'Topred-Delicious' apple trees increased the net photosynthesis of basal leaves by

36% eleven days after pruning and an additional 23% thirty-nine days after pruning compared to similar leaves on unpruned trees. These results were confirmed by Marini and Barden (1982b) on young trees, but Porpiglia and Barden (1981) did not observe increased photosynthesis as a result of summer pruning on mature bearing apple trees in spite of increased illumination of the tree canopy. Unlike apples, summer pruning did not increase light penetration with peach canopies, perhaps because the canopy at this time is quite open. However later in the season much less light penetrated into unpruned trees (Marini, 1985) (Figure 6.7).

Kanato and Nakaya (1968) asserted that severe dormant pruning of pear trees decreased the volume of respiring wood considerably; thus photosynthate loss due to respiration can be greatly decreased.

EFFECT OF PRUNING ON CARBOHYDRATE RESERVES

The previous discussion indicated that pruning effectively decreases tree size while increasing the new growth in the canopy. It may be expected that the change between various fractions of the canopy induced by pruning also changes the carbohydrate reserves within the tree. Studies on this subject are limited. Hooker (1924) reported that dormant pruning decreased starch and soluble sugar content in branches of apple trees. Cameron (1923) indicated that dormant pruned pear and peach trees started to accumulate starch and soluble sugars later in the season than unpruned trees. Aldrich and Grim (1938) found no change in carbohydrates in spurs of pears that remained on the tree after dormant pruning when they examined them during the spring. Also in long-term experiments Soczek et al. (1970) found that dormant pruning did not change carbohydrate reserves in leaves, shoots, and roots of young apple trees.

Taylor and Ferree (1981) reported that summer pruning of young apple trees of 'Topred-Delicious' decreased the dry weight of both basal stem sections and roots roughly in proportion to the amount of shoot removed, whereas the dry weight of regrowth shoots was influenced to a lesser degree. Although dry weight cannot be equated with carbohydrate reserves, it still can serve as an indicator of where such reserves may be deposited. Grochowska et al. (1977) studied the effect of dormant pruning and removing of buds from young apple trees on carbohydrate levels of shoots and roots. Total carbohydrate content was not significantly changed by the treatments. Carbohydrate metabolism, especially during the beginning of the growth season, was different in pruned and unpruned trees. Pruned trees were higher in soluble sugars and deposited less starch during the fast-growth period during the first 90 days after bloom. Starch deposition into the annual shoot of the pruned trees

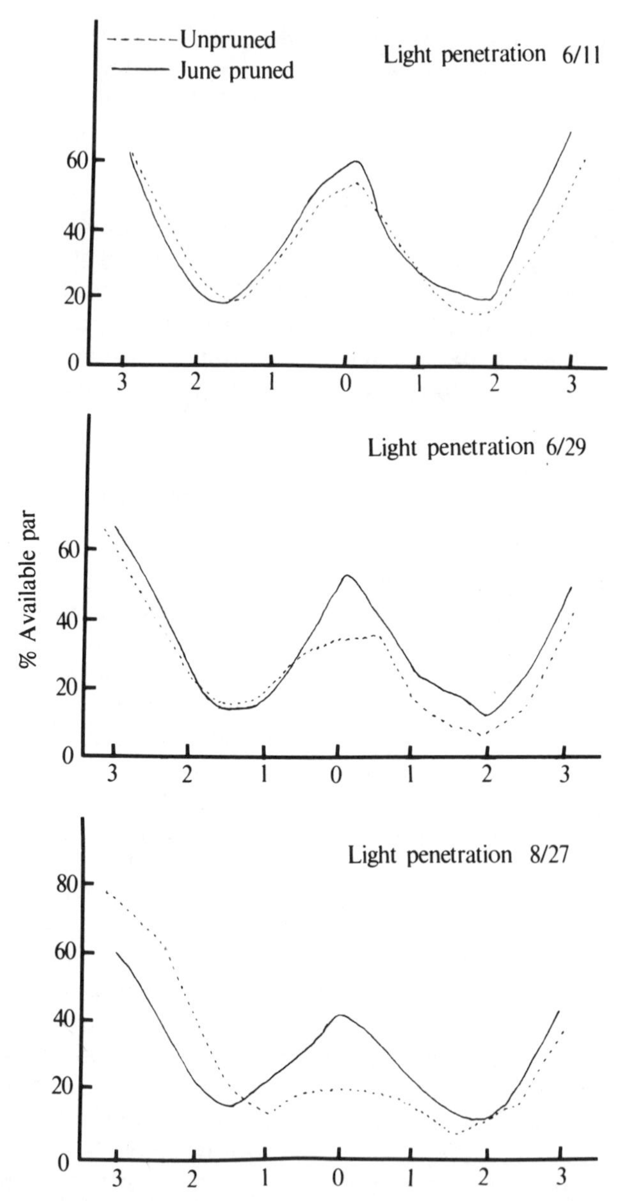

Figure 6.7 Light penetration into peach trees following summer topping at various times. Pruning has little effect early in the season, June 11, but by the end of the season, August 27, much less light penetrates into unpruned trees. (Reproduced by permission from Marini, 1985.)

started later, but by the end of the first 90 days of the growth period starch levels were the same.

EFFECT OF PRUNING ON FRUIT SET

Among the more important factors that should be kept in mind when pruning are light intensity (Jackson and Palmer, 1977) and tree vigor since these have a direct bearing on the maintenance of fruiting vigor. Consequently dormant-pruned, more vigorous trees are expected to fruit better. There are several reports that dormant pruning increased the percentage of blossoms that set fruit. (Aldrich, 1935; Aldrich and Grim, 1938; Aldrich and Work, 1935; Lalatta and Solaroli, 1970; Mika, 1982). Yet overly vigorous trees can become vegetative with a corresponding decrease in fruiting.

Chandler (1919) indicated that better fruit set of pruned trees might result from increased water and nitrogen supply to the tree. Aldrich and Grim (1938) found an increased N supply in fruiting spurs of pruned trees, concomitantly higher water content of leaves, and longer periods of stomatal opening. However, none of these factors seemed sufficient to explain the increased fruit set. Presently evidence is accumulating that fruit set is either controlled by hormonal levels or indirectly controlled by the available photosynthates. Growing shoots may compete for the available carbohydrates (Abbott, 1970; Grochowska and Karaszewska, 1978) However, the studies of Wilders et al. (1978) and Chalmers et al. (1978) indicate that the role of hormones in fruit set must be considered jointly with nutrient distribution.

Summer pinching (Sansavini, 1969) may also result in increased fruit set. Growing shoot tips have higher sink power than competing fruit sets (Abbott, 1970). Removal of growing shoot tips can improve fruit retention in apple (Quinlan and Preston, 1971), eliminating or partially eliminating competing sinks.

EFFECT OF PRUNING ON YIELD AND ANNUAL BEARING

Results of numerous experiments indicate that pruning influences yield and regularity of bearing. The results greatly depend on tree age and growing conditions, both of which exert a simultaneous influence on the tree at the time of pruning.

Mika (1986) reviewed the literature and concluded that as a rule dormant pruning of young trees suppresses cropping. He emphasized the point of view presented by Maggs (1965) that the main effects of pruning are on growth. Any pruning that stimulates the growth of young trees prolongs the vegetative phase and delays the fruiting process. In Chapter 4 conditions that push the tree toward the vegetative phase were discussed. It is easy to see why cropping of trees is particularly delayed if one or more of these conditions exist at the time pruning is applied.

Grubb (1931) concluded that young trees on very dwarfing rootstocks will fruit even when severely pruned, while similar trees on vigorous rootstock delay cropping as a result of pruning. This has been confirmed by many pruning experiments (Mika et al., 1983; Preston, 1957, 1960).

The type of pruning that stimulates growth will decrease fruiting. Dormant pruning decreases yield more than summer pruning (Mika et al., 1983). Heading cuts decrease yield more than thinning cuts (Heinicke, 1975; Soczek and Mrozowski, 1971), and many small cuts decrease fruiting more than a few larger cuts if the same amount of wood is removed (Preston, 1960). Because pruning of young trees delays fruiting, several investigators (Baldini, 1974; Crowe, 1980; Norton, 1980; Wertheim, 1980) recommend shoot bending rather than pruning to obtain the desired frame structure.

In older trees the number of fruit buds seldom limits yield unless the tree is in the 'off' year of a biennial bearing habit. Consequently, branches of older trees over the age of 6–8 years can be thinned quite severely without significantly affecting yield (Czynczyk et al., 1968; Mika et al., 1980). Greene and Lord, 1983; Pruning may prevent excessive yield of cultivars that tend to be biennial and promotes regular bearing (Grubb, 1929, 1937; Lazniewska and Mika, 1971; Preston, 1957, 1960, 1968). Pruning of such trees must be done in the winter of their 'on' year when fruit buds are numerous (Grubb, 1929). Pruning in an on year decreases fruit and stimulates growth in the on year, which in turn promotes more flowering in the off year. Hoblyn et al. (1936) reported that leader tipping had no effect on regularity of bearing but spear shortening early in the off year reduced biennial intensity compared with a similar treatment applied early in the on year. Although pruning may be helpful in relatively moderately bearing biennial cultivars, it could not correct irregular bearing of strongly biennial cultivars.

PRUNING OF PEACHES

Pruning of peaches has a very different purpose when it is compared to that of apples. As has been discussed before, the peach produces its flower buds on shoots while they are growing during the season (1-year-old wood). Therefore, it is essential that the tree is annually forced to grow. This is in contrast to the apple and similar types of trees that produce flower buds on several-year-old structures and annual growth is not required to produce the flower-bud-bearing surface. Another major

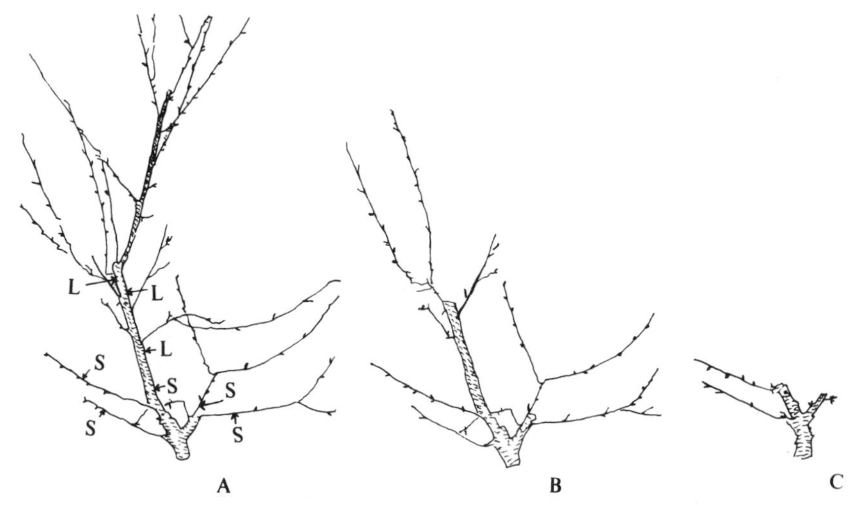

Figure 6.8 Pruning types of peach: A, unpruned; B, simplifying pruning; C, short pruning. Pruning cuts: L = for simplifying pruning; S = for short pruning.

difference between apple and peach is that the peach has much less apical dominance or, in the case of some cultivars such as the 'Compact Redhaven,' practically none. This leaves the choice to heading a peach shoot to its base buds to force it to grow or just severely 'simplify' the structure of the several-year-old branch, and the new shoots will develop without removing the top bud of each shoot. The choice between the so-called 'short pruning' and the 'simplifying' pruning is determined by the general growth rate of the tree. In dry climates, unirrigated trees need to be pruned short because the tree must be forced to grow at the maximum rate. In climates and under soil conditions where the peach tree has sufficient moisture and nutrition, the simplifying pruning is sufficient to induce growth on which flower buds will develop.

These two major pruning types are illustrated in Figure 6.8. Regardless of the scaffold system of the tree, the secondary branching almost always looks as illustrated in part A. The simplifying pruning is illustrated in part B and the short pruning in part C.

Scorza et al. (1986) compared the effect of simplifying pruning on regrowth of three types of peach trees: standard, semidwarf, and compact. He found that pruning of standard type of trees increased the current season wood fivefold in comparison to unpruned trees. In contrast, the effect of pruning was not significant on compact-type cultivars. The semidwarf trees, which he created by crossing the other two types, were intermediate in response to pruning. The number of current season shoots of compact types were much higher than standard types whether

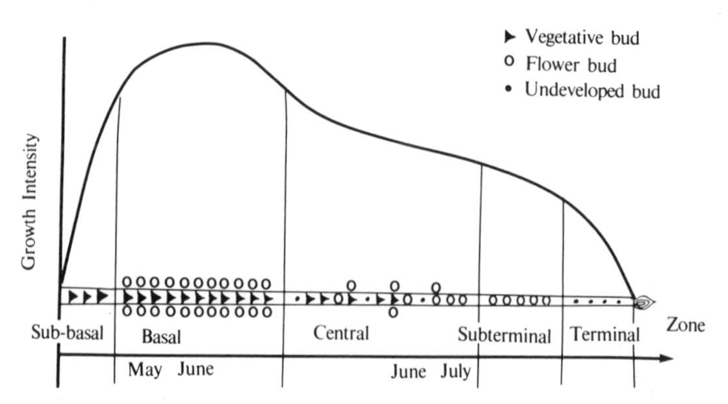

Figure 6.9 Growth intensity and flower bud development on peach shoots. (Reproduced by permission from Timon, 1976.)

they were pruned or not. These trees, as mentioned previously, have practically no apical dominance, and the buds will break whether or not pruning is applied. In contrast, the standard type of trees only break buds as a response to pruning even though only a portion of the total apical buds are removed and apical buds on remaining shoots are intact. By reducing the overall effect of a conglomerate of apical buds, the remaining buds will respond and grow. The standard and compact type of trees are illustrated in Figure 1.10.

The purpose of pruning peaches is to produce shoots at the 40–60 cm range because these are the shoots on which flower bud development is most numerous. Shoots of this size also develop the approximate thickness used to characterize peach shoots that develop flower buds. Timon (1976) classified flower bud development on peach shoots into five zones (Figure 6.9). In the *subbasal* zone, shoot growth rate is relatively weak, only partially developed vegetative buds can be found. In the *basal* zone, growth intensity reaches its maximum. The internodes are strongest here. Two flower and one vegetative buds develop at every node. The *central* zone is characterized by less growth intensity. The vegetative character of this zone can be observed by few vegetative buds and often solitary flower buds on the nodes. In the *subterminal* zone, the vegetative character is usually completely absent. Flower buds are solitary and single. The *terminal* zone is characterized by short internodes. Few undeveloped buds (flower or vegetative) are visible.

On shoots less than 30 cm, the subbasal, the upper portion of the central, the subterminal, and the terminal zones are apparent. The most valuable part of the shoot, the basal and lower part of the central zones, are missing. On shoots 10 cm or shorter, usually the basal, subterminal, and terminal zones are present. It appears that the growth intensity of shorter shoots is never strong enough to produce the intensity needed for bud development in the basal zone. Consequently on shoots less than 40 cm, the most valuable part is missing. Thus pruning should produce shoots of 40–60 cm in length. Development of peach flower buds related to the intensity of growth is illustrated in Figure 6.9.

EFFECT OF PRUNING ON FRUIT QUALITY

Fruit quality has been described by many parameters. Fruit size, color, and storage quality, which are directly related to Ca content, are the three quality factors that could be related to pruning.

The effect of dormant pruning on fruit size has been reviewed by Mika (1986). Dormant pruning decreases the number of fruits per tree and correspondingly increases the size of the individual fruit. Fruit size is especially increased by pruning on cultivars such as 'Golden Delicious' apple, 'Conference' pear, and 'Wangenheim' prune, which tend to bear abundant crops. Summer pruning, particularly summer hedging with mechanical hedgers, decreases fruit size on peach (Brown and Harris, 1958) and on apples (Ferree et al., 1984; Green and Lord, 1983; Sako, 1966; Upshall and Barkov, 1963; Wertheim and Lemmens, 1978; Lemmens and Spruit, 1980). Marini and Barden (1982a) reported a decrease in fruit size as a result of winter pruning in 'Stayman' apples but no influence on 'Golden Delicious' and 'Delicious' apples. In contrast, Lemmens (1980) and Taylor and Ferree (1984) obtained larger apples after summer pruning.

The effect of dormant pruning on fruit color is not always positive. In some studies, dormant pruning improved apple fruit color (Rogers and Preston, 1947); but in many other studies, fruit color was decreased (Christensen, 1969; Preston, 1968; Thorsrund, 1965).

Few large thinning cuts that enhance high penetration into the canopy but do not stimulate much growth usually enhance fruit color (Mika, 1982). In contrast, many small cuts such as inflicted by spur pruning decrease fruit color (Christensen, 1969; Preston, 1968). Poor color may be caused by shading resulting from intensive shoot growth induced by many cuts (Mika, 1969) or by increased N supply experienced in pruned trees. Summer pruning in most pruning trials increased color by increasing light penetration to the fruit (Lemmens, 1980; Lemmens and Spruit, 1980; Lord and Green, 1982; Preston and Perring, 1974; Tymoszuk et al., 1984). However, light alone cannot improve color unless other factors are also met.

The Ca content of fruit is directly connected with fruit quality. Directing Ca into the fruit is important. Any horticultural manipulation that achieves this purpose is desirable. Preston and Perring (1974) reported that summer pruning removes the competition for Ca by eliminating the strong sink created by growing shoot tips, and this increases Ca transport into the fruit. These results were confirmed by Lord et al. (1979) and Olszewki and Slowik (1982). Contradictory results also have been reported by Lord and Green (1982) and Lemmens (1982), who found no increase in Ca from similar treatments. Low-Ca apples develop bitter pit and other storage disorders. Elimination of this disorder by itself, even without measuring Ca content, is an important measure of storage quality. Less bitter pit incidence has been noted on summer-pruned trees (Borsboom, 1976; Lemmens, 1982; Lord et al., 1979; Marini and Barden, 1982; Preston and Perring, 1974; Schumacher and Frankhauser, 1974; Strukclec, 1981; Tepe, 1969).

In some experiments, pruning did not lower the incidence of bitter pit (Greene and Lord, 1983; Schumacher et al., 1974). It must be noted that Ca uptake requires carbohydrates transported into the root. Decreasing foliage may deprive the roots of the necessary carbohydrates. This results in decreased Ca uptake, and obviously pruning has no or an adverse effect on bitter pit of apples.

Summer pruning could, under certain conditions, decrease assimilates available for the fruit. If such conditions exist, the soluble solid content of the fruit may decrease (Lord and Green, 1982; Marini and Barden, 1982c; Strukclec, 1981).

EFFECT OF PRUNING ON HORMONAL PATTERN OF FRUIT TREES

Pruning removes plant parts that are sites of hormone production. It also reduces the upper portion of the tree without changing the roots. Thus the root-produced hormones are expected to dominate, and the balance of hormones is expected to change. In 'McIntosh' trees, Mika et al. (1983) and Grochowska et al. (1984) analyzed the growth and hormonal content of various parts of the tree by bioassays.

Mika et al. (1983) demonstrated that removing parts of apple trees that may produce the essential hormones greatly influences the growth and development of the tree. They compared dormant thinning, light and severe dormant heading, early summer pinching, late summer thinning, removing dormant buds, removing shoot apexes repeatedly, and removing new leaves repeatedly. Results are presented in Table 6.6. The heading, thinning, and summer pinching treatments gave the expected results. Shoot-apex removal and removal of new leaves repeatedly severely retarded growth and flower bud development. These results indicated that hormones produced by young leaves are essential for development of the tree. The various growth types resulting from the treatments are illustrated in Figure 6.10. Heavy dormant pruning induced vigorous

TABLE 6.6 Effect of Removal of Various Parts of the Tree on Tree Response in Apple*

Treatment[2]	Branch Spread (m^2)	Number of Spurs (cm^{-2} trunk Cross-sectional area)	Length of Shoot (cm average)	Total Shoot Growth (m)	Trunk Cross-sectional Area (cm^2)	Number of Flower Buds per Tree
1. Thinning third of dormant shoots removed	9 de†	11 dc†	40.4 c†	93 cfg†	38 bcd†	187 bc†
2. Dormant shoots headed by one-third	7 bcd	10 dc	38.8 c	126 h	44 cde	171 b
3. Dormant shoots headed by two-thirds	5 b	6 a	47.9 g	145 i	38 bcd	196 bcd
4. Pinching in June to 4–5 leaves	8 cd	15 fg	29.2 c	80 c–f	44 cde	265 cde
5. Pinching in July to 7 leaves	7 bc	12 fg	22.2 b	73 cde	49 c	210 bcd
6. One-third of shoots removed in August	8.4 cd	15 fg	39.2 e	90 d–g	42 cde	281 de
7. All dormant lateral buds removed	8.2 cd	8 bc	44.2 f	77 c–f	32 b	277 cde
8. One-half of dormant lateral buds removed	9.7 de	12 bf	34.7 d	106 gh	43 cde	209 bcd
9. Shoot apexes removed every 2 weeks	1.0 a	7 ab	15.8 a	19 a	18 a	48 a
10. New leaves removed every week	2.1 a	18 fg	27.7 c	52 b	19 a	66 a
11. Control, untreated	11.1 e	18 fg	39.7 e	96 fg	40 bc	388 c

* 'McIntosh' cultivar M26 rootstock.
† Separation within columns by Duncan's multiple range test (5% level). Numbers followed by different letters are significantly different. Reproduced by permission from Mika et al., 1983.

Figure 6.10 Tree responses of apple to removal of various plant parts in a northern producing area. Numbers correspond to treatments described in Table 6.6. Marker is 0.5 m. (Reproduced by permission from Mika et al., 1983.)

growth of apples, which was accompanied by higher levels of cytokinins, auxins, and gibberellins (Grochowska et al., 1984). The hormones were determined by bioassays, and the magnitude of change was higher in the trunk and scaffold branches than in leaves or annual shoots (Figure 6.11). As a result of pruning, hormone content increased in the resulting shoots, but whether the higher hormone content was the result of vigorous growth or the vigorous growth was the result of higher hormone content is not clear.

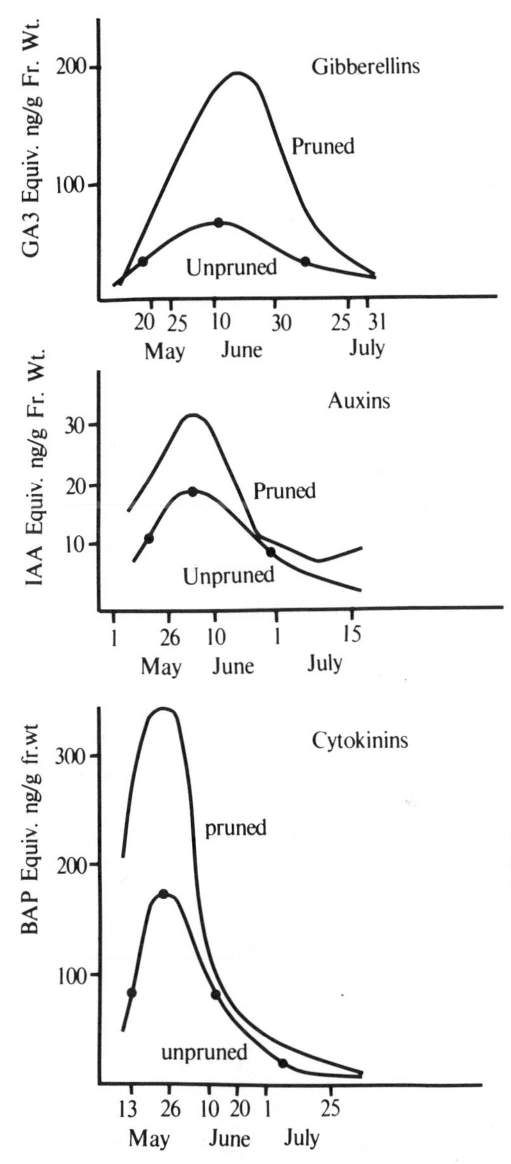

Figure 6.11 Hormone fluctuation in trunk and scaffold branches of apples due to pruning (for data points, see original paper). (Reproduced by permission from Grochowska et al., 1984.)

TABLE 6.7 Ethylene Production in Apple Shoots[a]

	μl/g fw
One-year-old wood	0.45
Two-year-old wood	0.50
Spur-forming flower bud	1.20
Stub 50 hr after Lorette-type pruning	1.00

[a] Reproduced by permission from Klein and Faust 1978.

Pruning inflicts wounds on the tissue at the point of branch removal. Plants usually produce ethylene in tissues surrounding a wound. Klein and Faust (1978) measured the wound ethylene production of the portion of shoot left on the tree after Lorette pruning. Ethylene increased temporarily in the wound tissue to a level that is probably sufficiently high to induce flower bud formation (Table 6.7). It is possible that one pruning is not enough to produce flower buds at the base because the developing shoot and its young leaves produce GA, which can counteract the flower bud formation process. Perhaps this is the reason that two or more prunings are needed in the Lorette system. Steps of the Lorette pruning system are illustrated in Figure 6.12.

Pulling the branches in a horizontal position also increases their ethylene content (Robitaille, 1975; Robitaille and Leopold, 1974), and it usually increases flower bud formation.

Pruning can be done not only to the shoots but also to the roots of the tree. Root pruning is presently done by tractor-mounted implements that cut into the orchard soil cutting away the roots in its path. Cutting depth is generally 30–35 cm, and the distance is 60–80 cm from the trunk depending on the vigor of the trees. The time for cutting in Switzerland is from December to March. Treatments in April did not produce the desired results (Schumacher, 1975). In England better results were obtained by winter root pruning. Cutting is done on both sides of hedgerows or on both sides of the row in relatively closely planted orchards. Schumacher (1975) remarked that it is difficult to determine the optimum distance of the cut from the tree. If too few roots are cut, excessive root development is induced and shoot growth will increase. If too many roots are cut, the trees may die, especially in dry years. If the correct cut is made, the amount of vigorous shoot growth is reduced, shoot growth terminates earlier, and flower bud initiation is encouraged (Geisler and Ferree, 1984).

Root growth produces a number of physiological changes in the tree. The reasons these changes occur in most cases are not known. The effects were reviewed by Geisler and Ferree (1984), and a summary is presented in Table 6.8.

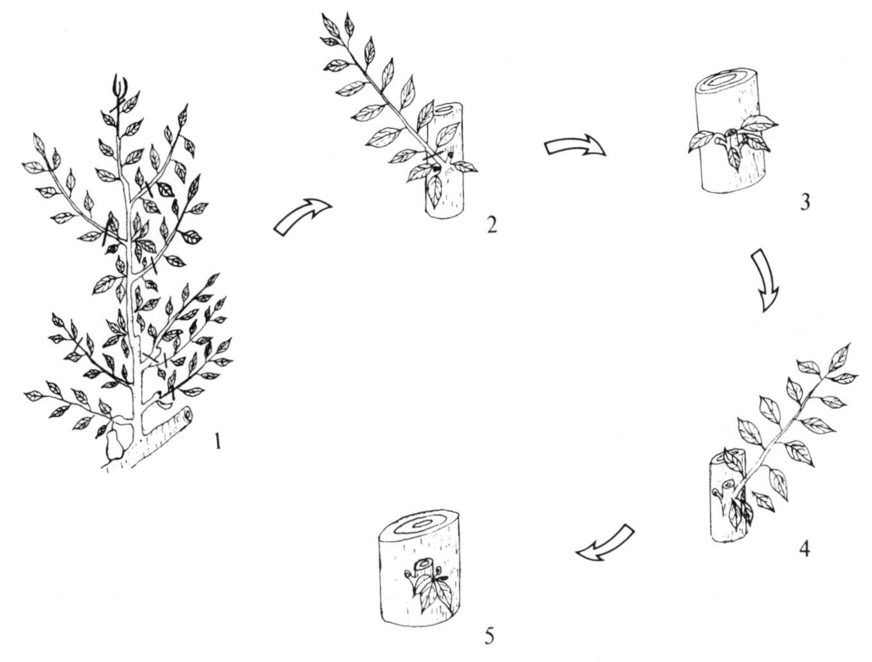

Figure 6.12 Lorette pruning. Step 1: Development of shoot after dormant pruning; mid-June cuts are made as indicated. Step 2: After regrowth of 20 cm, shoots are cut back. Step 3: Spars may develop (s) or spurs and shoots develop. Step 4: When spurs and shoots or only shoots develop, shoots are cut again. Step 5: Usually spurs develop.

TABLE 6.8 Effect of Root Pruning on Fruit Trees[a]

Effect	Importance
Reduced shoot growth	Tree is smaller
Root growth and root activity increases after moderate pruning	More hormone (cytokinin) and Ca uptake by roots
Root–shoot ratio is readjusted in favor of roots	Better balanced tree
Reduction of water uptake and lowered water potential (especially if cuts are repeated)	Smaller fruit; water stress, which may cause stomatal closing
Increased assimilation of CO_2 3–7 weeks after root pruning	More carbohydrates are available for tree functions
Induces or enhances flower bud development (pruning must be done in summer)	More flowers but not necessary more fruit
Increases fruit set (pruning must be done in winter)	Possible less competition from shoot growth

[a] Details for each point are reviewed by Geisler and Ferree (1984). Considerations pertinent to decreasing tree size are also presented in Chapter 5.

REFERENCES

Abbott, D. L. 1970. XVII Int. Hort. Congr. 1970:1–57.

Alderman, W. H., and E. C. Auchter. 1916. W. Virg. Agr. Exp. Sta. Bull. No. 158.

Aldrich, W. W. 1935. Proc. Amer. Soc. Hort. Sci. 32:107–114.

Aldrich, W. W., and J. H. Grim. 1938. Proc. Amer. Soc. Hort. Sci. 36:328–334.

Aldrich, W. W., and R. A. Work. 1935. Proc. Amer. Soc. Hort. Sci. 32:115–123.

Aselage, J., and R. F. Carlson. 1977. Compact Fruit Tree 10:77–85.

Baldini, E. 1974. XIX Inter. Hort. Congress, Warszawa 11–18 Sept. 1974, 3:115–124.

Barlow, H. W. B., and C. R. Hancock. 1960. Bot. Gaz. 121:208–215.

Barlow, H. W. B., and C. R. Hancock. 1962. Rep. E. Malling Res. Sta. 1961/1962:71–76.

Beakbane, A. B., and A. P. Preston. 1962. Rep. E. Malling Res. Stn. 1961/1962:57–60.

Blake, M. A. 1917. Proc. Amer. Soc. Hort. Sci., 14:14–23.

Borsboom, O. 1976. Fruitteelt, 66:784–785.

Brown, D. S., and R. W. Harris. 1958. Proc. Amer. Soc. Hort. Sci., 72:79–84.

Cameron, S. H. 1923. Proc. Amer. Soc. Hort. Sci., 20:98.

Chalmers, D. J., R. A. Wilders, I. R. Dann, and C. C. Hunter. 1978. XX Int. Hort. Congr., Sydney, Australia (Abstract 1492).

Chandler, W. H. 1919. Proc. Amer. Soc. Hort. Sci. 16:88–101.

Christensen, J. V. 1969. Tids skrift Planteavl 73:429–433.

Christiev, J. Z. 1970. Effect of puning on biology and physiology of apple. Ph.D. Dissertation, Fruit Research Institute, Plowdiw, Bulgaria.

Crabbe, J. J. 1984. Acta Hort. 146:113–120.

Crowe, A. D. 1980. Compact Fruit Tree. 13:91–96.

Czynczyk, A., A. Mika, and H. Domanska. 1968. Prace Inst. Sad. 12:236–242.

Emmerson, I. H., and R. A. Hayden. 1985. Amer. Fruit Grower. 105(10):10–11.

Engel, G. 1974. Erwerbsobstbau 16(3):47–48.

Ferree, D. C., S. C. Myers, C. R. Rom, and B. H. Taylor. 1984. Acta Hort. 146:243–252.

Ferree, D. C., and E. J. Stang. 1980. Ohio Agri. Res. Dev. Ctr., Res. Circ. 259:4–6.

Gardner, V. R., F. C. Bradford, and H. D. Hooker. 1922. The fundamentals of fruit production. McGraw-Hill, New York.

Geisler, D., and D. C. Ferree. 1984. Hort. Rev. 6:155–188.

Greene, D. W., and W. J. Lord. 1983. J. Amer. Soc. Hort. Sci., 108(4):590–595.

Grochowska, M. J., A. Karaszewska, and A. Mika. 1977. Fruit Sci. Rep. 4:7–13.

Grochowska, M. J., and A. Karaszewska. 1978. Acta Hort. 80:457–464.

Grochowska, M. J., A. Karaszewska, B. Jankowska, J. Maksymiuk, and M. W.

Williams, 1984. J. Amer. Soc. Hort. Sci. 109(3):312–318.

Grubb, N. H. 1929. Apple pruning. Ann. Rep. E. Malling Res. Stn., 1929/1930:62–69.

Grubb, N. H. 1931. Rep. E. Malling Res. Stn. 1931/1932:61–69.

Grubb, N. H. 1937. Rep. E. Malling Res. Stn. 1937/1938:237–241.

Gyuro, C. 1980. Müvelesi rendszerek es metszesmodok a modern gyümölcstermesztesben. Budapest: Mezögardasagi Kiado. pp. 356.

Heinicke, D. R. 1975. Agric. Res. Service USDA Agric. Handbook 458, pp. 1–34.

Hoblyn, T. N., N. H. Grunn, A. C. Painter, and B. L. Wates. 1936. J. Pomol. 14:39–76.

Hooker, H. D. 1924. Missouri Stat. Res. Bull. No. 72.

Jackson, J. F., and J. W. Palmer. 1977. J. Hort. Sci. 52(2):253–266.

Jonkers, H. 1962. Hort. Congr. Brussels 1962:441–443.

Jonkers, H. 1982. Scientia Hort. 16(3):209–215.

Kanato, Kitsuo, and Nakaya, Hidaji. 1968. Bull. Hort. Res. Sta. Japan, Ser. A., No. 7, 143–156.

Klein, J. D., and M. Faust. 1978. Hort. Sci. 13:164–166.

Knight, R. C. 1934. J. Pomol. 12:1–14.

Koopmann, K. 1896. Grundlehren des Obstbaumschnittes. Berlin: Paul Parey.

Lalatta, F., and Solaroli. 1970. Ann. Instit. Sparimentale Fruticol. Roma, 1(1):165–172.

Lazniewska, I., and A. Mika. 1971. Proc. Inst. Sad. 15:73–79.

Lemmens, J. J. 1980. Fruit Belge 48(391):205–210.

Lemmens, J. J. 1982. Fruitteelt 72(6):196–198.

Lemmens, J. J., and G. Spruit. 1980. Fruitteelt 70(32):988–990.

Lord, W. J., and D. W. Greene. 1982. Hort. Sci. 17:372–373.

Lord, W. J., D. W. Greene, W. S. Bramlage, and M. Drake. 1979. Compact Fruit Tree 12:23–29.

Lorette, L. 1926. La taille Lorette, 6th ed. Versailles: Bibliothéque de la Revue Jardinage.

Maggs, D. H. 1959. Ann. Bot. Lond. N.S. 23:319–330.

Maggs, D. H. 1965. J. Hort. Sci. 40(3):249–265.

Marini, R. P. 1985. J. Amer. Soc. Hort. Sci. 110:133–139.

Marini, R. P., and J. A. Barden. 1982a. J. Amer. Soc. Hort. Sci. 107:474–479.

Marini, R. P., and J. A. Barden. 1982b. J. Amer. Soc. Hort. Sci., 107:39–43.

Marini, R. P., and J. A. Barden. 1982c. J. Amer. Soc. Hort. Sci., 107:34–39.

Mika, A. 1971. Proc. Inst. Sad. 15:63–72.

Mika, A. 1975. Fruit Sci. Rep. 2(1):31–42.

Mika, A. 1982. Proc. XXI Inter. Hort. Congress, 29 August—4 September, Hamburg 1982. 1:209–221.

Mika, H. 1986. Hort. Rev. 8:337–378.

Mika, A., and R. Antoszewski. 1973. Biologia Plant. 15(3):202–207.

Mika, A., M. J. Grochowska, and A. Karaszewska. 1977. Fruit Sci. Rep. 4(3): 1–5.

Mika, A., A. Jakiewicz, and M. Potocka. 1980. Pr. Inst. Sadov. (Ser. A.) 22:25–31.

Mika, A., M. J. Grochowska, and A. Karaszewska. 1978. Abstract of XX Inter. Hort. Congr. Sydney, Australia, 15–23 August, Abstr. no. 1479.

Mika, A., M. J. Grochowska, A. Karaszewska, and M. Williams. 1983. J. Amer. Soc. Hort. Sci., 108(4):655–660.

Myers, S. C., and D. C. Ferree. 1983a. J. Amer. Soc. Hort. Sci. 108(1):4–9.

Myers, S. C., and D. C. Ferree. 1983b. J. Amer. Soc. Hort. Sci. 108(4):630–633.

Norton, R. L. 1980. Ann. Rep. State Hort. Soc., New York 110:35–47.

Olszewski, T., and K. Slowik. 1982. Abstracts XXI Int. Hort. Congr. Hamburg, 1, 1114.

Polikarpov, V. P., and M. M. Adaskalieij. 1973. Vestn. Skh. Nauki (Moskow) 18(8):84–87.

Porpiglia, P. J., and J. A. Barden. 1981. J. Amer. Soc. Hort. Sci. 106(6): 752–754.

Preston, A. P. 1957. J. Hort. Sci. 32:133–141.

Preston, A. P. 1960. J. Hort. Sci. 35:146–156.

Preston, A. P. 1968. J. Pomol. 12:110.

Preston, A. P., and M. A. Perring. 1974. J. Hort. Sci. 49:77–83.

Quinlan, J. D., A. P. Preston. 1971. J. Hort. Sci. 46(4):525–534.

Robitaille, H. A. 1975. J. Amer. Soc. Hort. Sci. 100:524–527.

Robitaille, H. A., and A. C. Leopold. 1974. Physiol. Plant 32:301–304.

Rogers, W. S., and A. P. Preston. 1947. Rep. E. Malling Res. Sta. 1946/1947: 49–54.

Rom, C. R. 1982. The influence of time and severity of summer pruning on peach tree growth and development. M.S. Thesis. Ohio State University.

Sako, J. 1966. Hedelmalehtr 13:30–31.

Sansavini, S. 1969. Universitá di Bologna, Public, 175:267–302.

Schumacher, R. 1975. Schweiz. Obst. Weinbau. 111:115–116.

Schumacher, R., and F. Frankhauser. 1974. Schweiz. Zeitschr. Obst. Weinbau. 108:243–251.

Schumacher, R., F. Frankhauser, and W. Stadler. 1974. Schweiz. Obst. Weinbau. 110(24):654–657.

Scoiza, R., Li Zailong, G. W. Lighner, and L. E. Gilreath. 1986. J. Amer. Soc. Hort. Sci. 111:541–545.

Soczek, Z., W. Klossowski, K. Slowik, A. Mika, and W. Reszczyk. 1970. Pr. Inst. Sadow. Skierniewicach 14:43–70.

Soczek, Z., and M. Mrozowski. 1971. Proc. Inst. Sad. 15:37–51.

Strukclec, A. 1981. Gartenbauwissenschaft. 46(6):268–276.

Taylor, B. H., and D. C. Ferree. 1981. J. Amer. Soc. Hort. Sci. 106(3):389–393.

Taylor, B. H., and D. C. Ferree. 1984. J. Amer. Soc. Hort. Sci. 109:19–24.

Tepe, W. 1969. Iber 1968 Hessische Lehr-u. Forschungsants. für Wein- Obst- u. Gartenb., Geisenheim. Rheingau, pp. 108–109.

Thorsrund, J. 1965. Yrksfruktdyrking 2:15–16.

Timon, B. 1976. Öszibarack (Peach). Budapest: Mezögazdasagi Kiado.

Tymoszuk, S., A. Mika, and R. Antoszewski. 1984. Fruit Sci. Rep., 11(4): 149–154.

Upshall, W. H., and J. Barkov. 1963. Rep. Ont. Hort. Exp. Sta. Prod. Lab. 16–19.

Wertheim, S. J. 1980. Acta Hort. 114:318–325.

Wertheim, S. J., and J. J. Lemmens. 1978. Ann. Rep. Res. Sta. Fruit Growing, Wilhelminadorp 11–18.

Wilders, R. A., I. R. Dan, D. J. Chalmers, and C. Hunter. 1978. XX Int. Hort. Congr., Sydney, Australia (Abstract 1941).

Zelev, I. 1972. Gradin. lozar. Nauka 9(15):13–18.

Zelev, I., and I. Filizence 1977. Gradin, lozar Nauka 14(1):3–9.

7

RESISTANCE OF FRUIT TREES TO COLD

Most temperate zone fruit trees are subject to periodic winter cold or spring frosts, which cause severe economic losses. Fruit-producing areas developed at geographic locations where cold-induced losses are minimized nevertheless occur occasionally. Growers are painfully aware of the consequences if the trees, for any reason, are not prepared for the winter. Thus understanding hardiness as a phenomenon and as an aid in avoiding cold injury is of major importance for horticulturists and physiologists alike. The early literature on cold hardiness of fruit trees was summarized by Chandler (1913, 1954); a recent general summary can be found in the book written by Sakai and Larcher (1987).

Cold hardiness means resistance to a wide array of types of injuries caused by low temperature. Quamme (1978) lists the types of injuries occurring in apple to illustrate this point. Late fall immature buds may be susceptible to fall frosts (Simons, 1972); severe freezes may even injure the trunk and the larger branches of the tree (Maney, 1940). Darkening of xylem, an oxidative browning of xylem cells, is a result of winter freezes and is called 'black heart' (Proebsting, 1963). Sunscald is a type of bark injury that occurs on the southwest side of the tree and is thought to be the result of radiant heating and cooling during periods of high radiation in winter (Levitt, 1972). Injury may occur to the roots, and occasionally the tree may be killed because of severe root injury (Cummins and Aldwinkle, 1974). The bark may split as a result of winter injury, and the unprotected xylem may dry out, or secondary rotting organisms may invade the tissue through the split (Faust, unpublished). Flower buds may be killed just before or during the period when they resume growth in spring. The array of injuries occurring in other fruit species is similar to that occurring in apple.

Susceptibility of tissues to cold injury may differ. Leaves of deciduous fruit trees acclimate little. Roots are consistently less hardy than overwintering stems (Weiser, 1970a). During fall the xylem and bark tissues have about equal hardiness, but during the winter the bark survives lower temperatures (Potter, 1939). The tissues of flowers differ in their cold resistance with the stage of their development (Westwood, 1978).

Cold hardiness is changeable during the winter and is under the control of a number of variables (Quamme, 1978). During active growth tissues have little cold tolerance. During the fall growth ceases and hardiness develops. The development of hardiness is called *acclimation*, and the loss of hardiness is *deacclimation*. Acclimation and deacclimation processes are under the control of environmental stimuli. Depending on the species and the acclimation process, fruit trees survive various temperatures. Approximate hardiness zones were established on the North American continent and are illustrated in Figure 7.1. Fruit trees are not expected to survive in zone 2 with temperatures below −40°C. In zone 3, in which temperatures range during the winter from −34.4 to −40°C, the trees survive but commercial production areas are confined to geographic locations where the local microclimate is somewhat tempered by the presence of large bodies of water such as Lake Michigan or Lake Ontario.

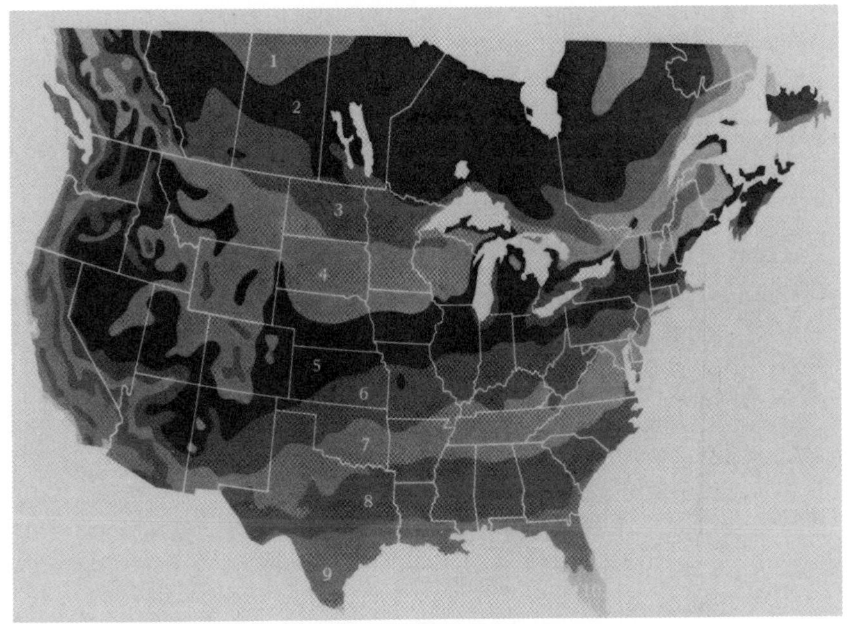

Figure 7.1 Winterhardiness zones in the North American continent. Zone 1, below −45.6°C. Zone 2, −45.6 to −40°C. Zone 3, −40 to −34.4°C. Zone 4, −34.4 to −28.9°C. Zone 5, −28.9 to −23.3°C. Zone 6, −23.3 to −17.9°C. Zone 7, −17.9 to −12.2°C. Zone 8, −12.2 to −6.7°C. Zone 9, −6.7 to −1.1°C. Zone 10, −1.1 to 4.4°C. (From Plant Hardiness Zone Map U.S. Department of Agriculture Miscellaneous Bulletin 814 1960.)

The hardiness zones were established with data points representing minimum temperatures and do not take into consideration the deacclimation processes. In general, during warm periods in winter deacclimation may occur. Fruit trees, even though they are 'hardy' if acclimated, are susceptible to severe cold injury in geographic areas where southern winter winds warm the air for brief periods of time and the trees loose acclimation. The central and mid-Atlantic areas of the United States are especially vulnerable to this type of cold injury. The changing temperature pattern over the mid-Atlantic area is illustrated in Figure 7.2. As high-pressure systems pass over the area from northwest to southeast, the clockwise circulation pattern of the wind changes and a southern, warm air flow is followed by a northern, cold air flow that could have devastating results even though the temperature of the air is only −10 to −15°C.

In geographic areas where the deacclimation process is important, trees that have the ability to remain dormant seem to be hardier than those known from the northern areas for their extreme hardiness. A good example of this is the comparison between the peach rootstocks 'Siberian C', a very hardy rootstock well adapted to very cold conditions, and

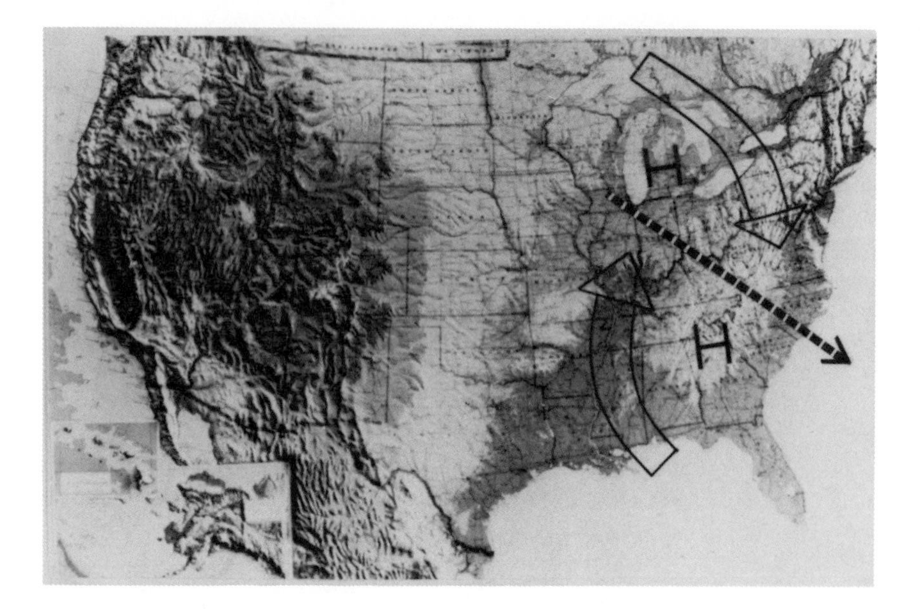

Figure 7.2 Schematic representation of major weather patterns governing air flow over the eastern United States: H, centers of high pressure with clockwise circulation patterns. High-pressure areas move from northwest to southeast, represented by dotted line.

'Boone County' or 'Bailey,' two rootstocks developed in the central United States. 'Boone County' and 'Bailey' are late blooming peaches, which extricate from dormancy very slowly. In the mid-Atlantic states, where dormancy is important, both are hardier than Siberian C, a rootstock where dormancy is shallow. In contrast, the opposite is true for areas of Southern Canada. There the dormancy is not important because temperature fluctuation is rare. Therefore, 'Siberian C', the hardy rootstock, has the clear advantage over 'Boone County' or 'Bailey.' Thus, depending on environmental conditions, different biochemical or physiological processes emerge as the most important to enable trees to survive cold stresses. In northern areas, the ability to develop a high degree of hardiness is most important for survival at low temperatures. In areas with fluctuating winters, even though the cold is relatively mild by northern standards, the ability to maintain dormancy and retain acclimation or the ability to reacclimate quickly are characteristics that assure survival.

Breeders studying the inheritance of cold hardiness to develop cultivars of fruit trees that are 'hardy' continually look toward the northern limits of fruit production in their aim to develop hardy cultivars (Quamme, 1978). It must be emphasized that fruit production has moved south during the past three decades. Thus in the future, the deacclimation component of hardiness will be more important than the ability to withstand very low temperatures. Furthermore, it must be noted that species or cultivars evolved or developed in geographic areas with prolonged cold winters usually deacclimate rapidly and resume growth as soon as temperatures reache the low threshold. Thus trees developed to withstand very low temperatures experienced in the north are not suitable for geographic areas where deacclimation is the most important feature of cold survival. Apricot is a good example for this. Apricot is indigenous in northwestern China and central Asia, areas of continental climates with −33 to −40°C temperatures and nonfluctuating winter temperatures (Bailey and Hough, 1975). Apricot can be grown in southern Canada, where winters are cold and fluctuation is minimal, or in California, where there are no winters. In between, especially in the mid-Atlantic states, apricots cannot survive. Therefore, the central idea is that the deacclimation process is and will be very important in hardiness.

EVALUATION OF HARDINESS

Cold hardiness can be evaluated in several ways: by visual observations after a natural freeze, freezing the trees or part of the trees in situ in portable freezing chambers, freezing excised parts of trees in the laboratory in freezing chambers, determining conductivity of plant extracts, and determining hardiness by exotherm analysis. Each of the methods has some unique features and a brief description of each is warranted.

Determination of Cold Hardiness following Natural Freezes

Useful information can be obtained by visually assessing survival after natural freezes. This method allows an opportunity for varietal evaluation and determining the nature and frequency of the injury. The degree of injury can be scored using a numerical rating system, and since the entire orchard is affected, examining a large number of samples is easily possible. The disadvantage of this method is that test winters may be infrequent and variable. For example, there were only 11 test winters in this century. Thus a test winter may occur on average only every 7–8 years (Childers, 1983).

Determination of Cold Hardiness by Freezing Plants In Situ in Portable Freezing Chambers

This method can be applied to intact trees or branches at any desired time during the season(s), thus determining the development of hardiness (Quamme, 1978). It is the closest technique to natural freezes because the trees are in an orchard setting throughout the process. The disadvantage of this method is that the size of the chamber is a limiting factor and the number of chambers are usually few because of the expenses involved. Thus only a limited number of experiments can be made in this way.

Determination of Cold Hardiness by Freezing Tree Parts in Laboratory Freezing Chambers

A common method to determine hardiness is to freeze the plant parts in the laboratory and evaluate the damage by tissue browning or one of the several objective methods measuring cell vitality. Usually, more precise control of temperature can be achieved in the laboratory than in the orchard. The laboratory procedures can be adapted to handle a large number of samples in a short period of time, and the laboratory techniques allow the determination of lethal doses of temperature, which can express absolute hardiness levels. The expression of cold hardiness is usually affected by freezing and thawing rates used in the test. Cooling rates most often used are in the range of 1–2°C per hour. Cooling rates above 8–15°C per hour usually overestimate the sensitivity of tissues (Proebsting and Sakai, 1979).

Commonly, tissue browning after incubation is a good estimate of injury. There are several objective methods that can be used in evaluating cold damage (Quamme, 1978), but all of them are more labor requiring than observing browning. Since the correlation between visual rating and 2,2,3-triphenyl tetrazolium chloride (TTC) determination and visual rating and conductivity measurements are good ($r = 0.85$ and $r = 0.98$, respectively), usually only the visual evaluation is used (Rajashekar et al., 1982).

The temperature at which 50% injury occurs, designated T_{50} or LT_{50}, is commonly used to estimate absolute hardiness. To determine flower bud hardiness, the values of LT_{10} or LT_{90} are used to determine the temperature at which injury occurs and the temperature at which few buds survive. More complex equipment and procedures are required to determine absolute hardiness and to relate absolute hardiness indices to environmental stimuli (Quamme, 1976). Those methods need not be discussed here.

Determination of Hardiness by Exotherm Analysis

Solute-containing water can be supercooled to relatively low temperatures if nucleation does not occur. Sudden freezing and release of the heat of fusion can be detected as a deflection on the time–temperature profile during constant cooling. Such deflections are called *exotherm(s)*. *Differential thermal analysis* (DTA) is a sensitive method for detecting

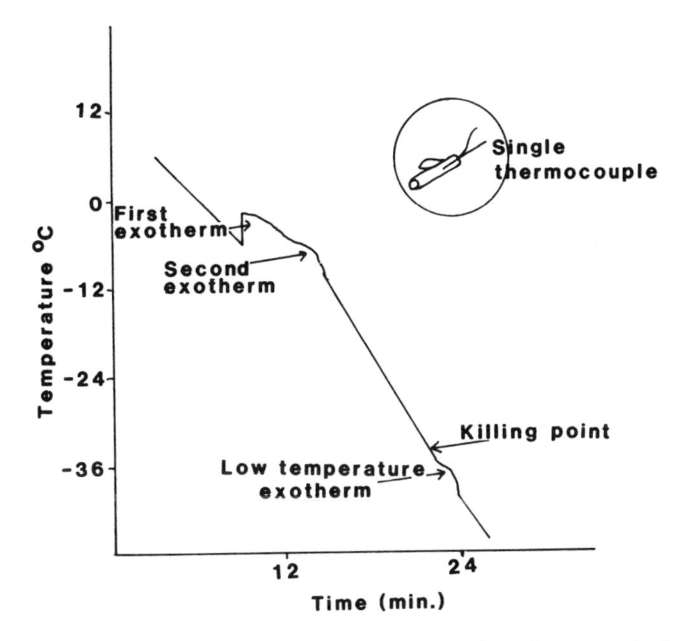

Figure 7.3 Typical record of tissue temperature (freezing curve) during the controlled freezing of an acclimated tissue. Inset: sample with single thermocouple represents the type of measurement used to obtain curve. (Redrawn by permission from Weiser, 1970.)

exotherms. A typical record of tissue temperature can be presented as recorded during the controlled freezing (Figure 7.3). The differential thermal analysis profile is developed by using the tissue temperature for one electrode and a dried sample for the other. The record shows the temperature difference (Figure 7.4).

As the tree cools, water freezes in the intercellular spaces a few degrees below the freezing point. In hardy plants ice crystals in the intercellular spaces do not harm the plant. The freezing of water creates an exotherm. The first exotherm is called the *high-temperature exotherm* (HTE). As the plant is cooled, one or more exotherms are observable. The last observable exotherm is the *low-temperature exotherm* (LTE) at which the remaining water in the protoplast is frozen or the last resistance to retain water in the protoplast is eliminated and water is drawn out and frozen. This is usually the death point for the tissue.

Tissues show LTE during DTA only when the tissue contains sufficient amount of water. The LTE does not develop if twigs of apple are freeze-dried below 8.5% water content per unit fresh weight, but LTE reappears when twigs are rehydrated to 20% water content (Quamme et

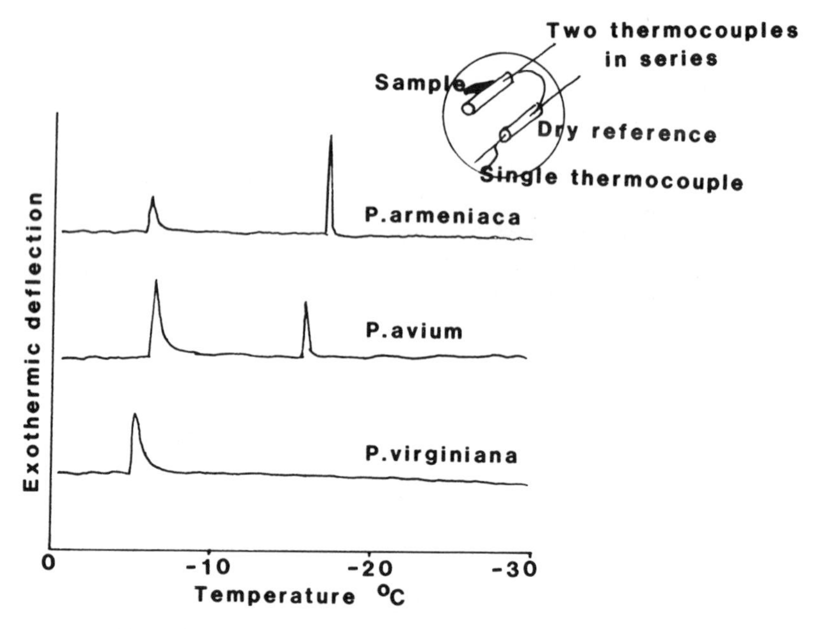

Figure 7.4 Record of exotherms established with differential thermal analysis. Inset: sample and dry reference represent the type of measurement used to obtain differential thermal record. Record was obtained in December with 5 degrees per hour cooling rate. (Redrawn by permission from Ashworth, 1984.)

al., 1973). The development of LTE also requires a certain, yet undefined, structural configuration. If, for example, dried twigs are ground before rehydration, LTE does not develop (Quamme et al., 1973).

Several correlations have been made within the temperature at which the LTE was exhibited and the tissue is damaged. Correlations are good between LTE and tissue death for tissues, buds, shoots, and bark of deciduous fruit trees in general and peach, apple, apricot, and pear in particular (Ashworth et al., 1983; Quamme, 1974, 1976, 1978; Rajashekar and Burke, 1978).

The size of the sample used for DTA has an influence on the temperature at which the LTE develops. Developing young fruit detached with its pedicel from woody tissue supercooled to a lower temperature than fruit attached to shoots 30 cm long. A similar difference occurs between detached and attached peach flower buds and developing flowers (Proebsting et al., 1982). The difference in the LTE between attached and detached samples may be as much as 3–6°C. Attached fruit shows a nucleation temperature of −2.2 to −3.8°C in comparison to detached fruit, which nucleates from −6.2 to −8.3°C including sweet cherry, peach, and per. Andrews and Proebsting (1986) estimated the rela-

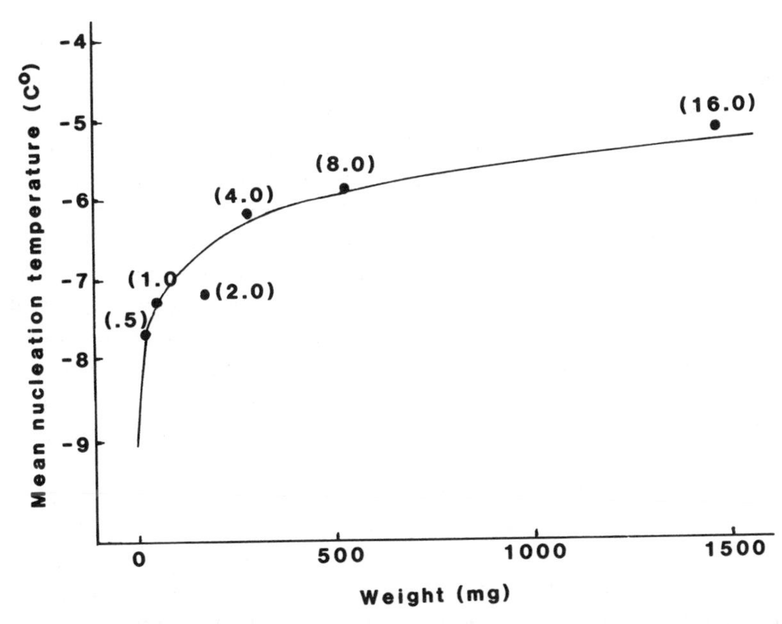

Figure 7.5 Relationship between sample size and nucleation temperature. Data were obtained March 28. Numbers in parentheses represent length of stem segments in centimeter. (Reprinted by permission from Andrews and Proebsting, 1986.)

tionship between the size of sample and the *mean ice nucleation temperature* (MNT) with DTA. The MNT slowly decreases as the sample size decreases from 1500 to 500 mg and rapidly decreases afterward to a few milligram sample size. The relationship is illustrated in Figure 7.5.

FREEZING INJURY

The freezing and death of plant tissues and cells largely depends upon the speed of cooling, the method of ice nucleation, the capacity of supercooling, and the permeability properties of cells (Sakai and Larcher, 1987).

Extracellular Freezing

Extracellular freezing is defined by Sakai and Larcher (1987) as ice formation on the surface of the cells or between the protoplast and the cell wall (extraprotoplasmic freezing). The water pressure of the cell water is higher than that of ice at the same temperature, and water will diffuse through the plasma membrane to the extracellular ice. The cells may plasmolize. If plasmolysis is not severe, after rewarming the cells can reabsorb water and regain turgor and normal metabolic activity. Of course, if plasmolysis is severe, they are injured and remain collapsed.

The rate of diffusion of water into the extracellular space is limited by the permeability of plasma membrane lipids. If the temperature drops rapidly, the diffusion to the extraplasmic ice must occur with sufficient speed to maintain an equilibrium and prevent lethal freezing. When water movement through the membranes is rapid, the extracellular ice formed is made of fern-shaped crystals that are too large to penetrate the protoplasmic membranes. Hardy cells have a high rate of permeability through the plasma membranes and are resistant to the penetration of ice crystals (Sakai and Larcher, 1987). The prerequisite for the initiation of extracellular freezing is the inoculation with ice crystals or other ice-nucleating agents.

Freezing of tissues is different from freezing of isolated cells. Under natural conditions, the intercellular spaces are filled with air, and extracellular freezing has to start from water films or droplets on the cell surfaces. Thus ice may be confined to regions where water diffuses from relatively distant unfrozen regions. Water droplets are commonly found around the vascular bundle sheets, guard cells, and some mesophilic cells around the stomatal cavity (Sakai and Larcher, 1987). Thus ice formation is likely to start in these tissues. Ice masses may form between the cambium and xylem of the basal stems of young trees. The ice forms from the water migrating from the unfrozen roots and causes splitting of the bark (Sakai and Larcher, 1987). Bark splitting is a major injury on dwarfing apple rootstock such as M9 and in apricots in warmer fruit-growing regions where soils are not frozen but cold air masses cause damage to the above-ground part of the tree.

Intracellular Freezing

Intracellular ice formation usually does not occur unless the cells are supercooled to at least −10°C. At the instant of intracellular freezing the cells are killed, probably due to the destruction of biomembranes resulting from the fast growing of ice crystals in the protoplast. Intracellular freezing of parenchymal cells could occur in two distinct ways characterized by flash or nonflash type of freezing. When sudden freezing occurs (flash type), the whole cell darkens instantaneously. When freezing is slow (nonflash type), ice growth in the protoplasm is clearly visible. A high cooling rate and a high degree of supercooling favor flash-type cell freezing. Under normal cooling conditions intracellular freezing kills the cells in the xylem ray parenchyma of fruit trees (Quamme et al., 1973) after deep supercooling. This is why xylem ray parenchyma is one of the most sensitive tissues of fruit trees to low winter temperatures.

Events Occurring during Slow Cooling

If the cooling rate is slow, which most often occurs in nature, ice forms outside of the protoplasts. Hardened cells of woody plants survive such extracellular freezing. However, the cells reach a point somewhere between −15 and −45°C when they are killed. Weiser's (1970b) summary of the events that occur during slow freezing death in hardened bark cells follows:

1. Tissues supercool to −2 to −8°C depending on the amount of solutes in the cells.
2. Water freezes between cells and nonliving xylem elements; ice is formed extracellularly.
3. Rapid propagation of ice occurs throughout the stem and results in a substantial release of heat of fusion (an exotherm), which raises the tissue temperature from the supercooling point of −2 to −8°C to a plateau at the freezing point of the free water in the stem of −0.3 to −1.0°C.
4. Tissues cool further after the readily available water is froze.
5. Water migrates from the protoplast to the extracellular ice nuclei in response to the extracellular vapor pressure deficit.
6. Additional exotherms may occur followed by decreasing tissue temperature and continuous slow movement of water out of the protoplast to the ice nuclei.
7. Protoplasts shrink, cells plasmolize, and solutes in the cells concentrate.
8. At one point a calorimetric lag indicates that freezing or water movement from the cell is arrested.
9. The last exotherm is observable; perhaps intracellular ice is formed.
10. Granulation of protoplasm occurs and death follows.

The number of exotherms may not be the same in every determinations. This will be discussed later.

The interpretation of events occurring around the death of the tissue is not clear. It has been proposed that water is trapped in the protoplast where it continues to supercool. As the temperature continues to decrease, the trapped water nucleates and intracellular ice formation causes death (Tumanov and Krasavtsev, 1959). Weiser (1970a) proposed that during the freezing process a point is reached when the only water that is left in the protoplasm is 'vital water' necessary for the hydration of macromolecules, and thus this water is 'necessary for life.' The removal of this water during the freezing process may be the key factor in the loss of viability. The protoplast may disintegrate by dehydration rather than by ice nucleation. Ashworth et al. (1983), working with peaches and apricot, noted that bark and xylem tissues show contrasting freezing patterns and mechanisms of freezing resistance. According to them, water in the xylem parenchyma cells supercools and tissue injury results from intracellular freezing of this supercooled water. In contrast, bark injury results from stresses that accompany extracellular ice formation and cellular dehydration.

The time–temperature interaction required to kill the tissues strongly underlines the point that the protoplast disintegrates by dehydration rather than by freezing. Apple roots showed considerably more injury after 17 hr at −8°C than after 4 hr (Potter, 1924), and apple twigs and buds were injured more after 12 hr at −30°C than after 3 hr (Hildreth, 1926). Chandler (1954) interpreted the damage caused by extended exposure to changes (dehydration) in the protoplasm rather than by freezing. Although his interpretation is from the period before the LTE was discovered, it is in line with modern interpretations.

ROLE OF DORMANCY IN HARDINESS

Before resistance to very low temperature can develop, active growth must cease (Weiser, 1970a). In fruit trees the cessation of growth by different parts of the tree is not synchronized (Proebsting, 1978). Terminal growth ceases in some shoots early, but vigorous shoots continue to grow until late summer or early fall. Cambial activity is believed to cease first in the periphery of the tree and last in the trunk and crotches (Proebsting, 1978). The age of the tree, soil fertility, soil moisture, growth-regulating chemicals, and fall temperatures can modify the time of cessation of growth. Early winter freezes tend to injure the tissues that are last to become dormant (Potter, 1939). These are all part of the acclimation process, which is presumed to develop in three stages. Stage 1 is initiated by shortening days; in stage 2 cold temperature triggers major metabolic changes, and in stage 3 further reversible increases in hardiness occur while the tissues are in the frozen state. The basic information for this has been developed with *Cornus* (Weiser, 1970b), but Proebsting (1978) considers them compatible with field observations of *Malus*, *Pyrus*, and *Prunus*. There is information for development of stage 2. Much less is known about events occurring in stages 1 and 3.

Flower buds are in a special situation as far as dormancy is concerned. They are formed during the summer. develop slowly with relatively little visible growth, and become 'dormant.' Yet there is ample evidence that the primordia within the bud continues to develop morphologically late into the fall or into early winter. During late fall the dormant flower buds gain a relatively high degree of resistance to cold. However, because of their double nature, the flower buds must also be considered developing organs. Therefore, they lose hardiness very quickly when temperatures warm to the threshold level.

Andrews and Proebsting (1986) investigated the effect of temperature on developing the hardiness in sweet cherry flower buds. They measured hardiness by the development of the capacity to deep supercool in the buds. Ambient temperatures on average were 20°C during the day and 11°C at night in the orchard during August when they conducted the experiments. Buds exposed on the tree, in controlled chambers, to 15°C during the day and 5°C at night advanced the development of LTE by 4 weeks, and a warm treatment of 30°C during the day and 20°C at night delayed hardening and the development of LTEs by 2 weeks compared with buds exposed to ambient temperatures. They concluded that in flower buds the development of the capacity to supercool is temperature dependent. However, the fact that the 30°C day–20°C night temperature regime did not prevent the development of capacity to supercool indicates that some aspects of acclimation of flower buds is independent of temperature.

There are differences among species in the development of hardiness during the fall. Sweet cherries develop the capacity for deep supercooling 1–2 weeks earlier than peaches (Andrews and Proebsting, 1986). Which stage(s) hasten development of hardiness is not known.

When hardening develops, a few degrees of temperature difference could mean whether the flower buds are damaged only to a slight degree (10% killed) or to a very severe extent (90% killed). Proebsting and Mills (1966) determined the percentage of survival of 'Elberta' peach flower buds and expressed the deviation from the temperature required to kill 50% of the buds (T_{50}). Only 4°C above the T_{50}, 90% of the flower buds survived and only 3°C lower temperature killed all the buds (Figure 7.6).

Once hardiness develops in buds, it does not stay constant. Loss of hardening is very rapid if buds are exposed to warm temperature. Cherry buds lost 6.1°C of hardiness in 4 hr when exposed to 24°C temperature (Proebsting and Mills, 1972). Dehardening is also rapid in apple bark. The bark of 'Haralson' apple lost as much as 15°C of hardiness during 1 day exposure to 20°C temperature (Howell and Weiser, 1970). Ketchie and Beeman (1973) determined the rate of dehardening in 'Delicious' apple shoots after the shoots had been exposed in the field for various temperatures for 7 days. During the 3 years of the study the rate of

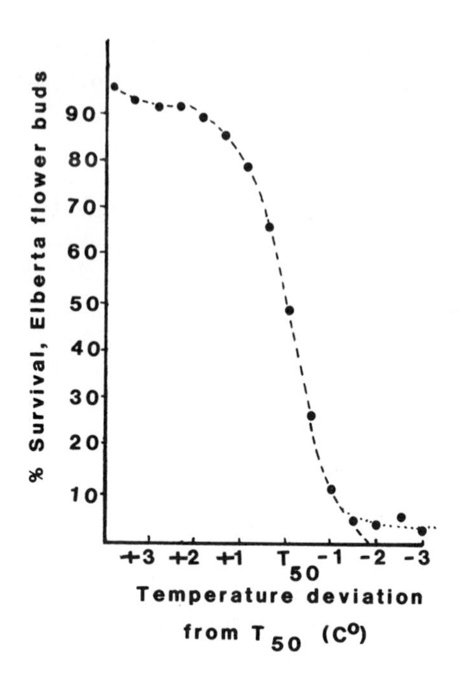

Figure 7.6 Effect of temperature around the T_{50} point on survival of 'Elberta' flower buds. Temperature is expressed as the deviation from that required to kill 50% of the buds. (Reprinted by permission from Proebsting and Mills, 1966.)

dehardening was remarkably the same as temperature increased (Figure 7.7). Exposure to +10°C temperature for 7 days resulted in a loss of almost 30°C in hardiness in 2 out of the 3 years and 20°C in the third year.

Both buds and bark can reharden if the temperature is sufficiently low. Cherry and peach buds reharden at a rate of 1.9°C per day for peaches and 2.8°C per day for cherries if the temperature is below −1.1 to −2.2°C (Proebsting and Mills, 1972). Dehardening usually occurs faster than rehardening. This is the case at least in peach (Proebsting, 1963). The dehardening–rehardening cycle is not completely reversible. Data obtained in controlled dehardening–rehardening experiments with apple bark indicate that with each successive day of dehardening less rehardening is possible. Hardened apple trees in containers exposed to about 20°C dehardening temperature for 2 days reharden within 3 days at −15°C to their original degree of hardiness. Trees dehardened at 20°C for 7 days reharden only to a small degree and are only slightly hardier than those not rehardened at all (Howell and Weiser, 1970) (Figure 7.8). Dehardening and rehardening also occur in orchards in natural conditions. During the natural dehardening during spring at least 5°C rehardening was re-

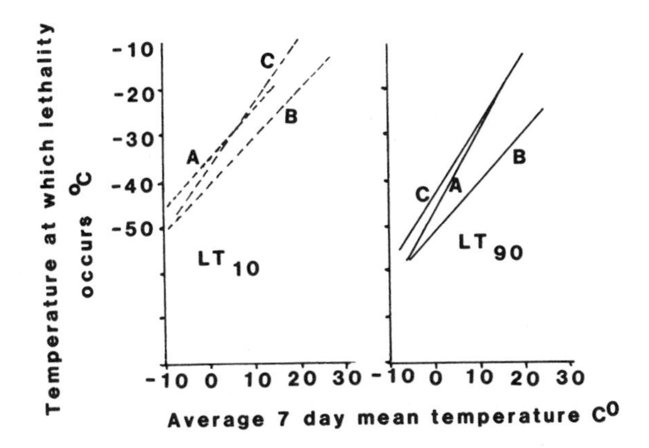

Figure 7.7 Rate of dehardening in 'Delicious' shoots LT_{10} and LT_{90} responses to mean temperatures during the 7 days preceding collection of samples during the winters of 1969–1970 (*A*), 1970–1971 (*B*), and 1971–1972 (*C*). Natural conditions. (Redrawn by permission from Ketchie and Beeman, 1973.)

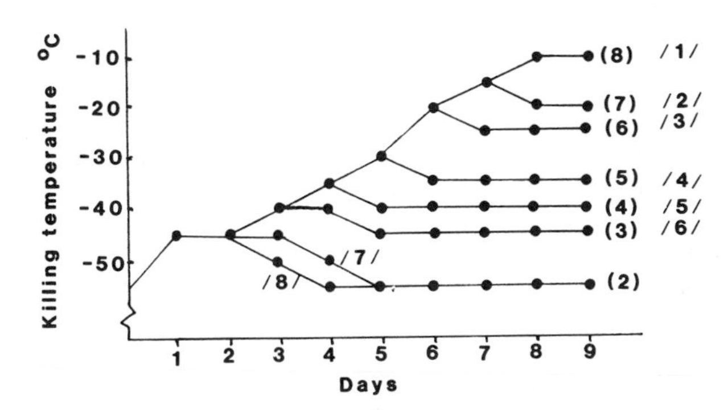

Figure 7.8 Rate of rehardening in 'Haralson' apple after approximate various durations of dehardening period at greenhouse temperature (~20°C) and rehardening at −15°C for various lengths of time. Numbers in parentheses ( ) are the days of dehardening and those in between slashes // are the days of rehardening at 15°C. (Redrawn by permission from Howell and Weiser, 1970.)

ported in several occasions when temperature decreased from the daily maximum 19°C–minimum 9°C to maximum 9°C–minimum 7°C (Howell and Weiser, 1970).

As time goes on, during the winter and especially during spring, trees are permanently dehardened (Howell and Weiser, 1970). Within 2 months beginning in early March, Haralson apples dehardened from −50

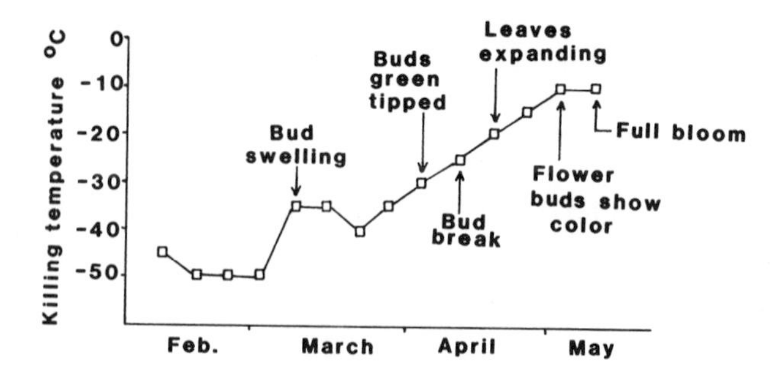

Figure 7.9 Rate of dehardening of flower buds of 'Haralson' apple during spring. (Redrawn by permission from Howell and Weiser, 1970.)

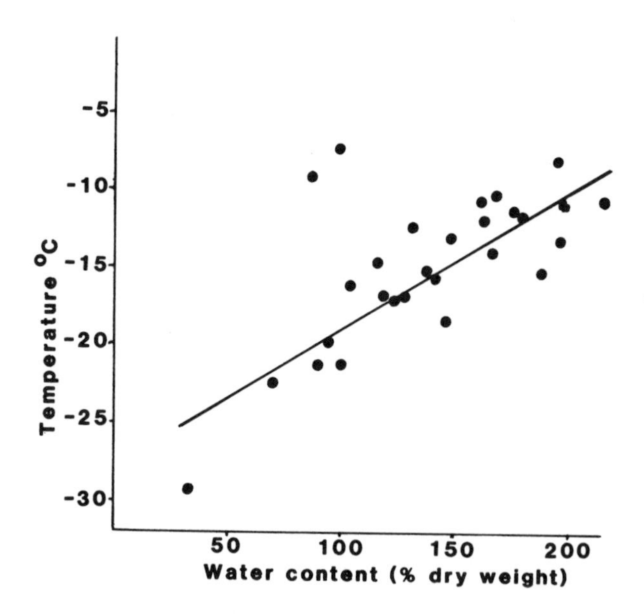

Figure 7.10 Relationship between the temperature at which the LTE is observable and the water content of flower primordia. (Reprinted by permission from Quamme, 1983.)

to −8°C (Figure 7.9). The dehardening, or deacclimation, in spring is markedly associated with the increase in water content of the flower primordia in peach and cherry (Quamme, 1983). The water content of the flower primordium and the vascular traces, not that of the whole flower bud, is critical to the level of supercooling attained (Quamme, 1983) (Figure 7.10). Andrews and Proebsting (1987) calculated linear regression

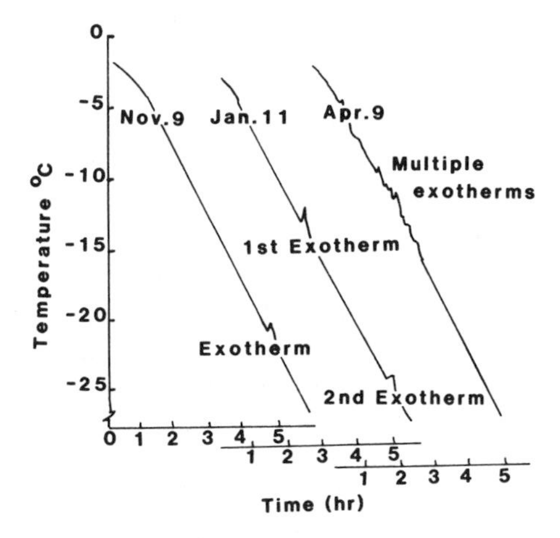

Figure 7.11 Typical time temperature profiles of 'Elberta' peach flower buds at three different stages of acclimation. (Redrawn by permission from Quamme, 1974.)

equations using data from controlled experiments to estimate bud hardiness from the water content of primordia. Their calculated value for LTE_{50} and the measured value of LTE_{50} were remarkably close. However, the constant in the equations changed as the season advanced from October to February. They proposed that the water content of flower buds may be useful for estimating the LTE_{50} of flower buds.

ICE NUCLEATING IN TISSUE OF FRUIT TREES

Even without dehardening there are considerable changes in hardiness of tissues of fruit trees during the winter. A close relationship between hardiness and LTE in peaches is illustrated in Figure 7.9. As the season changes, the LTE pattern also changes. Before December 12 only one exotherm is observable between −16 and −26°C. From this time until budswell two exotherms are exhibited on the profile, the first of which is between −5 and −14°C and a second from −10 to −27°C. After budswell there are multiple exotherms on the time–temperature profile (Figure 7.11) (Quamme, 1974). Ashworth (1984) examined the supercooling ability of apricot, cherry, sour cherry, peach, and plum (*P. salicina* and *P. sargentii*) during the winter. The buds had no ability to supercool in September and October; all buds supercooled in November and December. *Prunus sargentii* lost its ability to supercool in January, peach, cherry, apricot, and *P. salicina* lost their ability to supercool sometime between February and March, and sour cherry retained its ability to supercool the longest, to almost April. *Prunus serotina* and *P. virginiana* are two species in which the buds do not supercool. There are no xylem vessels within the primordia of the single- or double-flowering *Prunus* species through December. In January vascular differentiation resumes in *P. sargentii* and differentiation continues throughout the spring. Cherry and *P. salicina* establish xylem continuity between the primordia and adjacent tissues in February and March. Sour cherry is the last to undergo xylem differentiation. The vascular system in the racemous inflorescences of *P. serotina* and *P. virginiana* is different than that observable in other *Prunus* species. Xylem vessels run the length of the raceme, and vascular traces enter every primordia. Ashworth (1984) suggests that the lack of xylem continuity is important to enable the buds to supercool. As soon as xylem continuity develops, the buds lose their ability to supercool or those species that have xylem structures in their primordia cannot develop the ability to supercool. He reasoned that xylem continuity allows ice to propagate via the vascular system and nucleate the water within the primordia. Other members of the genus *Prunus* that have racemous flowers also do not supercool (Burke and Stushnoff, 1979).

In stems of trees there are usually more than one LTE resulting from the freezing of supercooled water in the tissue. In others the LTE has a very broad nondistinct peak. The stem contains at least two types of tissues. The bark tissue supercools differently than the xylem tissue, resulting in multiple exotherms in the DTA profile. In *Pyrus cordata* the analysis of area under the peaks in the DTA profile indicates that 24% of the water is in the bark and the rest is in the xylem parenchyma cells (Rajashekar et al., 1982).

Once initiated, ice formation spreads throughout the tissue. In peach trees in the field freezing is very rapid (Ashworth et al., 1985). On one occasion ice spread throughout the tree in 16 min. However, time intervals of more than 1 hr were also observed between initiation and complete ice formation throughout the tree. Ice formation usually occurs first in the current season's wood, subtending branches, and scaffold limbs. Ashworth et al. (1985) observed that the location of ice nucleation varied. Once the current season's wood cooled faster than larger branches and ice nucleation started in the shoots. On other occasions, the tissue temperature was more uniform, and freezing was initiated in the lower branches. Larger tissues, such as the trunk and scaffold limbs, remain 2–3°C warmer than air temperature for several hours after freezing is initiated.

As buds develop, they are more vulnerable to frost injury. The first stage of bud development is relatively frost tolerant, and −8 to −10°C is required to cause 50% injury. During the bloom or postbloom stages −2 to −5°C will injure the flowers or the young fruit.

TABLE 7.1 Mean Ice Nucleation Temperature for Attached Peach Flower Buds[a]

Stage of Development	MNT (°C)
First swelling	−6.0
Green calyx	−5.7
Red calyx	−5.8
First pink	−5.2
First bloom	−4.8
Full bloom	−2.4
Post bloom	−1.9

[a] Reprinted by permission from Proebsting et al., 1982.

As buds resume growth during spring and flower buds develop, the MNT increases (Proebsting et al., 1982). The MNT for various stages of peach flower buds is given in Table 7.1.

The most prominent ice-nucleating agents associated with plants are bacteria (Lindow et al., 1982). Strains of *Pseudomonas syringae* van Hall, *Erwinia herbicola* (Lohnis) Dye, *Pseudomonas viridiflava* strain W1, and *Pseudomonas fluorescence* Migula have been identified as effective ice nucleators (Anderson and Ashworth, 1986; Lindow et al., 1978, Maki et al., 1974). Such bacteria are called *ice nucleation active* (INA) bacteria. *Pseudomonas syringae* pv. syringae strain B301D originally isolated from pear flowers expressed a frequency of ice nucleation of 3×10^2 cells/−5°C ice nucleus (Gross et al., 1984). Ice-nucleating bacteria are relatively short lived in nature (Gross et al., 1984), and few bacteria survive the winters. In a survey in Washington state less than 10 CFU/g (colony-forming unit) were found in orchards during the dormant period. This is insufficient to cause ice nucleation in flower buds. Upon budbreak flower buds often harbor a small population of ice-nucleation active (INA) bacteria that presumably originated from those undetected during dormancy. In contrast, in the southeastern United States, *P. syringae* is quite common on peach trees during the winter. It causes bacterial canker on the trees. There is more bacteria on trees that are otherwise under stress. On a *peach tree short-life* (PTSL) soil 90% of trees had considerable bacterial populations, whereas trees on new peach soil had very few (Weaver et al., 1974).

Inoculation studies (Gross et al., 1984) could not detect differences in supercooling between inoculated flower organs or fruit and those free of bacteria. They presumed an intrinsic nonbacterial source of ice nucleation active at temperatures similar to INA bacteria that limited the supercooling of the tissues. Ashworth et al. (1985) came to the same conclusion about the formation of ice in peach shoots. The INA seems to be a constitutive component of mature wood and is stable under a range of chemical treatments. It appears that even in the absence of INA bacteria, no supercooling of water occurs in flowers and young shoots of fruit trees because the intrinsic INA component acts as an ice nucleator.

CRITERIA FOR ESTABLISHING HARDINESS IN FRUIT TREES

Since there are many aspects of hardiness, it seems necessary to list the criteria that can be used for establishing varietal differences in hardiness of fruit trees. As has been discussed, *hardiness* is an all-inclusive term. Thus at any given location only part of the listed criteria is applicable. Proebsting and Mills (1972) assembled such a list applicable to cherries and peaches. An abbreviated list is presented here.

The factors that influence the hardiness level of fruit trees are:

1. The rate of hardening in response to low temperature during autumn.
2. The minimum hardiness level the tree is capable of reaching.
3. The time that the minimum hardiness level remains constant during winter.
4. The pivotal temperature separating hardening from dehardening during dormancy.
5. The rate of hardiness loss of hardened buds in response to warm temperatures.
6. The time when capability for rehardening is lost.
7. The shape of the temperature survival curve.

EFFECT OF CULTURAL PRACTICES OF HARDINESS OF FRUIT TREES

Most of the acclimation of deciduous fruit trees is preordained by the genetic constitution of the tree and the normal environments the tree responds (Proebsting, 1978). Cultural practices usually can only add a very small increment to hardiness compared with the natural progression of acclimation. However, because of the steep response curve that causes the injury, a small increment can appear as a major advantage at the point of critical temperature. Here only practices that modify hardiness are presented. Those that modify temperature can be found in production-related books and need not be repeated here.

Practices Connected with Hastening Dormancy

Any practice that extends growth into fall decreases the hardiness of tissues. Among these 'late' N application and irrigation are foremost. Nitrogen application between July and September is considered late. However N application either as soil application or spray application after the days are shortened and shoot growth permanently ceases is not only not considered detrimental but also beneficial. As discussed previously, N application after shoot growth ceases keeps the photosynthetic activity of the leaves longer, which increases the reserve materials of the tree and promotes hardiness (Heinicke, 1934). In regions with low rainfall, irrigation may be used as an additional control of cold resistance. Late summer and early fall reduced irrigation helps the tree stop growth. Near the time of defoliation, when dormancy develops, the trees need irrigation to keep roots moist through the winter. This final irrigation slows the rate of decrease of soil temperature below freezing and helps prevent cold injury to the roots (Proebsting, 1978). It also assures normal growth during the spring. The late fall irrigation is somewhat controversial because there is some evidence that flower buds are more hardy if soil moisture is low during the winter (Proebsting, 1978). Therefore, growers must judge which is the greater risk in their region, root injury or flower bud injury.

Photosynthesis

Acclimation is an active metabolic process that requires a product of photosynthesis (Chandler, 1954). Foliage should be in good condition in late fall to produce the maximum photosynthate possible. Early thinning of fruit, which reduces the total carbohydrate requirement for fruit, increases cold resistance (Edgerton, 1948), presumably by allowing more photosynthate for the woody part of the tree. Early maturing cultivars often survive low fall temperatures better than cultivars with fruit still on the tree or with the crop harvested only before the frost (Potter, 1939).

Pruning

Pruning prior to the occurrence of low-temperature injury tends to increase injury. Thus in several areas pruning is delayed until after the threat of damaging low temperature is passed (Burkholder, 1936; Prince, 1966).

Chemical Control of Hardiness

Proebsting (1978) reviewed the existing literature on chemicals that promote or inhibit the development of hardiness. The earliest attempts were through control of dormancy and rest. Then cryoprotectants were used. Finally plant growth regulators were tried. Since Proebsting's review several additional investigations have been reported. None of the treatments are satisfactory either in the degree of response they can elicit from the tree or the consistency of their effect from year to year. Much of the promotion of hardiness depends on environmental parameters and on coordinating measures to induce the tree to develop the highest degree of hardiness possible. Thus a chemical treatment is unlikely to succeed.

Evaporative Cooling

To keep trees from dehardening, evaporative cooling has been tried to lower the temperature of the wood after the rest was completed (Bauer et al., 1976). Sprinklers were turned on automatically whenever air temperature was above 7°C and were left on for 5 min every 15 min from January 22 until April 18 (full bloom of nonsprinkled trees). Although sprinkling delayed full bloom by 15 days, it did not influence hardiness. Hardiness was measured by exosmosis and electrical conductance tests. The results are not unexpected. Sprinkling lowered temperature by 6.5°C when air temperature was above 16°C. Even the lowered wood temperature of the sprinkled trees was above the temperature threshold that causes rapid dehardening. Thus the rate of temperature reduction by which sprinkling is able to lower temperature is not satisfactory to retain hardiness.

Peach Tree Short Life

In the southeastern United States peach trees decline or are killed by winter freezes. A complex set of conditions predisposes peach trees so that they lose hardiness and become vulnerable to relatively mild cold. Cold injury usually occurs in late winter after the chilling requirement of the trees has been met. Trees exhibit the classical signs of cold injury. Symptoms include separation of the bark from the wood and browning of the cambium, xylem, and phloem. Trunk, crotches, and major scaffold branches are most often affected, with more severe damage on the sunward side of these plant parts. Some of the most severe losses occurred when temperatures in late winter reached 22°C or more for several days in succession followed by a rapid decrease to −6°C or below (Neshmith and Dowler, 1976). It is obvious that the environmental conditions favored losing the hardiness as described. Several cultural methods are known to hasten or lessen the development of cold injury. It is likely that the cultural manipulations known to affect PTSL either influence the depth of dormancy or inhibit the dehardening process, both of which lessen susceptibility to cold.

Fall pruning before December usually is detrimental and induces the peach tree short life (PTSL) symptoms (Prince and Horton, 1972). Pruning wounds usually partially heal after early pruning, indicating that cambium is active midwinter (Neshmith and Dowler, 1976). When cambial activity occurs, peach trees are very susceptible to cold injury (Alden and Hermann, 1971). Fumigation of peach soils before planting with, 1,2-dibromo-3-chloropropane (DBCP) greatly decreased the mortality of trees. The effect is through elimination nematode populations in the soil and preventing the infection of peach roots with nematodes (Neshmith and Dowler, 1976). Nematode infection may increase root metabolic activity, which in turn produces cytokinins that hasten dehardening or spring activity in the top of the tree, rendering it more vulnerable to cold injury. Nitrogen applications near leaf fall reduce the cold hardiness of trees (Neshmith and Dowler, 1976). When fumigation and N is applied together, the N application eliminates the beneficial effect of fumigation (Neshmith and Dowler, 1976).

The rootstock on which peach trees are grafted affects greatly the development of PTSL symptoms. Among trees grafted on 'Nemagard,' the losses are much greater than among those grafted on 'Lovell' rootstock (Zehr et al., 1976). Rootstock and nematode infection interact. Ring nematode (*Criconemoides xenoplax*) populations usually damage trees grafted on 'Nemagard' to a much greater extent than those grafted on 'Lovell' roots. Trees in the PTSL sites, where roots are infected with ring nematodes, show a higher level of cytokinins in the leaves during the growing season, and the leaves of such trees remain greener longer during autumn than those not infected with nematodes (Riley, personal communication). The possibility exists that the nematode-infected roots, having a higher metabolic activity, produce higher levels of cytokinins. The higher cytokinin levels in turn either allow only a shallower dormancy or aid in the dehardening process by shortening dormancy and eventually predisposing the tree to a higher level of cold damage. Soils on old peach sites, where the PTSL is the biggest problem, are usually very acid, having a pH of 4.5 or less. On such soils aluminum is usually taken up by the plant and blocks the uptake of Ca. In turn the Ca needed for regulatory action in cells and the high Ca concentration in tissues of peaches are considered essential to avoid the development of PTSL (Edwards and Horton, 1977).

Finally the bacterial infection of trees and the effect of the infection on hardiness need to be considered. *Pseudomonas syringae* and *P. morsprunorum* attack trees after defoliation, and tissue damage occurs during the dormant season. Symptoms of bacterial canker and cold injury are indistinguishable if the inspections are made in April or early May. However, some differences can be observed if inspection is made in March (Weaver et al., 1974). Bacterial infection is definitely a contributing factor in lowering the hardiness of peach trees, but the similarity of symptoms makes evaluation difficult. Bacterial infection interacts with planting site, nematode infection, and fumigation (Weaver et al., 1974).

In PTSL the predominant influence is environmental, but hardiness is greatly influenced by planting site, nematode infection, soil pH, Ca and Al concentrations, bacterial infection of trees, time of pruning, and rootstock. All these factors can interact, but not all are necessarily present in any given situation. The growers are urged to follow a 10-point plan, taking the preceding factors into consideration, to minimize damage.

REFERENCES

Alden, J., and R. K. Hermann. 1971. Bot. Rev. 37:142–171.

Anderson, J. A., and E. N. Ashworth. 1986. Plant Physiol. 80:956–960.

Andrews, P. K., and E. L. Proebsting. 1986. J. Amer. Soc. Hort. Sci. 111: 232–236.

Andrews, P. K., and E. L. Proebsting. 1987. J. Amer. Soc. Hort. Sci. 112: 334–340.

Ashworth, E. N. 1984. Plant Physiol. 74:862–865.

Ashworth, E. N., J. A. Anderson, and G. A. Davis. 1985. J. Amer. Soc. Hort. Sci. 110:287–291.

Ashworth, E. N., D. J. Rowse, and L. A. Billmyer. 1983. J. Amer. Soc. Hort. Sci. 108:299–303.

Bailey, C. H., and L. F. Hough. 1975. In Advances in fruit breeding. J. Janick and J. N. Moore (eds.). West Lafayette, IN: Purdue University Press, pp. 367–383.

Bauer, M., C. E. Chaplin, G. W. Schneider, B. J. Barfield, and G. M. White. 1976. J. Amer. Soc. Hort. Sci. 101:452–454.

Burke, M. J., and C. Stushnoff. 1979. In Stress physiology in crop plants. H. Mussel and R. Staples (eds.). New York: Wiley, pp. 197–225.

Burkholder, C. L. 1936. Proc. Amer. Soc. Hort. Sci. 34:49–51.

Chandler, W. H. 1913. Missouri Agr. Exp. Sta. Research Bull. 8:1–309.

Chandler, W. H. 1954. Proc. Amer. Soc. Hort. Sci. 64:552–572.

Childers, N. F. 1983. Modern fruit science. Gainesville, FL: Horticultural Publications.

Cummins, J. N., and H. S. Aldwinkle. 1974. Hort. Sci. 9:367–372.

Edgerton, L. J. 1948. Proc. Amer. Soc. Hort. Sci. 52:112–115.

Edwards, J. H., and B. D. Horton. 1977. J. Amer. Soc. Hort. Sci. 102:459–461.

Gross, D. C., E. L. Proebsting, and P. K. Andrews. 1984. J. Amer. Soc. Hort. Sci. 109:375–380.

Heinicke, A. J. 1934. Proc. Amer. Soc. Hort. Sci. 32:78–80.

Hildreth, H. C. 1926. Univ. Minnesota Agr. Exp. Sta. Techn. Bull. 42.

Howell, G. S., and C. J. Weiser. 1970. J. Amer. Soc. Hort. Sci. 95:190–192.

Ketchie, D. O., and C. H. Beeman. 1973. J. Amer. Soc. Hort. Sci. 98:257–261.

Levitt, J. 1972. Responses of plants to environmental stresses. Academic Press, New York.

Lindow, S. E., D. C. Arny, and C. D. Upper. 1978. Phytopathology 68:523–527.

Lindow, S. E., D. C. Arny, and C. D. Upper. 1982. Plant Physiol. 70:1090–1092.

Maki, L. R., E. L. Galyan, M. Chang-Chien, and D. R. Caldwell. 1974. App. Microbiol. 28:456–459.

Maney, T. J. 1940. Proc. Amer. Soc. Hort. Sci. 40:215–219.

Neshmith, W. C., and W. M. Dowler. 1976. J. Amer. Soc. Hort. Sci. 101: 116–119.

Potter, G. F. 1924. New Hampshire Agr. Exp. Sta. Techn. Bull. 27.

Potter, G. F. 1939. Proc. Amer. Soc. Hort. Sci. 36:184–189.

Prince, V. E. 1966. Proc. Amer. Soc. Hort. Sci. 88:190–195.

Prince, V. E., and B. D. Horton. 1972. J. Amer. Soc. Hort. Sci. 97:303–305.

Proebsting, E. L. 1963. Proc. Amer. Soc. Hort. Sci 83:259–269.

Proebsting, E. L. 1978. In Plant cold hardiness and freezing stress. P. H. Li and A. Sakai (eds.). New York: Academic Press, pp. 267–280.

Proebsting, E. L., and H. H. Mills. 1966. Proc. Amer. Soc. Hort. Sci. 89:85–90.

Proebsting, E. L., and H. H. Mills. 1972. J. Amer. Soc. Hort. Sci. 97:802–806.

Proebsting, E. L., and A. Sakai. 1979. Hort. Sci. 14:597–598.

Proebsting, E. L., P. K. Andrews, and D. Gross. 1982. Hort. Sci. 17:67–68.

Quamme, H. A. 1974. J. Amer. Soc. Hort. Sci. 99:315–318.

Quamme, H., C. J. Weiser, and C. Stushnoff. 1973. Plant Physiol. 51:273–277.

Quamme, H. A. 1976. Canad. J. Plant Sci. 56:493–501.

Quamme, H. A. 1978. In Plant cold hardiness and freezing stress. P. H. Li and A. Sakai (eds.). New York: Academic Press, pp. 313–332.

Quamme, H. A. 1983. J. Amer. Soc. Hort. Sci. 108:697–701.

Rajashekar, C., and M. J. Burke. 1978. In Plant cold hardiness and freezing stress. P. H. Li and A. Sakai (eds.). New York: Academic Press, pp. 213–226.

Rajashekar, C., M. N. Westwood, and M. J. Burke. 1982. J. Amer. Soc. Hort. Sci. 107:968–972.

Sakai, A., and W. Larcher. 1987. Frost survival of plants. Berlin: Spinger-Verlag.

Simons, R. 1972. Hort. Sci. 7:401–402.

Tumanov, I., and O. Krasavtsev. 1959. Sovjet Plant Physiol. 6:663–673.

Weaver, D. J., E. J. Wehunt, and W. M. Dowler. 1974. Plant Disease Reporter 58:76–79.

Weiser, C. J. 1970a. Hort. Sci. 5:403–408.

Weiser, C. J. 1970b. Science 169:1269.

Westwood, M. 1978. Temperate zone pomology. San Francisco: Freeman.

Zehr, E. I., R. W. Miller, and F. H. Smith. 1976. Phytopathology 66:689–694.

INDEX